de Gruyter Studies in Mathematics 36

de Gruyter Studies in Mathematics

1 Riemannian Geometry, 2nd rev. ed., *Wilhelm P. A. Klingenberg*
2 Semimartingales, *Michel Métivier*
3 Holomorphic Functions of Several Variables, *Ludger Kaup and Burchard Kaup*
4 Spaces of Measures, *Corneliu Constantinescu*
5 Knots, 2nd rev. and ext. ed., *Gerhard Burde and Heiner Zieschang*
6 Ergodic Theorems, *Ulrich Krengel*
7 Mathematical Theory of Statistics, *Helmut Strasser*
8 Transformation Groups, *Tammo tom Dieck*
9 Gibbs Measures and Phase Transitions, *Hans-Otto Georgii*
10 Analyticity in Infinite Dimensional Spaces, *Michel Hervé*
11 Elementary Geometry in Hyperbolic Space, *Werner Fenchel*
12 Transcendental Numbers, *Andrei B. Shidlovskii*
13 Ordinary Differential Equations, *Herbert Amann*
14 Dirichlet Forms and Analysis on Wiener Space, *Nicolas Bouleau and Francis Hirsch*
15 Nevanlinna Theory and Complex Differential Equations, *Ilpo Laine*
16 Rational Iteration, *Norbert Steinmetz*
17 Korovkin-type Approximation Theory and its Applications, *Francesco Altomare and Michele Campiti*
18 Quantum Invariants of Knots and 3-Manifolds, *Vladimir G. Turaev*
19 Dirichlet Forms and Symmetric Markov Processes, *Masatoshi Fukushima, Yoichi Oshima and Masayoshi Takeda*
20 Harmonic Analysis of Probability Measures on Hypergroups, *Walter R. Bloom and Herbert Heyer*
21 Potential Theory on Infinite-Dimensional Abelian Groups, *Alexander Bendikov*
22 Methods of Noncommutative Analysis, *Vladimir E. Nazaikinskii, Victor E. Shatalov and Boris Yu. Sternin*
23 Probability Theory, *Heinz Bauer*
24 Variational Methods for Potential Operator Equations, *Jan Chabrowski*
25 The Structure of Compact Groups, 2nd rev. and aug. ed., *Karl H. Hofmann and Sidney A. Morris*
26 Measure and Integration Theory, *Heinz Bauer*
27 Stochastic Finance, 2nd rev. and ext. ed., *Hans Föllmer and Alexander Schied*
28 Painlevé Differential Equations in the Complex Plane, *Valerii I. Gromak, Ilpo Laine and Shun Shimomura*
29 Discontinuous Groups of Isometries in the Hyperbolic Plane, *Werner Fenchel and Jakob Nielsen*
30 The Reidemeister Torsion of 3-Manifolds, *Liviu I. Nicolaescu*
31 Elliptic Curves, *Susanne Schmitt and Horst G. Zimmer*
32 Circle-valued Morse Theory, *Andrei V. Pajitnov*
33 Computer Arithmetic and Validity, *Ulrich Kulisch*
34 Feynman-Kac-Type Theorems and Gibbs Measures on Path Space, *József Lörinczi, Fumio Hiroshima and Volker Betz*
35 Integral Representation Theory, *Jaroslas Lukeš, Jan Malý, Ivan Netuka and Jiri Spurný*

Gerrit van Dijk

Introduction to Harmonic Analysis and Generalized Gelfand Pairs

Walter de Gruyter
Berlin · New York

Author

Gerrit van Dijk
Mathematical Institute
Leiden University
2300 RA Leiden, Netherlands
E-Mail: dijk@math.leidenuniv.nl

Mathematics Subject Classification 2000: 43-01, 43-02, 43A85, 22Bxx, 22Dxx

Keywords: Fourier theory, harmonic analysis, Gelfand pairs, locally compact groups, Haar measures

∞ Printed on acid-free paper which falls within the guidelines of the ANSI
to ensure permanence and durability.

Bibliographic information published by the Deutsche Nationalbibliothek

The Deutsche Nationalbibliothek lists this publication in the Deutsche Nationalbibliografie; detailed bibliographic data are available in the Internet at http://dnb.d-nb.de.

ISBN 978-3-11-022019-3

Printed in Germany.
Cover design: Martin Zech, Bremen.
Typeset using the author's LaTeX files: Kay Dimler, Müncheberg.
Printing and binding: Hubert & Co. GmbH & Co. KG, Göttingen.

Preface

This book is intended as an introduction to harmonic analysis, and especially to my favorite topic, generalized Gelfand pairs. It is aimed at advanced undergraduates or beginning graduate students. The scope of the book is limited, with the aim of enabling students to reach a level suitable for starting PhD-research. It is based on lectures I have given in several places, most recently at Kyushu University in Fukuoka, Japan (2008). Student input has strongly influenced the writing, and I hope that this book will help students to share my enthusiasm for the beautiful topics discussed.

Starting with the elementary theory of Fourier series and Fourier integrals, I proceed to abstract harmonic analysis on locally compact abelian groups. Here I follow the classical paper [7] of H. Cartan and R. Godement. It turns out that the technique they developed works as well for Gelfand pairs (G, K), where G is a not necessarily abelian locally compact group, and K is a compact subgroup of G. This approach is based on my thesis [51]. Finally I develop part of the theory of generalized Gelfand pairs (G, H) where H is a closed, possibly noncompact, subgroup of G. The basic ideas are due to E. G. F. Thomas, see for example [49], and several applications are from my own work [52]. I also draw on papers by J. Faraut and V. F. Molchanov to deal with examples related to the generalized Lorentz group.

There is relatively little expository literature on generalized Gelfand pairs, and there is no standard reference. For further reading, I recommend my Kyushu Lecture Note [53].

The main prerequisites for the book are elementary real, complex and functional analysis. In the later chapters we shall assume familiarity with some more advanced functional analysis, in particular with the spectral theory of (unbounded) self-adjoint operators on a Hilbert space. Some knowledge of distribution theory and Lie theory is also assumed. References to these topics are given in the text.

For terminology and notations we generally follow N. Bourbaki. Proofs following theorems, propositions and lemmas are written in small print. The index will be helpful to trace important notions defined in the text.

Thanks are due to my colleagues and students in several countries for their remarks and suggestions. Especial thanks are due to Dr. J. D. Stegeman (Utrecht) whose help in developing the final version of the manuscript has greatly improved the presentation.

Leiden, September 2009
Gerrit van Dijk

The picture shows Snellius' quadrant. Snellius (1580–1626), best known for Snell's law on the breaking of light, used this quadrant to measure the earth. It is now on display in the Museum Boerhaave at Leiden. The building of the Leiden Mathematical Institute is named after Snellius. A replica of the quadrant was offered to me on the occasion of my retirement and placed in the hall of the Snellius building.

Contents

Chapter 1
Fourier Series

Literature: [12], [25], [61].

1.1 Definition and elementary properties

We denote by $\mathbb{T}$ the unit circle in the complex plane. Functions on $\mathbb{T}$ can be identified with periodic functions on the real line $\mathbb{R}$, for example with period 1, via the mapping $x \mapsto e^{2\pi i x}$ $(x \in \mathbb{R})$. All functions are supposed to be complex valued. Set

- $L^1(\mathbb{T})$: the space of Lebesgue measurable periodic functions f on $\mathbb{R}$ with period 1, satisfying $\|f\|_1 = \int_0^1 |f(x)|dx < \infty$,
- $L^2(\mathbb{T})$: the space of Lebesgue measurable periodic functions f on $\mathbb{R}$ with period 1, satisfying $\|f\|_2 = (\int_0^1 |f(x)|^2 dx)^{1/2} < \infty$.

Notice that $L^2(\mathbb{T}) \subset L^1(\mathbb{T})$.

Let $f \in L^1(\mathbb{T})$. The nth *Fourier coefficient* a_n of f is defined by

$$a_n = \int_0^1 f(x)\, e^{-2\pi i n x} dx$$

where $n \in \mathbb{Z}$, the set of integers. We also write $a_n = a_n(f)$. The series $\sum_{n=-\infty}^{\infty} a_n\, e^{2\pi i n x}$ is called the *Fourier series* of f. The series does not need to converge for all x, and, moreover, if convergence takes place, the sum is not necessarily equal to $f(x)$.

Here are some *elementary properties*:

$$a_n(\alpha f + \beta g) = \alpha\, a_n(f) + \beta\, a_n(g) \text{ for all } f, g \in L^1(\mathbb{T}), \quad \alpha, \beta \in \mathbb{C} \text{ and } n \in \mathbb{Z}, \tag{1.1.1}$$

$$|a_n(f)| \le \|f\|_1 \text{ for all } f \in L^1(\mathbb{T}) \text{ and } n \in \mathbb{Z}. \tag{1.1.2}$$

If f is any function, integrable on the interval $[a, b]$, then one has

$$\lim_{|n|\to\infty} \int_a^b f(x)\, e^{-2\pi i n x} dx = 0.$$

This is easily shown by approximating f with continuously differentiable functions on $[a, b]$ and then applying partial integration. In particular we thus obtain

$$\lim_{|n|\to\infty} a_n(f) = 0 \text{ for all } f \in L^1(\mathbb{T}). \tag{1.1.3}$$

Furthermore we have:

$$\begin{aligned}&\text{If } f \in L^2(\mathbb{T}), \text{ then } \textstyle\sum_{n=-\infty}^{\infty} |a_n|^2 \text{ is convergent}\\ &\text{and one has } \textstyle\sum_{n=-\infty}^{\infty} |a_n|^2 \le \|f\|_2^2.\end{aligned} \tag{1.1.4}$$

The latter result is nothing but the *Bessel inequality* in the Hilbert space $L^2(\mathbb{T})$ with respect to the orthonormal system $\{e^{2\pi i n x}\}_{n\in\mathbb{Z}}$. In fact, as we will see later in Section 1.5, we have equality.

1.2 Convergence

In this section we start finding criteria for the convergence of the Fourier series of f.

Let $f \in L^1(\mathbb{T})$ and define $S_N(f, x) = \sum_{n=-N}^{N} a_n(f)\, e^{2\pi i n x}$. We may write

$$\begin{aligned} S_N(f,x) &= \sum_{n=-N}^{N} \int_0^1 f(t)\, e^{-2\pi i n(t-x)}\, dt \qquad (1.2.1)\\ &= \int_{x-\frac{1}{2}}^{x+\frac{1}{2}} f(t) \frac{\sin(2N+1)\pi(t-x)}{\sin\pi(t-x)}\, dt\\ &= \int_0^{1/2} \{f(x+y) + f(x-y)\} \frac{\sin(2N+1)\pi y}{\sin\pi y}\, dy. \end{aligned}$$

The function $y \mapsto \frac{\sin(2N+1)\pi y}{\sin \pi y}$ is called the *Dirichlet kernel* of S_N. If we take $f(x) = 1$, then we get $\int_0^{1/2} \frac{\sin(2N+1)\pi y}{\sin\pi y}\, dy = \frac{1}{2}$. Fourier already posed the following question. Does $S_N(f, x)$ converge to $f(x)$ at each point x? In almost all points x? In at least one point x?

We shall say that a function f on $\mathbb{R}$ has a *discontinuity of the first kind* at x if it is discontinuous at x and both one-sided limits $f(x+0) = \lim_{h\downarrow 0} f(x+h)$ and $f(x-0) = \lim_{h\uparrow 0} f(x+h)$ exist and are finite.

We have the following result:

Theorem 1.2.1. *If $f(x)$ is a periodic function with period 1, which is continuous on every finite interval, except for finitely many discontinuities of the first kind, then $S_N(f, x)$ converges for all x where the one-sided limits $f'(x+0) = \lim_{h\downarrow 0} \frac{f(x+h)-f(x+0)}{h}$ and $f'(x-0) = \lim_{h\uparrow 0} \frac{f(x+h)-f(x-0)}{h}$ exist and are* *finite to* $\frac{1}{2}\{f(x+0)+f(x-0)\}$. *In particular,* $S_N(f,x)$ *converges to* $f(x)$ *at all points where* f *is differentiable.*

From the conditions on f it follows that f is bounded and $f \in L^1(\mathbb{T})$. Then we have for all x, applying (1.2.1),

$$S_N(f,x) - \frac{f(x+0)+f(x-0)}{2} = \int_0^{1/2} \phi(y) \sin(2N+1)\pi y \, dy,$$

where

$$\phi(y) = \frac{\{f(x+y)-f(x+0)\}+\{f(x-y)-f(x-0)\}}{\sin \pi y}.$$

For every x for which $f'(x+0)$ and $f'(x-0)$ exist and are finite, the function ϕ (as a function of y) is continuous on $[0,\frac{1}{2}]$, except for at most finitely many discontinuities of the first kind. Hence ϕ is bounded on $[0,\frac{1}{2}]$ and integrable. The theorem now follows from the remarks just preceding (1.1.3).

Examples

1. $f(x) = x - \frac{1}{2}$ $(0 < x < 1)$, $f(0) = f(1) = 0$ and f is periodic with period 1. We obtain $a_0 = 0$, $a_n = -\frac{1}{2\pi i n}$ $(n \neq 0)$. Applying Theorem 1.2.1 gives $x - \frac{1}{2} = -\frac{1}{\pi}\sum_{n=1}^{\infty} \frac{\sin 2\pi n x}{n}$ for $0 < x < 1$. Taking $x = \frac{1}{4}$, we get
$$\frac{\pi}{4} = 1 - \frac{1}{3} + \frac{1}{5} - \frac{1}{7} + \frac{1}{9} - \cdots.$$
2. The "saw": $f(x) = |x - \frac{1}{2}|$ $(0 \leq x \leq 1)$ and periodic with period 1. We obtain: $a_0 = \frac{1}{4}$, $a_n = \frac{1-(-1)^n}{2\pi^2 n^2}$ $(n \neq 0)$. Applying Theorem 1.2.1 gives
$$\left|x - \frac{1}{2}\right| = \frac{1}{4} + \frac{2}{\pi^2}\sum_{k=0}^{\infty} \frac{\cos 2\pi(2k+1)x}{(2k+1)^2} \quad \text{for } 0 \leq x \leq 1.$$
Taking $x = 1$ we get $\sum_{k=0}^{\infty} \frac{1}{(2k+1)^2} = \frac{\pi^2}{8}$. Observe that the Fourier series converges uniformly to $f(x)$ in this case.

1.3 Uniform convergence

We now investigate the (uniform) convergence of the Fourier series itself.

Theorem 1.3.1. *If* f *is a periodic function with period* 1 *and Fourier coefficients* a_n, *continuous on each finite interval except for at most finitely many discontinuities of the first kind, then we can integrate* f *by integrating its Fourier series term* *by term. For all x we have*

$$\int_0^x f(t)\,dt = a_0 x + \sum_{n\neq 0} \frac{a_n}{2\pi i n}\,(\mathrm{e}^{2\pi i n x} - 1).$$

Let for $0 \le x \le 1$, $F(x) = \int_0^x \{f(t) - a_0\}dt$, so that $F(0) = F(1) = 0$ and $F'(x-0) = f(x-0) - a_0$, $F'(x+0) = f(x+0) - a_0$ for all x. Extend F periodically to $\mathbb{R}$.

Denote the Fourier coefficients of F by A_n. By partial integration we find, for $n \neq 0$,

$$A_n = \int_0^1 F(x)\,\mathrm{e}^{-2\pi i n x}\,dx = \frac{1}{2\pi i n}\int_0^1 f(x)\,\mathrm{e}^{-2\pi i n x}\,dx = \frac{a_n}{2\pi i n}.$$

From the inequality

$$\sum_{n=p}^{q}\left|\frac{a_n}{2\pi i n}\right| \le \left(\sum_{n=p}^{q}|a_n|^2\right)^{1/2}\left(\sum_{n=p}^{q}\frac{1}{4\pi^2 n^2}\right)^{1/2}$$

($0 < p \le q$ or $q \le p < 0$) it follows, taking into account that $f \in L^2(\mathbb{T})$ and (1.1.4), that the Fourier series of F converges absolutely and uniformly on $\mathbb{R}$. From Theorem 1.2.1 now it follows that $F(x) = \sum_{n=-\infty}^{\infty} A_n \mathrm{e}^{2\pi i n x}$ for all x. Hence, in particular, $A_0 + \sum_{n\neq 0} A_n = F(0) = 0$, so $A_0 = -\sum_{n\neq 0}\frac{a_n}{2\pi i n}$. For every x with $0 \le x \le 1$ we thus have

$$\int_0^x f(t)dt = F(x) + a_0 x = a_0 x + \sum_{n\neq 0}\frac{a_n}{2\pi i n}(\mathrm{e}^{2\pi i n x} - 1).$$

A function f on $\mathbb{R}$ is said to be *smooth* if:

(i) f is continuous,

(ii) f is, in addition, continuously differentiable except at a set of points of which each bounded interval contains only finitely many, and

(iii) at these points the left and right limit of the derivative f' exist.

Theorem 1.3.2. *Let f be periodic with period* 1 *and smooth. Then the Fourier series of f converges uniformly to $f(x)$.*

We shall apply Theorem 1.3.1. One clearly has $f(x) = \int_0^x f'(t)dt + f(0)$. Notice that f' fulfills the requirements of Theorem 1.3.1. So applying this theorem we get for all x with $0 \le x \le 1$,

$$f(x) - f(0) = a_0 x + \sum_{n\neq 0}\frac{a_n}{2\pi i n}\,(\mathrm{e}^{2\pi i n x} - 1).$$

Here a_n is the nth Fourier coefficient of f'. In particular we have $a_0 = \int_0^1 f'(t)dt = 0$. So

$$\begin{aligned} f(x) &= f(0) + \sum_{n\neq 0}\frac{a_n}{2\pi i n}\,(\mathrm{e}^{2\pi i n x} - 1) \\ &= \left(f(0) - \sum_{n\neq 0}\frac{a_n}{2\pi i n}\right) + \sum_{n\neq 0}\frac{a_n}{2\pi i n}\,\mathrm{e}^{2\pi i n x}. \end{aligned}$$

This series, with sum $f(x)$, converges absolutely and uniformly (see proof of Theorem 1.3.1) and is the Fourier series of f.

Example 2 (Section 1.2) provides a nice application of Theorem 1.3.2.

Remark 1.3.3. Alternatively, Theorem 1.3.2 can be proved without using Theorem 1.3.1 (which is of independent interest). Indeed, by Theorem 1.2.1 we have $f(x) = \lim_{N\to\infty} \sum_{n=-N}^{N} a_n e^{2\pi i n x}$ for all x. The series $\sum_{n=-\infty}^{\infty} |a_n|$ is clearly convergent, since for $n \neq 0$, $a_n = \frac{a_n(f')}{2\pi i n}$ (see above). Hence we have $f(x) = \sum_{n=-\infty}^{\infty} a_n e^{2\pi i n x}$, the series being absolutely and uniformly convergent.

1.4 Some facts about convergence of Fourier series

Throughout this section f is periodic with period 1.

1. There exists a continuous function f whose Fourier series diverges at one given point only (Du Bois–Reymond 1876, Fejér 1911).
2. There is a locally integrable function f for which the Fourier series diverges almost everywhere (Kolmogoroff 1923).
3. There is a locally integrable function f for which the Fourier series diverges everywhere (Kolmogoroff 1926).
4. Let E be a set of Lebesgue measure zero. There exists a *continuous* function f for which the Fourier series diverges on E only (Kahane, Katznelson 1965).
5. If $f \in L^2(\mathbb{T})$, then $S_N(f,x)$ converges a.e. to $f(x)$ (Carleson 1966), cf. [48, Chapter 7].
6. If $f \in L^p(\mathbb{T})$ $(p > 1)$, then $S_N(f,x)$ converges a.e. to $f(x)$ (Hunt 1968). A new proof has been given by Fefferman (1973), see [16].

1.5 Parseval's theorem

Let f be a continuously differentiable function on $\mathbb{R}$ with period 1. By Theorem 1.3.2 the Fourier series of f converges uniformly to $f(x)$. In particular we have:

Lemma 1.5.1. *Every periodic continuously differentiable function f can be uniformly approximated by trigonometric polynomials $\sum_{n=-M}^{N} a_n e^{2\pi i n x}$.*

Corollary 1.5.2. *The mapping $f \mapsto (a_n(f))_{n=-\infty}^{\infty}$ from $L^1(\mathbb{T})$ to the space $\mathbf{c}_0$ consisting of all sequences of complex numbers c_n with $\lim_{|n|\to\infty} c_n = 0$ is injective.*

This mapping is far from being surjective (see [12, p. 37]).

Corollary 1.5.3. *Let f be continuous and periodic with period 1 and Fourier coefficients a_n. If $\sum_{n=-\infty}^{\infty} |a_n| < \infty$, then $f(x) = \sum_{n=-\infty}^{\infty} a_n e^{2\pi i n x}$.*

We now turn to functions in $L^2(\mathbb{T})$. By Lemma 1.5.1, every $f \in L^2(\mathbb{T})$ can be approximated in L^2-norm by finite linear combinations $\sum_{n=-M}^{N} a_n e^{2\pi i n x}$. Therefore $\{e^{2\pi i n x}\}_{n\in\mathbb{Z}}$ is an orthonormal basis of the Hilbert space $L^2(\mathbb{T})$. This gives:

Theorem 1.5.4 (Parseval). *For every $f \in L^2(\mathbb{T})$ its Fourier series converges to f in L^2-norm. In particular*

$$\sum_{n\in\mathbb{Z}} |a_n(f)|^2 = \|f\|_2^2. \tag{1.5.1}$$

Example. Let $f(x) = x - \frac{1}{2}$ $(0 < x < 1)$, $f(0) = f(1) = 0$ and f periodic with period 1. Relation (1.5.1) then yields $\sum_{n=1}^{\infty} \frac{1}{n^2} = \frac{\pi^2}{6}$. Similarly, taking $f(x)^2$, we obtain $\sum_{n=1}^{\infty} \frac{1}{n^4} = \frac{\pi^4}{90}$, etc.

1.6 Generalization

The previous sections can be generalized to functions on $\mathbb{T}^n$, or, in other words, to functions $f = f(x_1, \dots, x_n)$ on $\mathbb{R}^n$ with the property $f(x_1 + k_1, \dots, x_n + k_n) = f(x_1, \dots, x_n)$ for all $k_1, \dots, k_n \in \mathbb{Z}$. Writing for $f \in L^1(\mathbb{T}^n)$

$$\int_{\mathbb{T}^n} f(x) dx = \int_0^1 \cdots \int_0^1 f(x_1, \dots, x_n)\, dx_1 \dots dx_n$$

and using the notation $(x, y) = \sum_{i=1}^{n} x_i y_i$ if $x = (x_1, \dots, x_n)$, $y = (y_1, \dots, y_n)$, the Fourier coefficients of $f \in L^1(\mathbb{T}^n)$ are given by

$$a_k = \int_{\mathbb{T}^n} f(x)\, e^{-2\pi i\,(x,k)} dx$$

where $k = (k_1, \dots, k_n) \in \mathbb{Z}^n$.

Let $\Delta = \frac{\partial^2}{\partial x_1^2} + \cdots + \frac{\partial^2}{\partial x_n^2}$ be the Laplace operator and assume that $\Delta^m f$ exists and is a continuous function for some $f \in L^1(\mathbb{T}^n)$. Then the Fourier series $\sum_{k\in\mathbb{Z}^n} a_k e^{2\pi i\,(x,k)}$ of f converges uniformly to $f(x)$ as soon as $m \geq [\frac{n}{2}]$. We leave this as an exercise.

Chapter 2
Fourier Integrals

Literature: [1], [6], [19], [44], [38], [25].

2.1 The convolution product

We first devote some words to the *convolution product* of functions on $\mathbb{R}$. All functions that we consider are again complex-valued.

Let f and g be two Lebesgue integrable functions on $\mathbb{R}$. Then the function $(x, y) \mapsto f(y)\, g(x-y)$ is measurable on $\mathbb{R}^2$ and one has

$$\int_{-\infty}^{\infty}\int_{-\infty}^{\infty} |f(y)\, g(x-y)| dx dy = \int_{-\infty}^{\infty} |f(y)| dy \int_{-\infty}^{\infty} |g(x)| dx = \|f\|_1 \|g\|_1.$$

By Fubini's theorem the integral

$$\int_{-\infty}^{\infty} f(y)\, g(x-y)\, dy \tag{2.1.1}$$

exists then for almost all x. We write

$$(f * g)\,(x) = \int_{-\infty}^{\infty} f(y)\, g(x-y)\, dy,$$

and call it the convolution product of f and g.

Also, by Fubini, $f * g$ is an integrable function and one has

$$\|f * g\|_1 \leq \|f\|_1 \, \|g\|_1.$$

From the well-known Hölder inequalities we conclude that the convolution product $f * g$, defined by the same formula (2.1.1), exists *everywhere* if $f \in L^p$ and $g \in L^q$ where $\frac{1}{p} + \frac{1}{q} = 1,\ p \geq 1$. Furthermore we have, in summary,

Theorem 2.1.1. (i) *If $f, g \in L^1(\mathbb{R})$, then $f * g \in L^1(\mathbb{R})$ and $\|f * g\|_1 \leq \|f\|_1 \|g\|_1$.*

(ii) *If $f \in L^p(\mathbb{R})$, $g \in L^q(\mathbb{R})$ ($\frac{1}{p} + \frac{1}{q} = 1,\ p > 1$), then $f * g$ is a continuous function which vanishes at $\pm\infty$. In addition we have $\|f * g\|_\infty \leq \|f\|_p \|g\|_q$.*

(iii) *If* $f \in L^1(\mathbb{R})$ *and* $g \in L^\infty(\mathbb{R})$, *then* $f * g$ *is a bounded continuous function. One has* $\|f * g\|_\infty \le \|f\|_1 \|g\|_\infty$.

The result in (ii) can be proved by approximating f and g by continuous functions with compact support. In (iii) we approximate f by such functions. In both cases one applies that continuous functions with compact support are uniformly continuous.

Remark 2.1.2. In the same way, by applying Minkowski's inequality, one shows: if $f \in L^1$ and $g \in L^p$ $(p \ge 1)$, then $f * g \in L^p$ and $\|f * g\|_p \le \|f\|_1 \, \|g\|_p$.

Notice that the convolution product, if it exists, is commutative. Let us define $\widetilde{f}(x) = \overline{f(-x)}$ $(x \in \mathbb{R})$. Then $L^1(\mathbb{R})$ is, with the addition, the L^1-norm, the convolution product and the *involution* $f \mapsto \widetilde{f}$, a standard example of a commutative *Banach algebra with involution.*

2.2 Elementary properties of the Fourier integral

We now introduce the *Fourier integral* and present some of its properties.

For $f \in L^1(\mathbb{R})$ we define

$$\widehat{f}(y) = \int_{-\infty}^{\infty} f(x)\,\mathrm{e}^{-2\pi i x y}\,dx \qquad (y \in \mathbb{R}). \tag{2.2.1}$$

We call $\widehat{f}$ the *Fourier transform* of f, also denoted by $\widehat{f} = \mathcal{F} f$.

We list some *elementary properties* (without proof):

(a) $(c_1 f_1 + c_2 f_2)\widehat{\ } = c_1 \widehat{f}_1 + c_2 \widehat{f}_2$ for $c_1, c_2 \in \mathbb{C}$ and $f_1, f_2 \in L^1(\mathbb{R})$.

(b) $|\widehat{f}(y)| \le \|f\|_1$, $\widehat{f}$ is continuous and $\lim_{|y| \to \infty} \widehat{f}(y) = 0$ for all $f \in L^1(\mathbb{R})$ (compare this with (1.1.3)). The latter result is often called the Riemann–Lebesgue lemma.

(c) $(f * g)\widehat{\ } = \widehat{f} \cdot \widehat{g}$ for all $f, g \in L^1(\mathbb{R})$.

(d) $(\widetilde{f})\widehat{\ } = \overline{\widehat{f}}$ for all $f \in L^1(\mathbb{R})$.

(e) Let $(L_t f)(x) = f(x - t)$, $(M_\rho f)(x) = f(\rho x)$, $\rho > 0$. Then

$$(L_t f)\widehat{\ }(y) = \mathrm{e}^{-2\pi i t y} \widehat{f}(y),$$

$$\left[\mathrm{e}^{2\pi i t x} f(x)\right]\widehat{\ }(y) = (L_t \widehat{f})(y),$$

$$(M_\rho f)\widehat{\ }(y) = \frac{1}{\rho} \widehat{f}\left(\frac{y}{\rho}\right)$$

for all $f \in L^1(\mathbb{R})$.

Examples

1. Let $A > 0$ and Φ_A the characteristic function of the closed interval $[-A, A]$. Then $\widehat{\Phi}_A(y) = \frac{\sin 2\pi A y}{\pi y}$ $(y \neq 0)$, $\widehat{\Phi}_A(0) = 2A$.
2. Set
$$\Delta(x) = \begin{cases} 1 - |x| & \text{for } |x| \leq 1, \\ 0 & \text{for } |x| > 1, \end{cases}$$
(triangle function). Then $\Delta = \Phi_{1/2} * \Phi_{1/2}$, so $\widehat{\Delta}(y) = \left(\frac{\sin \pi y}{\pi y}\right)^2$ $(y \neq 0)$, $\Delta(0) = 1$.
3. Let T be the trapezoidal function:
$$T(x) = \begin{cases} 1 & \text{if } |x| \leq 1, \\ 2 - |x| & \text{if } 1 < |x| \leq 2, \\ 0 & \text{if } |x| > 2. \end{cases}$$
Then $T = \Phi_{1/2} * \Phi_{3/2}$, hence $\widehat{T}(y) = \frac{\sin 3\pi y \, \sin \pi y}{(\pi y)^2}$ $(y \neq 0)$ and $\widehat{T}(0) = 3$.
4. Let $f(x) = \mathrm{e}^{-a|x|}, a > 0$. Then $\widehat{f}(y) = \frac{2a}{a^2 + 4\pi^2 y^2}$.
5. Let $f(x) = \mathrm{e}^{-ax^2}, a > 0$. Then we get by complex integration $\widehat{f}(y) = \sqrt{\frac{\pi}{a}}\, \mathrm{e}^{-\pi^2 y^2/a}$. In particular $\mathcal{F}(\mathrm{e}^{-\pi x^2}) = \mathrm{e}^{-\pi y^2}$.

For a proof we might refer to [41].

2.3 The inversion theorem

Theorem 2.3.1. *Let $f \in L^1(\mathbb{R})$ and x be a point where $f(x+0)$ and $f(x-0)$ exist. Then one has*
$$\lim_{\alpha \downarrow 0} \int_{-\infty}^{\infty} \widehat{f}(y)\, \mathrm{e}^{2\pi i x y}\, \mathrm{e}^{-4\pi^2 \alpha y^2}\, dy = \frac{f(x+0) + f(x-0)}{2}.$$

Step 1. One has
$$\begin{aligned}
&\int_{-\infty}^{\infty} \widehat{f}(y)\, \mathrm{e}^{2\pi i x y}\, \mathrm{e}^{-4\pi^2 \alpha y^2}\, dy \\
&\quad = \int_{-\infty}^{\infty} \int_{-\infty}^{\infty} f(t)\, \mathrm{e}^{2\pi i (x-t) y}\, \mathrm{e}^{-4\pi^2 \alpha y^2}\, dt\, dy \\
&\quad = \int_{-\infty}^{\infty} f(t) \left[\int_{-\infty}^{\infty} \mathrm{e}^{-4\pi^2 \alpha y^2}\, \mathrm{e}^{2\pi i (x-t) y}\, dy \right] dt \quad \text{by Fubini's theorem} \\
&\quad = \frac{1}{2\sqrt{\pi\alpha}} \int_{-\infty}^{\infty} f(t)\, \mathrm{e}^{(x-t)^2/4\alpha}\, dt = \frac{1}{2\sqrt{\pi\alpha}} \int_{-\infty}^{\infty} f(x+t)\, \mathrm{e}^{-t^2/4\alpha}\, dt.
\end{aligned}$$

Step 2. We now get

$$\begin{aligned}
&\int_{-\infty}^{\infty} \widehat{f}(y)\, e^{2\pi ixy}\, e^{-4\pi^2\alpha y^2} dy - \left\{\frac{f(x+0)+f(x-0)}{2}\right\} \\
&= \frac{1}{2\sqrt{\pi\alpha}} \int_0^{\infty} [\{f(x+t)-f(x-0)\} + \{f(x-t)-f(x-0)\}]\, e^{-t^2/4\alpha} dt \\
&= \frac{1}{2\sqrt{\pi\alpha}} \int_0^{\delta} \cdots + \frac{1}{2\sqrt{\pi\alpha}} \int_{\delta}^{\infty} \cdots = I_1 + I_2,
\end{aligned}$$

where $\delta > 0$ still has to be chosen.

Step 3. Set $\phi(t) = |f(x+t) - f(x+0) + f(x-t) - f(x-0)|$. Let $\varepsilon > 0$ be given. Choose δ such that $\phi(t) < \varepsilon/2$ for $|t| < \delta$, $t > 0$. This is possible, since $\lim_{t\downarrow 0} \phi(t) = 0$. Then we have $|I_1| \leq \varepsilon/2$.

Step 4. We now estimate I_2. We have

$$|I_2| \leq \frac{1}{2\sqrt{\pi\alpha}} \int_{\delta}^{\infty} \phi(t)\, e^{-t^2/4\alpha} dt \leq \frac{2\sqrt{\alpha}}{\sqrt{\pi}} \int_{\delta}^{\infty} \frac{\phi(t)}{t^2} dt.$$

Now $\int_{\delta}^{\infty} \frac{\phi(t)}{t^2} dt$ is bounded, because

$$\begin{aligned}
\int_{\delta}^{\infty} \frac{\phi(t)}{t^2} dt &\leq \int_{\delta}^{\infty} \frac{|f(x+t)|}{t^2} dt + \int_{\delta}^{\infty} \frac{|f(x-t)|}{t^2} dt + \int_{\delta}^{\infty} \frac{|f(x+0)-f(x-0)|}{t^2} dt \\
&\leq \frac{2}{\delta^2} \|f\|_1 + \frac{1}{\delta} \{|f(x+0)-f(x-0)|\}.
\end{aligned}$$

Hence $\lim_{\alpha\downarrow 0} I_2 = 0$, so $|I_2| < \varepsilon/2$ as soon as α is small, say $0 < \alpha < \eta$.

Step 5. Summarizing: if $0 < \alpha < \eta$, then

$$\left| \int_{-\infty}^{\infty} \widehat{f}(y)\, e^{2\pi ixy}\, e^{-4\pi^2\alpha y^2} dy - \frac{f(x+0)+f(x-0)}{2} \right| < \varepsilon.$$

Remark 2.3.2. (a) Under the conditions of Theorem 2.3.1 one shows in a similar way

$$\lim_{\alpha\downarrow 0} \int_{-\infty}^{\infty} \widehat{f}(t)\, e^{2\pi ixt}\, e^{-\alpha|t|} dt = \frac{f(x+0)+f(x-0)}{2},$$

and also

$$\lim_{\alpha\to\infty} \int_{-\alpha}^{\alpha} \left(1 - \frac{|t|}{\alpha}\right) \widehat{f}(t)\, e^{2\pi ixt} dt = \frac{f(x+0)+f(x-0)}{2}.$$

(b) One can show, by careful reconsidering our proof, that for every $f \in L^1$ the three above limits converge almost everywhere to $f(x)$. One uses the so-called *Lebesgue set* of f. See [19], [44].

(c) If $f \in L^1$ and if x is a point where $f(x+0)$, $f(x-0)$, $f'(x+0)$ and $f'(x-0)$ exist, then one can show, almost in the same way as in the proof of Theorem 1.2.1, that

$$\lim_{R\to\infty} \int_{-R}^{R} \widehat{f}(t)\, e^{2\pi ixt}\, dt = \frac{1}{2}\,[f(x+0) + f(x-0)].$$

Corollary 2.3.3. *If both f and $\widehat{f}$ belong to L^1 and f is continuous in x, then*

$$\int_{-\infty}^{\infty} \widehat{f}(y)\, e^{2\pi ixy}\, dy = f(x).$$

Corollary 2.3.4. *Let $f \in L^1$ and $\widehat{f} \geq 0$. If f is continuous in $x = 0$, then one has $\widehat{f} \in L^1$ and therefore*

$$f(0) = \int_{-\infty}^{\infty} \widehat{f}(t)\, dt.$$

Corollary 2.3.3 is easily proved by applying Lebesgue's theorem on dominated convergence; Corollary 2.3.4 is a consequence of Fatou's lemma.

Remark 2.3.5. When does $\widehat{f}$ belong to L^1? Here differentiability of f plays a role (as in the case of Fourier series). One has for example:

If $f \in L^1$, *f continuously differentiable*, $f' \in L^1 \cap L^2$, then $\widehat{f'}(y) = (2\pi iy)\,\widehat{f}(y)$, hence $\widehat{f} \in L^1$ (use Theorem 2.4.1 for example).

Applying this to $\varphi \in C_c^{(1)}(\mathbb{R})$, we easily get $\widehat{\varphi} \in L^1$ and, using Corollary 2.3.3 for φ,

$$\int_{-\infty}^{\infty} f(x)\,\varphi(x)dx = \int_{-\infty}^{\infty} \widehat{f}(-x)\,\widehat{\varphi}(x)dx$$

for all $f \in L^1$. Hence, if $\widehat{f} = 0$, then $f = 0$ almost everywhere. Consequently:

Corollary 2.3.6. *The mapping $\mathcal{F} : L^1 \to \mathcal{C}_0(\mathbb{R})$ is injective.*

Here $\mathcal{C}_0(\mathbb{R})$ is the complex vector space of continuous functions f on $\mathbb{R}$ satisfying $\lim_{|x|\to\infty} f(x) = 0$.

2.4 Plancherel's theorem

Theorem 2.4.1. *If $f \in L^1 \cap L^2$, then $\widehat{f} \in L^2$ and $\|f\|_2 = \|\widehat{f}\|_2$.*

Let f be a function in $L^1 \cap L^2$. Then $f * \widetilde{f}$ is in L^1 and is a continuous function. Moreover $(f * \widetilde{f})\widehat{\ } = |f|^2$. According to Corollary 2.3.4 we have $|\widehat{f}|^2 \in L^1$, hence $\widehat{f} \in L^2$. Furthermore

$$f * \widetilde{f}(0) = \int_{-\infty}^{\infty} |f(x)|^2 dx = \int_{-\infty}^{\infty} |\widehat{f}(y)|^2 dy.$$

So $\|f\|_2 = \|\widehat{f}\|_2$.

The Fourier transform is therefore an *isometric linear mapping*, defined on the dense linear subspace $L^1 \cap L^2$ of L^2. Let us extend this mapping $\mathcal{F}$ to L^2, for instance by

$$\mathcal{F} f(x) = \lim_{k\to\infty} \int_{-k}^{k} f(t)\, e^{-2\pi i x t}\, dt \qquad (L^2\text{-convergence}).$$

We then have:

Theorem 2.4.2 (Plancherel's theorem). *The Fourier transform is a unitary operator on L^2.*

For any $f, g \in L^1 \cap L^2$ one has

$$\int_{-\infty}^{\infty} f(x)\widehat{g}(x)dx = \int_{-\infty}^{\infty} \widehat{f}(x)g(x)dx.$$

Since both $\int_{-\infty}^{\infty} f(x)(\mathcal{F} g)(x)dx$ and $\int_{-\infty}^{\infty} (\mathcal{F} f)(x)g(x)dx$ are defined for $f, g \in L^2$, and because

$$\left|\int_{-\infty}^{\infty} f(x)(\mathcal{F} g)(x)dx\right| \le \|f\|_2\|\mathcal{F} g\|_2 = \|f\|_2\|g\|_2,$$

$$\left|\int_{-\infty}^{\infty} (\mathcal{F} f)(x)g(x)dx\right| \le \|\mathcal{F} f\|_2\|g\|_2 = \|f\|_2\|g\|_2,$$

we have for any $f, g \in L^2$

$$\int_{-\infty}^{\infty} f(x)(\mathcal{F} g)(x)dx = \int_{-\infty}^{\infty} (\mathcal{F} f)(x)g(x)dx.$$

Notice that $\mathcal{F}(L^2) \subset L^2$ is a closed linear subspace. We have to show that $\mathcal{F}(L^2) = L^2$. Let $g \in L^2$, g orthogonal to $\mathcal{F}(L^2)$, so $\int_{-\infty}^{\infty} (\mathcal{F} f)(x)g(x)dx = 0$ for all $f \in L^2$. Then, by the above arguments, $\int_{-\infty}^{\infty} f(x)(\mathcal{F} g)(x)dx = 0$ for all $f \in L^2$, hence $\mathcal{F}(g) = 0$. Since $\|g\|_2 = \|\mathcal{F} g\|_2$, we get $g = 0$. So $\mathcal{F}(L^2) = L^2$.

2.5 The Poisson summation formula

This formula, which is of independent interest, plays an important role in algebraic number theory.

Theorem 2.5.1. *Let f be a continuous L^1-function such that*

(i) $\sum_{n\in\mathbb{Z}} |f(x+n)|$ *is uniformly convergent on* $[0,1]$,

(ii) $\sum_{n\in\mathbb{Z}} |\widehat{f}(n)|$ *is convergent.*

Then one has $\sum_{n\in\mathbb{Z}} f(n) = \sum_{n\in\mathbb{Z}} \widehat{f}(n)$.

The function $\phi(x) = \sum_{n\in\mathbb{Z}} f(x+n)$ is periodic with period 1, and continuous (by (i)). Its Fourier coefficients are

$$a_n = \int_0^1 \phi(t)\,\mathrm{e}^{-2\pi i n t}\,dt = \sum_{k\in\mathbb{Z}} \int_0^1 f(t+k)\,\mathrm{e}^{-2\pi i n t}\,dt$$

$$= \sum_{k\in\mathbb{Z}} \int_k^{k+1} f(t)\,\mathrm{e}^{-2\pi i n t}\,dt = \widehat{f}(n).$$

By (ii) we have $\sum |a_n| < \infty$; hence, since ϕ is continuous,

$$\phi(x) = \sum_{n\in\mathbb{Z}} a_n\,\mathrm{e}^{2\pi i n x} \quad \text{(by Corollary 1.5.3).}$$

So we get

$$\sum_{n\in\mathbb{Z}} \widehat{f}(n)\,\mathrm{e}^{2\pi i n x} = \phi(x) = \sum_{n\in\mathbb{Z}} f(x+n).$$

The theorem now follows by taking $x = 0$.

Example. Take $f(x) = \mathrm{e}^{-ax^2}$ with $a > 0$. We get

$$\sum_{n\in\mathbb{Z}} \mathrm{e}^{-an^2} = \sqrt{\frac{\pi}{a}} \sum_{n\in\mathbb{Z}} \mathrm{e}^{-\pi^2 n^2/a}.$$

This gives a well-known identity for *theta* functions in algebraic number theory.

2.6 The Riemann–Stieltjes integral and functions of bounded variation

We introduce the *Riemann–Stieltjes integral* for functions from $\mathbb{R}$ to $\mathbb{R}$. Our approach resembles that in [59, Chapter I].

We begin with the Riemann–Stieltjes integral on a closed interval $[a,b]$. From now on g is an increasing function on $[a,b]$. Let f be a *bounded* function on $[a,b]$.

We denote by V a partition of $[a, b]$:

$$a = x_0 < x_1 < \cdots < x_n = b.$$

Define

$$S_V = \sum_{i=1}^{n} (\sup_i f)\,(g(x_i) - g(x_{i-1})),$$

$$s_V = \sum_{i=1}^{n} (\inf_i f)\,(g(x_i) - g(x_{i-1})).$$

Here

$$\sup_i f = \sup\{f(x) : x_{i-1} \le x \le x_i\}$$

and

$$\inf_i f = \inf\{f(x) : x_{i-1} \le x \le x_i\}.$$

Set $S = \inf_V S_V$, $s = \sup_V s_V$. The function f is called Riemann–Stieltjes (R-S) integrable with respect to g if $S = s$. The common value is denoted by

$$\int_a^b f(x)\,dg(x) \quad \text{or} \quad \int_a^b f\,dg.$$

The function g is called a *weight function*.

The following theorem can be proved similarly to the case $g(x) = x$, the case of the Riemann integral.

Theorem 2.6.1. (i) *The function f is R-S integrable with respect to g if and only if for any $\varepsilon > 0$ there is a partition V of $[a, b]$ such that $S_V - s_V < \varepsilon$.*

(ii) *If f_1 and f_2 are R-S integrable with respect to g, then so is $\alpha_1 f_1 + \alpha_2 f_2$ for all $\alpha_1, \alpha_2 \in \mathbb{R}$. Moreover*

$$\int_a^b (\alpha_1 f_1 + \alpha_2 f_2)\,dg = \alpha_1 \int_a^b f_1\,dg + \alpha_2 \int_a^b f_2\,dg.$$

(iii) *If f_1, f_2 are R-S integrable with respect to g and $f_1 \le f_2$ then $\int_a^b f_1\,dg \le \int_a^b f_2\,dg$.*

(iv) *If f is R-S integrable with respect to g, then so is $|f|$ and*

$$\left| \int_a^b f\,dg \right| \le \int_a^b |f|\,dg.$$

(v) *Any continuous function on $[a, b]$ is R-S integrable on $[a, b]$.*

Here are some more properties.

Theorem 2.6.2. (i) *Let g be a continuously differentiable weight function. Then for all continuous functions f on $[a,b]$ one has*

$$\int_a^b f\,dg = \int_a^b fg'\,dx.$$

(ii) *Let g_1 and g_2 be two weight functions on $[a,b]$ and α_1, α_2 positive real numbers. Then $\alpha_1 g_1 + \alpha_2 g_2$ is a weight function on $[a,b]$ and*

$$\int_a^b f\,d(\alpha_1 g_1 + \alpha_2 g_2) = \alpha_1 \int_a^b f\,dg_1 + \alpha_2 \int_a^b f\,dg_2,$$

for all continuous functions f.

(iii) *Let g be a weight function on $[a,b]$. The set of points of discontinuity of g is countable.*

(iv) *Let g and h be two weight functions on $[a,b]$. Let $g(x) = h(x)$ at all points x where g is continuous. Then one has*

$$\int_a^b f\,dg - \int_a^b f\,dh = f(b)\,[g(b) - h(b)] - f(a)\,[g(a) - h(a)],$$

for all continuous functions f.

(v) *Let g be a weight function on $[a,b]$. Then $\int_a^b f\,dg = 0$ for all continuous functions f if and only if $g(x) =$* constant.

Clearly, (i) is easily shown applying the mean value theorem. For (iii) one considers the union of the sets $V_n = \{x : g(x+0) - g(x-0) \geq 1/n\}$. The proof of the remaining properties is easy and is left to the reader.

Clearly, one may assume that weight functions g are normalized by $g(a) = 0$ and g left continuous, applying the above properties (iv) and (v). For such weight functions g_1, g_2 one has

$$\int_a^b f\,dg_1 = \int_a^b f\,dg_2 \text{ for all continuous } f \text{ if and only if } g_1 = g_2. \tag{2.6.1}$$

A linear form μ on $C[a,b]$, the space of real-valued continuous functions on $[a,b]$, is called positive if $\mu(f) \geq 0$ when $f \geq 0$.

Theorem 2.6.3 (Representation theorem of F. Riesz, without proof). *Let μ be a positive linear form on $C[a,b]$. Then there exists a weight function g on $[a,b]$ such that*

$$\mu(f) = \int_a^b f\,dg$$

for $f \in C[a,b]$.

Definition 2.6.4. Let g be a real function on $[a,b]$.

(i) The *total variation* $V_a^b(g)$ of g on $[a,b]$ is defined by

$$V_a^b(g) = \sup\left\{\sum_{i=1}^{n} |g(x_i)-g(x_{i-1})| : a = x_0 < x_1 < \cdots < x_n = b,\ n \geq 1\right\}.$$

(ii) The function g is said to be of *bounded variation* (b.v.) if $V_a^b(g) < \infty$.

Here are some properties of functions of bounded variation:

(a) Any monotone function $g : [a,b] \to \mathbb{R}$ is of b.v. and $V_a^b(g) = |g(b) - g(a)|$.

(b) A continuously differentiable function g is of b.v. and $V_a^b(g) \leq M(b-a)$, where $M = \sup\{|g'(x)| : x \in [a,b]\}$.

(c) Not every continuous function is of b.v. Consider for example on $[0,1]$ the function

$$g(x) = x \sin(1/x) \text{ if } x \neq 0,\ g(0) = 0.$$

(d) If $g : [a,b] \to \mathbb{R}$ is of b.v., then g is bounded.

(e) If $g : [a,b] \to \mathbb{R}$ is of b.v., then g is Riemann-integrable.

(f) If $g : [a,b] \to \mathbb{R}$ is of b.v., then for all $c \in [a,b]$,

$$V_a^c(g) + V_c^b(g) = V_a^b(g).$$

Lemma 2.6.5. *A function $g : [a,b] \to \mathbb{R}$ is of b.v. if and only if there are weight functions g_1 and g_2 on $[a,b]$ such that $g = g_1 - g_2$.*

By property (a) above, weight functions are of b.v. So if g_1 and g_2 are weight functions, $g = g_1 - g_2$ is of b.v. Conversely, let g be of b.v. Then choosing $g_1(x) = V_a^x(g)$ and $g_2 = g_1 - g$, g_1 and g_2 are weight functions.

Let now g be of b.v. and select weight functions g_1 and g_2 such that $g = g_1 - g_2$. For continuous functions $f : [a,b] \to \mathbb{R}$ one defines

$$\int_a^b f\,dg = \int_a^b f\,dg_1 - \int_a^b f\,dg_2. \tag{2.6.2}$$

This definition is clearly independent of the selected splitting $g = g_1 - g_2$ of g.

In particular one may assume $g(a) = 0$ and g left continuous.

One can even write $g(x) = g_1(x) - g_2(x)$ with weight functions g_1 and g_2 satisfying $V_a^x(g) = g_1(x) + g_2(x)$. Indeed, choose

$$g_1(x) = \frac{1}{2}[V_a^x(g) + g(x)],$$

$$g_2(x) = \frac{1}{2}[V_a^x(g) - g(x)].$$

Then we immediately obtain $V_a^b(g) = V_a^b(g_1) + V_a^b(g_2)$. In particular

$$\left| \int_a^b f dg \right| \leq \|f\|_\infty \cdot V_a^b(g) \tag{2.6.3}$$

for all $f \in C[a,b]$.

We are now going to define the Riemann–Stieltjes integral on the line $\mathbb{R}$.

Let g be a function on $\mathbb{R}$, which is of b.v. on every closed bounded interval $[a,b]$. Denote by $C_c(\mathbb{R})$ the linear space of all real-valued continuous functions on $\mathbb{R}$ with compact support. For any $f \in C_c(\mathbb{R})$ the integral

$$\int_{-\infty}^{\infty} f\,dg \quad \left(= \int_a^b f\,dg \quad \text{if } \operatorname{Supp} f \subset [a,b]\right)$$

exists and one has

$$\left| \int_a^b f\,dg \right| \leq \|f\|_\infty V_a^b(g)$$

if $\operatorname{Supp} f \subset [a,b]$. Observe that $\int_{-\infty}^{\infty} f\,dg$ is well-defined. The function g is said to be *of bounded variation on* $\mathbb{R}$ if

$$\sup_{a,b} V_a^b(g) < \infty.$$

Clearly, such a function g is bounded. Again we can write g as a difference of two bounded weight functions. Moreover one can then easily define

$$\int_{-\infty}^{\infty} f\,dg$$

for any *bounded* continuous function f.

The number $V(g) = \sup_{a,b} V_a^b(g)$ is called the *total variation* of g. Again one has

$$\left| \int_{-\infty}^{\infty} f\,dg \right| \leq \|f\|_\infty V(g).$$

Here are some results (which we state without proof).

Theorem 2.6.6. (i) *Let μ be a real continuous linear form on $C[a,b]$, provided with the supremum norm. There exists a function g of b.v. on $C[a,b]$ such that $\mu(f) = \int_a^b f\,dg$ for all $f \in C[a,b]$. Furthermore, $\|\mu\| = V_a^b(g)$.*

(ii) *Let μ be a continuous linear form on $C_c(\mathbb{R})$, provided with the supremum norm. There exists a function g of b.v. on $\mathbb{R}$ such that $\mu(f) = \int_{-\infty}^{\infty} f\,dg$ for all $f \in C_c(\mathbb{R})$. Furthermore, $\|\mu\| = V(g)$.*

Remark 2.6.7. The above theory has motivated Bourbaki to give the following definition, see [5, Chapter IV] and also Chapter 4.

- Let X be a locally compact topological space and $C_c(X)$ the linear space of the real-valued continuous functions on X with compact support.
- A *measure* μ on X is a real linear form on $C_c(X)$ such that for any compact subset $K \subset X$,
$$|\mu(f)| \le C_K \|f\|_\infty$$
for all f with $\operatorname{Supp} f \subset K$. Here C_K is a constant, only depending on the choice of K.
- The measure is said to be *bounded* if C_K can be chosen independently of K, hence if μ is a continuous linear form on $C_c(X)$ with respect to the supremum norm.

Up till now we have been considering integration of real-valued functions. In an evident way one can extend the theory to complex-valued functions.

Finally we give two important theorems (see [59]).

Theorem 2.6.8. *Let $g_1, g_2, \dots$ be left continuous weight functions on $\mathbb{R}$ such that $|g_n(t)| \le M$ for all n and t ($M > 0$, a constant). There exists a subsequence $n_1 < n_2 < n_3 < \cdots$ and a weight function g such that*
$$\lim_{k\to\infty} g_{n_k}(t) = g(t) \quad (t \in \mathbb{R}).$$
Moreover,
$$\lim_{k\to\infty} \int_{-\infty}^{\infty} f\,dg_{n_k} = \int_{-\infty}^{\infty} f\,dg$$
for all $f \in C_c(\mathbb{R})$.

Theorem 2.6.9. *Let $f, f_1, f_2, \dots$ be continuous functions on $\mathbb{R}$, $|f_n(t)| \le M$ for some $M > 0$ and all n, t and let $\lim_{n\to\infty} f_n(t) = f(t)$ for all $t \in \mathbb{R}$. Let g be a function of b.v. on $\mathbb{R}$. Then*
$$\lim_{n\to\infty} \int_{-\infty}^{\infty} f_n(t)\,dg(t) = \int_{-\infty}^{\infty} f(t)\,dg(t).$$

Theorem 2.6.8 is a consequence of a theorem from functional analysis saying that the unit ball in the dual of a normed space is sequentially compact in the weak topology. Theorem 2.6.9 is a variant of Lebesgue's theorem on dominated convergence applied to the measure corresponding to g (i.e. to dg_1 and dg_2 if $g = g_1 - g_2$).

2.7 Bochner's theorem

Definition 2.7.1. Let φ be a complex-valued function on $\mathbb{R}$. It is said to be positive-definite if for any n and any n-tuple of real numbers $x_1, \dots, x_n$ and any n-tuple of complex numbers $\lambda_1, \dots, \lambda_n$ the following inequality holds:

$$\sum_{i,j=1}^{n} \lambda_i \overline{\lambda_j}\, \varphi(x_i - x_j) \geq 0.$$

We shall denote by **P** the set of all positive-definite (p.d.) functions φ on $\mathbb{R}$.

Here are some examples of functions in **P**: $\varphi = 1$, $\varphi = u * \widetilde{u}$ ($u \in L^2$), $\varphi_t(x) = \mathrm{e}^{-2\pi i x t}$ ($x \in \mathbb{R}, t \in \mathbb{R}$).

Some properties. Let $\varphi \in \mathbf{P}$. Then

(a) $\varphi(0) \geq 0$,

(b) $|\varphi(x)| \leq \varphi(0)$ for all x,

(c) $\varphi = \widetilde{\varphi}$.

To prove these properties, consider the positive-definite matrix

$$\begin{pmatrix} \varphi(0) & \varphi(x) \\ \varphi(-x) & \varphi(0) \end{pmatrix}.$$

Let g be a bounded weight function on $\mathbb{R}$. Then the integral

$$\varphi(x) = \int_{-\infty}^{\infty} \mathrm{e}^{-2\pi i x t}\, dg(t)$$

exists for all $x \in \mathbb{R}$ and φ is continuous (apply Theorem 2.6.9).

Theorem 2.7.2. *Let g be a bounded weight function on $\mathbb{R}$. Then*

$$\varphi(x) = \int_{-\infty}^{\infty} \mathrm{e}^{-2\pi i x t}\, dg(t)$$

is a continuous function in **P**.

We only have to show that φ is p.d. So let $x_1,\dots,x_n$ in $\mathbb{R}$, $\lambda_1,\dots,\lambda_n$ in $\mathbb{C}$. Then

$$\begin{aligned}\sum_{j,k=1}^{n} \lambda_j\overline{\lambda_k}\,\varphi(x_j-x_k) &= \sum_{j,k=1}^{n} \lambda_j\overline{\lambda_k}\int_{-\infty}^{\infty} \mathrm{e}^{-2\pi i x_j t}\mathrm{e}^{2\pi i x_k t}\,dg(t)\\ &= \int_{-\infty}^{\infty}\left(\sum_{j=1}^{n}\lambda_j \mathrm{e}^{-2\pi i x_j t}\right)\left(\sum_{k=1}^{n}\overline{\lambda_k}\mathrm{e}^{2\pi i x_k t}\right)dg(t)\\ &= \int_{-\infty}^{\infty}\left|\sum_{j=1}^{n}\lambda_j \mathrm{e}^{2\pi i x_j t}\right|^2 dg(t)\ge 0.\end{aligned}$$

This is one part of Bochner's theorem. The second part says:

Theorem 2.7.3. *Let $\varphi \in \mathbf{P}$ be measurable (so $\varphi \in L^\infty$). Then there exists a bounded weight function g on $\mathbb{R}$ such that*

$$\varphi(x) = \int_{-\infty}^{\infty} \mathrm{e}^{-2\pi i x t}\,dg(t) \quad (a.e.).$$

Bochner himself assumed φ continuous and showed that the equality holds everywhere. F. Riesz weakened the condition on φ in the form as stated in the theorem.

Corollary 2.7.4. *Any p.d. function $\varphi \in L^\infty \cap \mathbf{P}$ coincides a.e. with a continuous p.d. function.*

We shall show Theorem 2.7.3. We begin with a few lemmas.

Lemma 2.7.5. *Let $\varphi \in \mathbf{P}$. For any $\varepsilon > 0$ the function ψ given by $\psi(x) = \mathrm{e}^{-\varepsilon x^2}\varphi(x)$ $(x \in \mathbb{R})$ is in $\mathbf{P}$.*

It is well known that

$$\mathrm{e}^{-\varepsilon x^2} = \sqrt{\frac{\pi}{\varepsilon}}\int_{-\infty}^{\infty} \mathrm{e}^{-\pi^2 t^2/\varepsilon}\,\mathrm{e}^{2\pi i x t}\,dt.$$

Hence

$$\begin{aligned}&\sqrt{\frac{\varepsilon}{\pi}}\sum_{j,k}\lambda_j\overline{\lambda_k}\psi(x_j-x_k)\\ &\quad= \sum_{j,k}\lambda_j\overline{\lambda_k}\varphi(x_j-x_k)\cdot\int_{-\infty}^{\infty}\mathrm{e}^{-\pi^2t^2/\varepsilon}\,\mathrm{e}^{2\pi i(x_j-x_k)t}\,dt\\ &\quad= \int_{-\infty}^{\infty}\mathrm{e}^{-\pi^2t^2/\varepsilon}\left\{\sum_{j,k}\left(\lambda_j\mathrm{e}^{2\pi i x_j t}\right)\left(\overline{\lambda_k}\mathrm{e}^{-2\pi i x_k t}\right)\varphi(x_j-x_k)\right\}dt \ge 0.\end{aligned}$$

Lemma 2.7.6. *Let $\varphi \in L^1 \cap \mathbf{P}$. There exists $\psi \in L^2$ such that*

(a) $\varphi = \mathcal{F}\psi$,

(b) $\psi(x) = \widehat{\varphi}(-x)$ *a.e.,*

(c) $\psi(x) \geq 0$.

Since $|\varphi|^2 \leq \varphi(0)\,|\varphi|$, the function φ belongs to $L^2 \cap L^1$. By Parseval's theorem there is $\psi \in L^2$ with $\varphi = \mathcal{F}\psi$. Property (b) is now immediately seen by applying the inverse Fourier transform $\overline{\mathcal{F}}$: $\overline{\mathcal{F}}(\mathcal{F})\psi(x) = \psi(x) = \overline{\mathcal{F}}\varphi(x) = \widehat{\varphi}(-x)$ as functions in L^2, so a.e. It remains to show property (c): $\widehat{\varphi}(x) \geq 0$ for all x. We know that

$$\sum_{j,k=1}^{n} \lambda_j \overline{\lambda_k}\, \varphi(x_j - x_k) \geq 0 \quad \text{and hence} \quad \sum_{j,k=1}^{n} \mathrm{e}^{-2\pi i t(x_j - x_k)} \varphi(x_j - x_k) \geq 0.$$

Integrating this expression with respect to $x_1, \dots, x_n$ from 0 to N yields:

- the terms with $j = k$ give $nN^n\varphi(0)$,
- the terms with $j \neq k$ give $n(n-1)N^{n-2} \int_0^N \int_0^N \mathrm{e}^{-2\pi i t(x-u)} \varphi(x-u)\, dx du$.

Together we get

$$n\varphi(0)N^n + n(n-1)N^{n-2} \int_0^N \int_0^N \mathrm{e}^{-2\pi i t(x-u)} \varphi(x-u)\, dx du \geq 0.$$

Divide this expression by $n(n-1)N^{n-2}$ and let $n \to \infty$; we obtain

$$\frac{1}{N} \int_0^N \int_0^N \mathrm{e}^{-2\pi i t(x-u)} \varphi(x-u)\, dx du \geq 0.$$

We shall compute this expression. We have

$$\frac{1}{N} \int_0^N du \int_0^N \mathrm{e}^{-2\pi i t(x-u)} \varphi(x-u)\, dx = \frac{1}{N} \int_0^N du \int_{-u}^{N-u} \mathrm{e}^{-2\pi i t x} \varphi(x)\, dx.$$

Changing the order of integration in the latter expression yields

$$\begin{aligned}
&\frac{1}{N} \int_0^N \mathrm{e}^{-2\pi i t x} \varphi(x)\, dx \cdot \int_0^{N-x} du + \frac{1}{N} \int_{-N}^0 \mathrm{e}^{-2\pi i t x} \varphi(x)\, dx \cdot \int_{-x}^N du \\
&= \int_0^N \mathrm{e}^{-2\pi i t x} \varphi(x) \Big(1 - \frac{x}{N}\Big)\, dx + \int_{-N}^0 \mathrm{e}^{-2\pi i t x} \varphi(x) \Big(1 - \frac{x}{N}\Big)\, dx \\
&= \int_{-N}^N \mathrm{e}^{-2\pi i x t} \varphi(x) \Big(1 - \frac{|x|}{N}\Big)\, dx \\
&= \int_{-\infty}^{\infty} \mathrm{e}^{-2\pi i t x} \varphi(x) \Delta_N(x)\, dx,
\end{aligned}$$

where $\Delta_N(x) = \Delta(\frac{x}{n})$ (for the definition of the triangle function Δ, see Section 2.2). Therefore

$$\int_{-\infty}^{\infty} \mathrm{e}^{-2\pi itx}\varphi(x)\Delta_N(x)\,dx \geq 0 \quad \text{for all } N > 0.$$

Now $\lim_{N\to\infty}\Delta_N(x) = 1$ for all x. By Lebesgue's theorem on dominated convergence one obtains, since $\varphi \in L^1$,

$$\int_{-\infty}^{\infty} \mathrm{e}^{-2\pi itx}\varphi(x)\,dx \geq 0 \quad \text{for all } t \in \mathbb{R}.$$

Lemma 2.7.7. *Set $\delta_N = \widehat{\Delta}_N$. Then for any $\varphi \in L^\infty$ one has*

$$\lim_{N\to\infty} \delta_N * \varphi(x) = \varphi(x) \text{ a.e.}$$

Recall that $\delta_N(x) = \frac{1}{N}\left(\frac{\sin \pi N x}{\pi x}\right)^2$ if $x \neq 0$, $\delta_N(0) = N$.

(a) Assume firstly that $\varphi \in L^1$. Then one has

$$\begin{aligned}\lim_{N\to\infty} \delta_N * \varphi(x) &= \lim_{N\to\infty} \widehat{\Delta}_N * \varphi(x)\\ &= \lim_{N\to\infty} \int_{-N}^{N}\int_{-\infty}^{\infty}\left(1-\frac{|t|}{N}\right)\mathrm{e}^{-2\pi iyt}\varphi(x-y)\,dt\,dy.\end{aligned}$$

Changing the order of integration (Fubini's theorem) gives

$$\lim_{N\to\infty} \delta_N * \varphi(x) = \lim_{N\to\infty}\int_{-N}^{N}\left(1-\frac{|t|}{N}\right)\mathrm{e}^{2\pi ixt}\widehat{\varphi}(t)\,dt = \varphi(x) \text{ a.e.}$$

by Remarks 2.3.2 (a) and (b).

(b) Let now $\varphi \in L^\infty$. Set for $s > 0$,

$$\varphi_s(t) = \begin{cases}\varphi(t) & \text{if } |t| \leq s,\\ 0 & \text{if } |t| > s.\end{cases}$$

Then $\varphi_s \in L^1$. Hence, according to (a),

$$\lim_{N\to\infty}\int_{-s}^{s}\delta_N(x-t)\,\varphi(t)\,dt = \varphi(x) \quad \text{for } |x| \leq s, \text{ a.e.}$$

Furthermore we have

$$\begin{aligned}\left|\int_{|t|\geq s}\delta_N(x-t)\,\varphi(t)\,dt\right| &\leq \frac{1}{N\pi^2}\int_{|t|\geq s}\frac{|\varphi(t)}{(x-t)^2}\,dt\\ &\leq \frac{1}{N\pi^2}\|\varphi\|_\infty\int_{|t|\geq s}\frac{dt}{(x-t)^2} \qquad \text{for } |x| < s.\end{aligned}$$

So for $|x| < s$ we obtain $\lim_{N\to\infty}\int_{|t|\geq s}\delta_N(x-t)\,\varphi(t)\,dt = 0$. Hence

$$\lim_{N\to\infty}\int_{-\infty}^{\infty}\delta_N(x-t)\,\varphi(t)\,dt = \varphi(x) \text{ a.e.}$$

We now arrive at the proof of Theorem 2.7.3.

For $n = 1, 2, \dots$ let $\varphi_n(x) = \mathrm{e}^{-x^2/n}\varphi(x)$. Then clearly $\varphi_n \in L^1 \cap \mathbf{P}$ by Lemma 2.7.5, and by Lemma 2.7.6 there exist $\psi_n \in L^2$ with

$$\varphi_n = \mathcal{F}\psi_n, \quad \psi_n(x) = \widehat{\varphi}_n(-x) \text{ a.e.}, \quad \psi_n(x) \geq 0.$$

We claim that $\psi_n \in L^1$. One has

$$\int_{-\infty}^{\infty} \delta_N(x-t)\,\varphi_n(t)\,dt = \int_{-\infty}^{\infty} \Delta_N(t)\,\mathrm{e}^{-2\pi i x t}\psi_n(t)\,dt \tag{2.7.1}$$

for all $x \in \mathbb{R}$.

For $x = 0$ this gives

$$\int_{-\infty}^{\infty} \delta_N(-t)\varphi_n(t)\,dt = \int_{-\infty}^{\infty} \Delta_N(t)\psi_n(t)\,dt.$$

Therefore

$$\int_{-\infty}^{\infty} \Delta_N(t)\,\psi_n(t)\,dt \leq \int_{-\infty}^{\infty} |\varphi_n(t)|\,\delta_N(-t)\,dt \leq \varphi_n(0)\|\delta_N\|_1 = \varphi(0).$$

By Fatou's lemma ($N \to \infty$), $\int_{-\infty}^{\infty} \psi_n(t)\,dt$ converges, so $\psi_n \in L^1$ and $\|\psi_n\|_1 \leq \varphi(0)$ for all n.

Now define $g_n(u) = \int_{-\infty}^{u} \psi_n(t)\,dt$ ($u \in \mathbb{R}$). Observe that we may assume ψ_n to be continuous. The functions g_n are bounded weight functions satisfying $0 \leq g_n(u) \leq \varphi(0)$ for all $n = 1, 2, \dots$.

Furthermore, by (2.7.1),

$$\int_{-\infty}^{\infty} \delta_N(x-t)\,\varphi_n(t)\,dt = \int_{-\infty}^{\infty} \mathrm{e}^{-2\pi i x t}\Delta_N(t)\,dg_n(t).$$

By Lebesgue's theorem ($|\varphi_n(t)| \leq \varphi(0)$) we now have

$$\lim_{n\to\infty} \int_{-\infty}^{\infty} \delta_N(x-t)\,\varphi_n(t)\,dt = \int_{-\infty}^{\infty} \delta_N(x-t)\,\varphi(t)\,dt.$$

Since $|g_n(t)| \leq \varphi(0)$ for all $t \in \mathbb{R}$, by Theorem 2.6.8 there exists a sequence $n_1 < n_2 < \cdots$ and a weight function g such that

$$\lim_{n_k\to\infty} \int_{-\infty}^{\infty} \Delta_N(t)\,\mathrm{e}^{-2\pi i x t}\,dg_{n_k}(t) = \int_{-\infty}^{\infty} \Delta_N(t)\,\mathrm{e}^{-2\pi i x t}\,dg(t).$$

Hence

$$\int_{-\infty}^{\infty} \delta_N(x-t)\,\varphi(t)\,dt = \int_{-\infty}^{\infty} \Delta_N(t)\,\mathrm{e}^{-2\pi i x t}\,dg(t).$$

Let now N tend to infinity and apply Lemma 2.7.7 and Theorem 2.6.9. We obtain

$$\varphi(x) = \int_{-\infty}^{\infty} \mathrm{e}^{-2\pi i x t}\,dg(t) \text{ a.e.}$$

2.8 Extension to $\mathbb{R}^n$

One easily can extend the definition of the Fourier transform $\widehat{f}$ to functions $f \in L^1(\mathbb{R}^n)$ by

$$\widehat{f}(x) = \int_{\mathbb{R}^n} f(y)\,\mathrm{e}^{-2\pi i(x,y)}\,dy \qquad (x \in \mathbb{R}^n)$$

where $(x, y) = x_1y_1 + \cdots + x_ny_n$ if $x = (x_1, \ldots, x_n)$, $y = (y_1, \ldots, y_n)$.

Similar theorems hold as in the case $n = 1$, but the proofs might be more involved. In particular the proof of Bochner's theorem seems difficult to generalize. It is done by Bochner in [1, Author's Supplement], and by F. Riesz and E. Hopf, see [58, pp. 122–123] for precise references. In Chapter 5, Section 5.4, we shall prove this theorem by putting the theory of the Fourier transform in a new context.

Chapter 3
Locally Compact Groups

Literature: [3], [29], [33], [34], [35], [38], [58].

3.1 Groups

We recall a few simple facts from elementary algebra. Let G be a group with *unit element* e. We denote the inverse of x by x^{-1}. The group G is said to be *abelian* or commutative if $xy = yx$ for all $x, y \in G$. We assume that the reader is familiar with the notions of homomorphism, isomorphism, kernel of a homomorphism, denoted by $\ker f$, isomorphic groups, subgroup H of G, coset spaces G/H and $H\backslash G$, canonical projection, normal subgroup. The coset space G/H is a group as soon as H is normal, and then called the quotient group or factor group. If f is a homomorphism of G *onto* G', then $G' \simeq G/\ker f$.

Let G_α ($\alpha \in I$) be an indexed family of groups. The set $G = \prod_\alpha G_\alpha$ is a group if the product is defined by $(x_\alpha).(y_\alpha) = (x_\alpha y_\alpha)$. G is called the *direct product* of the G_α. The product of two groups G_1 and G_2 is commonly denoted by $G_1 \times G_2$.

3.2 Topological spaces

We recall some elementary notions.

- A *topological space* X is a set X wherein one has selected certain subsets, which are called *open*, such that

  (a) the empty set $\emptyset$ and X are open,

  (b) if $(U_\alpha)_{\alpha \in I}$ are open, then $\bigcup_{\alpha \in I} U_\alpha$ is open,

  (c) if U_1 and U_2 are open, then $U_1 \cap U_2$ is open.

- A collection $\mathcal{B}$ of open sets is called a *basis* of the topology of X if every open set can be written as a union of elements of $\mathcal{B}$.

- A collection $\mathcal{C}$ of subsets of a set X can serve as a basis for a topology as soon as $\mathcal{C}$ is closed under taking finite intersections, $\emptyset \in \mathcal{C}$ and $\bigcup_{C \in \mathcal{C}} C = X$ (take arbitrary unions as elements of $\mathcal{B}$).

- The complement of an open set is called a *closed* set. A *neighbourhood* of a point $x \in X$ is any subset which has x as *interior point*, i.e. any subset which contains an open set U with $x \in U$.

- The topological space X is said to be *Hausdorff* if for any pair of points $x, y \in X$ there are neighbourhoods U of x and V of y with $U \cap V = \emptyset$.

- A collection of neighbourhoods of x is called a *neighbourhood basis* of x if any neighbourhood of x contains a neighbourhood from the collection.

- We say that X satisfies the *first axiom of countability* if every point of X has a countable neighbourhood basis.

- The space X satisfies the *second axiom of countability* if X has a countable basis for its topology.

- The space X is called *discrete* if each subset of X is open.

- By $\prod_{\alpha \in I} X_\alpha$ we denote the *product* of a collection of topological spaces $(X_\alpha)_{\alpha \in I}$. A basis for the topology is given by the products $\prod_{\alpha \in I} Y_\alpha$, Y_α open in X_α, $Y_\alpha = X_\alpha$ for all but a finite number of α. The product of two topological spaces X_1 and X_2 is commonly denoted by $X_1 \times X_2$.

- Let Y be a subset of X. Call $V \subset Y$ open if V is of the form $V = Y \cap U$ with U open in X. Thus Y becomes itself a topological space, a *topological subspace* of X, with the *induced topology*.

- A topological space X is called *compact* if X is a Hausdorff space and satisfies the following property: any open covering of X contains a finite sub-covering.

- A subset Y of X is called compact if Y is compact in the induced topology.

- The space X is called *locally compact* if X is a Hausdorff space in which every point has a compact neighbourhood (equivalently: X is a Hausdorff space in which every point has a neighbourhood basis consisting of compact neighbourhoods).

- A locally compact space X can be *compactified* (made compact) by adding one point. This is called the *one-point compactification* or *Alexandrov compactification*.

- Call this compactification X' and set $X' = X \cup (\infty)$. The open neighbourhoods of ∞ are given by the complements of the compact subsets of X; the topology of X induced by the topology of X' coincides with the original topology of X.

- Let X_α $(\alpha \in I)$ be compact spaces, then $\prod_{\alpha \in I} X_\alpha$ is also compact (Tychonov's theorem).

- The product of *finitely* many locally compact spaces is again locally compact.

- A topological space is said to be *countable at infinity* if X is the union of countably many compact subsets.

- A mapping $f : X \to X'$ is called *continuous* if the inverse image of any open (closed) subset of X' is open (closed) in X. The mapping f is said to be an *open* mapping if the image of any open set in X is open in X'.

- The spaces X and X' are said to be *homeomorphic* if there exists a bijection $f : X \to X'$ such that f and f^{-1} are continuous.

We conclude this section with a technical result without proof.

Theorem 3.2.1. *Let X be locally compact, $K \subset X$ a compact subset, U open in X with $K \subset U$. There exists a real-valued continuous function f on X with $0 \leq f \leq 1$ and such that $f(x) = 1$ on K, $f(x) = 0$ outside U.*

The proof of this theorem is based on Tietze–Urysohn's extension theorem (see, e.g., [34, pp. 43–44]).

3.3 Topological groups

See [3, Chapter 3] and [33].

Let G be a group, whose underlying space is a Hausdorff topological space. The group G is called a *topological group* if both maps $(x, y) \mapsto xy$ $(G \times G \to G)$ and $x \mapsto x^{-1}$ $(G \to G)$ are continuous.

Examples. $\mathbb{R}$, $\mathbb{R}^*$, $\mathrm{GL}_n(\mathbb{R})$, $\mathrm{GL}_n(\mathbb{C})$, $\mathbb{T}$, $\mathbb{Z}$, $\mathbb{C}$, $\mathbb{H}$, SU(2), any Lie group, the field of p-adic numbers $\mathbb{Q}_p$, the ring of adèles $\mathcal{A}_k$, the ring of idèles $\mathcal{I}_k$ (k an algebraic number field), any finite group; every (abstract) group is a topological group with the discrete topology.

The mappings $x \mapsto xy$ and $x \mapsto yx$ ($y \in G$, fixed) are homeomorphisms of G. So neighbourhoods of y are of the form Uy or yU, where U is a neighbourhood of e. The mapping $x \mapsto x^{-1}$ is also a homeomorphism between G and G. With U also U^{-1} is therefore a neighbourhood of e.

Some properties

(i) *Every neighbourhood U of e contains a symmetric neighbourhood V, i.e. a neighbourhood satisfying $V = V^{-1}$.*

Take $V = U \cap U^{-1}$.

(ii) *Every neighbourhood U of e contains a neighbourhood V of e with $V^2 \subset U$.*

There exist neighbourhoods V_1, V_2 with $V_1 V_2 \subset U$, since xy depends continuously on (x, y). Take now $V = V_1 \cap V_2$.

Similarly we have

(ii′) *For every n there is a neighbourhood V of e with $V^n \subset U$.*

(iii) *Let $K \subset G$ be compact, U open and $K \subset U$. There exists a neighbourhood W of e with $WK \subset U$.*

For any $k \in K$ there is a neighbourhood V of e with $Vk \subset U$ and a neighbourhood W of e with $W^2 \subset V$. Since K is compact, there exist $k_1, \dots, k_n$ and $W_1, \dots, W_n$ such that $K \subset_{i=1}^n W_i k_i$. Set $W = \bigcap_{i=1}^n W_i$. Then we have

$$WK \subset \bigcup_{i=1}^{n} W W_i k_i \subset \bigcup_{i=1}^{n} W_i^2 k_i \subset U.$$

(iv) *If K_1 and K_2 are compact subsets of G, then so is $K_1 K_2$.*

$K_1 K_2$ is the continuous image of $K_1 \times K_2$ under the mapping $(x, y) \mapsto xy$.

(v) *The product of two open subsets of G is open. The product of two closed subsets of G (even of two closed subgroups) need not be closed. Take for example in $\mathbb{R}$ (the additive group of real numbers) $F_1 = \mathbb{Z}$, $F_2 = \mathbb{Z}\sqrt{2}$. But if F is closed and K compact, then FK is closed.*

If $x \notin FK$, then $F^{-1}x \cap K = \emptyset$. Hence there is a neighbourhood V of e with $F^{-1}x \cap KV^{-1} = \emptyset$ by property (iii). So $F^{-1}xV \cap K = \emptyset$, or $xV \cap FK = \emptyset$.

(vi) *A subgroup H of a topological group is a topological group itself with the induced topology.*

A *homomorphism* from G_1 to G_2 is a continuous algebraic homomorphism. Two topological groups G_1 and G_2 are called *isomorphic* if there is a bi-continuous algebraic isomorphism from G_1 to G_2.

A topological group is called *locally compact* if the underlying space is locally compact. In a similar way we speak about compact, discrete,… groups. One easily shows:

The group G is locally compact if and only if $e \in G$ has a compact neighbourhood.

A locally compact group is said to be *compactly generated* if G (as a group) is generated by a compact subset of G. Clearly such a group is countable at infinity.

We finally mention here two results without proof.

(vii) *Any locally compact group with a countable neighbourhood basis of e is metrizable. Every locally compact group is complete.* (See [3], Chapter 9, §3, no. 1 and Chapter 3, §3, no. 3.)

3.4 Quotient spaces and quotient groups

Let G be a topological group and H a subgroup of G. The left cosets of H form the points of a new topological space, the quotient space G/H, with topology defined as follows. Let $\pi_H : G \to G/H$ be the canonical projection $\pi_H(x) = \dot{x} = xH$. A subset $E \subset G/H$ is said to be open if $\pi_H^{-1}(E)$ is open in G. In other words: the open sets of G/H correspond to the subsets $UH \subset G$ with U open in G.

One easily shows that G/H is a Hausdorff space if and only if H is a closed subgroup of G.

If H is a closed *normal* subgroup of G, then G/H is a topological group, the quotient group G/H.

Let H be a closed subgroup of G. Then G/H is locally compact if G is locally compact. More generally (without proof), let G be a topological group and H a closed subgroup. Then G *is locally compact if and only if* H *and* G/H *are locally compact.*

The canonical projection π_H is not only continuous, but also *open*.

Observe that any open subgroup of a topological group G is closed.

Let G be a topological group, f a (continuous) homomorphism of G *onto* G'. The kernel of f, $\ker f$, is a closed normal subgroup of G and $G/\ker f$ is algebraically isomorphic to G'. In order that this isomorphism is also a topological one, f must be an open mapping. This holds under specific conditions, for example if G and G' are locally compact, while G satisfies the second axiom of countability and G' satisfies the first one. It is a consequence of *Baire's theorem*:

If a complete metric space is the countable union of closed subsets, then at least one of these subsets contains a non-empty open subset.

According to Section 3.3, property (vii), G and G' are metrizable. Choose a dense sequence (x_n) in G and let U be a compact neighbourhood of e and V a symmetric compact neighbourhood with $V^2 \subset U$. The compact sets $x_n V$ cover G. Hence the compacts sets $f(x_n)\,f(V)$ cover G'. Now G' is a complete metric space. According to Baire's theorem at least one of the sets $f(x_n)\,f(V)$ is a neighbourhood of, let us say, $f(x_n)\,f(v_0) = f(x_n v_0)$. Therefore $f(v_0^{-1}x_n^{-1})\,f(x_n)f(V) = f(v_0^{-1})f(V)$ is a neighbourhood of $f(e)$, and also $f(v_0^{-1})f(V) \subset f(V^2) \subset f(U)$. It is now clear that the image of any open subset of G is open in G'. So f is an open mapping.

Proof of Baire's theorem. Let (M,d) be a complete metric space, V_n closed in M ($n = 1,2,\ldots$) and $M = \bigcup_{n=1}^{\infty} V_n$. Suppose no V_n has an interior point. Then $V_1 \neq M$ and $M\backslash V_1$ is open. There is $x_1 \in M\backslash V_1$ and a ball $B(x_1,\varepsilon_1) \subset M\backslash V_1$ with $0 < \varepsilon_1 < 1/2$. Now the ball $B(x_1,\varepsilon_1)$ does not belong to V_2, hence $M\backslash V_2 \cap B(x_1,\varepsilon_1)$ contains a point x_2 and a ball $B(x_2,\varepsilon_2)$ with $0 < \varepsilon_2 < 1/4$. In this way we get a sequence of balls $B(x_n,\varepsilon_n)$ with

$$B(x_1,\varepsilon_1) \supset B(x_2,\varepsilon_2) \supset \cdots; \quad 0 < \varepsilon_n < \frac{1}{2^n}, \quad B_n \cap V_n = \emptyset.$$

For $n < m$ one has $d(x_n,x_m) < 1/2^n$, which tends to zero if $n,m \to \infty$. The Cauchy sequence $\{x_n\}$ has a limit $x \in M$ since M is complete. But

$$d(x_n,x) \leq d(x_n,x_m) + d(x_m,x) < \varepsilon_n + d(x_m,x) \to \varepsilon_n \qquad (m \to \infty).$$

So $x \in B(x_n,\varepsilon_n)$ for all n. Hence $x \notin V_n$ for all n, so $x \notin M$, a contradiction.

3.5 Some useful facts

(i) *Let G be a locally compact group, H a closed subgroup of G. For every compact set $\dot{K} \subset G/H$ there exists a compact set $K \subset G$ with $\pi_H(K) = \dot{K}$.*

Let V be a compact neighbourhood of e in G. Because $\dot{K}$ is compact, one can find finitely many $s_i \in G$ such that $\dot{K}$ is contained in the image of $K' = \bigcup_{i=1}^{n} s_i V$ under π_H. Now K' is compact. Take finally $K = K' \cap \pi_H^{-1}(\dot{K})$.

(ii) *The product of arbitrary many topological groups is again a topological group.*

Observe that $\prod_{\alpha\in I} G_\alpha$, G_α discrete, is not necessarily discrete as soon as $\mathrm{Card}(I) = \infty$ and $\mathrm{Card}(G_\alpha) \geq 2$ for all $\alpha \in I$ (it is compact if all G_α are finite).

(iii) *The mapping $(x,y) \mapsto xy$ from $G \times G$ to G is open as is $(x,y) \mapsto xy^{-1}$ and $(x,y) \mapsto x$.*

This follows from the definition of the product topology.

(iv) *Any locally compact group is the union of open (and hence closed) subgroups which are compactly generated.*

Indeed, take $G_V = \bigcup_{n\geq 1} V^n$ with V an open symmetric neighbourhood of e with compact closure. Clearly G is the union of all such G_V, which are compactly generated, since clearly also $G_V = \bigcup_{n\geq 1} \overline{V}^n$, where $\overline{V}$ is the closure of V.

(v) *A locally compact group or quotient space, which has countably many elements, is discrete.*

This follows from Baire's theorem and Section 3.3, property (vii).

3.6 Functions on locally compact groups

(i) Let f be a real- or complex-valued function on the locally compact space X (or even with values in a complex vector space) and set $V_f = \{x : f(x) \neq 0\}$. The closure of V_f is usually called the *support* of f, denoted by $\operatorname{Supp} f$. By Theorem 3.2.1 there exist non-zero complex-valued functions on X with *compact* support.

Let us denote by $C_c(G)$ the set of all continuous complex-valued functions with compact support on the locally compact group G. A complex-valued function f on G is called right (left) *uniformly continuous* if, given $\varepsilon > 0$, there is a neighbourhood U of e such that

$$|f(yx) - f(x)| < \varepsilon \qquad (|f(xy) - f(x)| < \varepsilon)$$

for all $y \in U$ and $x \in G$.

Any complex-valued continuous function f on G with compact support is right and left uniformly continuous.

Let $K = \operatorname{Supp} f$. Choose $\varepsilon > 0$ and a compact symmetric neighbourhood U of e. For any point $a \in KU$ there is a neighbourhood V_a of e such that $|f(ya) - f(a)| < \varepsilon/2$ for $y \in V_a$. Select neighbourhoods W_a of a with $W_a^2 \subset V_a$. The W_a cover KU, hence there are $a_1, \dots, a_n$ such that $KU \subset \bigcup_{i=1}^n W_{a_i} a_i$. Set $W = \bigcap_{i=1}^n W_{a_i}$. For $x \in G$ and $y \in W \cap U$ we now have

$$|f(yx) - f(x)| = 0 \quad \text{for } x \notin UK.$$

For $x \in UK$ we get, if $x \in W_{a_i} a_i$,

$$|f(yx) - f(x)| \leq |f(yx - f(a_i)| + |f(a_i) - f(x)| \leq \varepsilon/2 + \varepsilon/2 = \varepsilon$$

for $y \in W \cap U$. A similar proof applies for the left uniform continuity.

The same result holds for the functions in $\mathcal{C}_0(G)$, the space of continuous complex-valued functions f on G which vanish at infinity, i.e. the functions f satisfying: for any $\varepsilon > 0$ there is a compact subset $K_\varepsilon \subset G$ such that $|f(x)| < \varepsilon$ for all $x \notin K_\varepsilon$.

(ii) Let H be a closed subgroup of G. A complex-valued function F on G is of the form $F' \circ \pi_H$, where F' is a function on G/H if and only if F is left H-invariant, i.e. $F(xh) = F(x)$ for all $x \in G, h \in H$. The function F is continuous if and only if F' is continuous. This follows immediately from the definition of the topology of G/H.

Chapter 4
Haar Measures

Literature: [5], [38], [58].

4.1 Measures

We follow the approach to measure theory of A. Weil and N. Bourbaki.

(i) Let X be a locally compact space and set as before $C_c(X)$ for the space of complex-valued continuous functions on X with compact support, and denote by $\mathcal{C}_0(X)$ the space of complex-valued continuous functions on X which vanish at infinity.

Observe that $\mathcal{C}_0(X)$ is the completion of $C_c(X)$ with respect to the supremum norm $\|f\|_\infty = \sup|f(x)|$.

Let $C_c(X;\mathbb{R})$ be the subspace of real-valued functions in $C_c(X)$.

A real (complex) *measure* μ on X is a real (complex) *linear* functional on $C_c(X;\mathbb{R})$ ($C_c(X)$) such that for any compact set $K \subset X$ there exists a constant $M_K > 0$ satisfying

$$|\mu(f)| \leq M_K \|f\|_\infty$$

for all $f \in C_c(X;\mathbb{R})$ ($C_c(X)$) with $\operatorname{Supp} f \subset K$.

The measure μ is called *positive* if $\mu(f) \geq 0$ as soon as $f \geq 0$, for all $f \in C_c(X)$. Clearly μ is a real measure in this case. We shall also write

$$\mu(f) = \int_X f(x)d\mu(x) = \int_X f d\mu.$$

Sometimes we shall delete the suffix "X" if the space is clear from the context.

The connection between our approach and the set-theoretic one is as follows (see, e.g., [22, §9]):

Theorem 4.1.1 (Riesz representation theorem). *If $I : C_c(X) \to \mathbb{R}$ is a positive measure (in our approach), then there exists a unique regular Borel measure μ on X such that for all $f \in C_c(X;\mathbb{R})$*

$$I(f) = \int f d\mu.$$

(ii) A complex measure is completely determined by its restriction to the space $C_c(X;\mathbb{R})$. Any real measure can be uniquely extended to a complex measure. Let μ be a complex measure. Then also $\overline{\mu}$ defined by $\overline{\mu}(f) = \overline{\mu(\overline{f})}$ $(f \in C_c(X))$ is a complex measure. Set $\mu_1 = \frac{\mu+\overline{\mu}}{2}$ and $\mu_2 = \frac{\mu-\overline{\mu}}{2i}$. Then μ_1 and μ_2 are real measures and $\mu = \mu_1 + i\mu_2$ (where μ_1 and μ_2 are being considered as complex measures).

A real measure is the difference of two positive measures.

Let μ be a real measure. Define $\nu(f) = \sup_{0\le g\le f} \mu(g)$ for $f \in C_c(X)$, $f \ge 0$. This functional ν gives, after extending it to $C_c(X;\mathbb{R})$, a positive measure, and $\mu = \nu-(\nu-\mu)$.

(iii) *If μ is a positive measure, then*

$$\left|\int f(x)\, d\mu(x)\right| \le \int |f(x)| d\mu(x) \qquad (f \in C_c(X)).$$

Assume $\mu(f) \ne 0$, $|\mu(f)| = \lambda\mu(f)$. Then $\mu(\lambda f)$ is a real number and $0 < \mu(\lambda f) = \mu(\mathrm{Re}(\lambda f)) \le \mu(|\lambda f|) = \mu(|f|)$, since $|\lambda| = 1$.

(iv) *Every positive linear form on $C_c(X)$ (i.e. every complex linear form μ on $C_c(X)$ with $\mu(f) \ge 0$ if $f \ge 0$) is a positive measure.*

One has, similar to (iii), with the same proof, $|\mu(f)| \le \mu(|f|)$ for all $f \in C_c(X)$. Let $K \subset X$ be a compact set and $g_K \in C_c(X)$ such that $g_K(x) = 1$ on K, $g_K \ge 0$. Then one has for all $f \in C_c(X)$ with $\mathrm{Supp}\, f \subset K$: $f = fg_K$ and

$$|\mu(f)| \le \mu(|f|g_K) \le \|f\|_\infty\, \mu(g_K).$$

(v) A *bounded measure* μ is a measure for which $M > 0$ exists with $|\mu(f)| \le M\|f\|_\infty$ for all $f \in C_c(X)$. Hence μ is a continuous linear form on $C_c(X)$ (or $\mathcal{C}_0(X)$) with respect to the sup-norm. The norm of μ is defined by $\|\mu\| = \sup_{\|f\|_\infty\le 1} |\mu(f)|$. The bounded measures form a Banach space, denoted by $M^1(X)$. Clearly $M^1(X)$ is the dual of $\mathcal{C}_0(X)$.

(vi) *Support of a measure.* Let O be an open subset of X such that $\mu(f) = 0$ for all $f \in C_c(X)$ with $\mathrm{Supp}\, f \subset O$. If the O_α $(\alpha \in I)$ have this property, then also $\bigcup_{\alpha\in I} O_\alpha$ has. This follows easily from the lemma stated below. Let U be the largest open subset with this property. The complement of U, which is a closed subset of X, is called the *support of* μ: $\mathrm{Supp}\, \mu$. The measures with compact support form a dense subset in $M^1(X)$.

Lemma 4.1.2 (Partition of unity). *Let X be a locally compact space. Suppose that $O_1, \dots, O_m$ are open subsets of X and let K be a compact subset such that*

$K \subset \bigcup_{i=1}^{m} O_i$. There exist functions $\phi_i \in C_c(X)$ with $\mathrm{Supp}(\phi_i) \subset O_i$ and such that $\phi_i \geq 0$, $\sum_{i=1}^{m} \phi_i \leq 1$, $\sum_{i=1}^{m} \phi_i = 1$ in a neighbourhood of K.

Select compact sets $K_i \subset O_i$ such that $K \subset \bigcup_{i=1}^{m} K_i$. According to Theorem 3.2.1 there exist functions $\psi_i \in C_c(X)$ with $\mathrm{Supp}\,\psi_i \subset O_i$, $\psi_i = 1$ on a neighbourhood of K_i, $0 \leq \psi_i \leq 1$. Now set $\phi_1 = \psi_1$ and

$$\phi_i = \psi_i(1-\psi_1)\cdots(1-\psi_{i-1}) \quad \text{for } i = 2, \ldots, m.$$

Then $\phi_1, \ldots, \phi_m$ satisfy the requirements. Observe that

$$\sum_{i=1}^{m} \phi_i = 1 - (1-\psi_1)(1-\psi_2)\cdots(1-\psi_m).$$

Proposition 4.1.3. *Let μ be a measure on X. Then $\mu(f) = 0$ for all $f \in C_c(X)$ satisfying $f(x) = 0$ for $x \in \mathrm{Supp}\,\mu$.*

Let $f \in C_c(X)$ vanish on the support of μ and let K be an open set with compact closure such that $\mathrm{Supp}\, f \subset K$. Then $|\mu(\varphi)| \leq M_K \|\varphi\|_\infty$ for some positive constant M_K and all $\varphi \in C_c(X)$ with $\mathrm{Supp}\,\varphi \subset K$. Choose for each $\varepsilon > 0$ a function $\varphi_\varepsilon \in C_c(X)$ such that $\varphi_\varepsilon = 1$ near $\mathrm{Supp}\,\mu \cap K$ and such that $\|\varphi_\varepsilon f\|_\infty < \varepsilon$. Then

$$|\mu(f)| = |\mu(\varphi_\varepsilon f)| \leq M_K \varepsilon$$

for all $\varepsilon > 0$, so $\mu(f) = 0$.

4.2 Invariant measures

Let G be a locally compact group. A measure μ on G is said to be *left-(right-) invariant* if

$$\int_G f(a^{-1}x)d\mu(x) = \int_G f(x)d\mu(x)$$

($\int_G f(xa)d\mu(x) = \int_G f(x)d\mu(x)$) for all $f \in C_c(X)$ and all $a \in G$. Notice that if μ is a left-invariant measure, then $\check{\mu}$ defined by

$$\check{\mu}(f) = \int_G f(x^{-1})\, d\mu(x) \qquad (f \in C_c(G))$$

is a right-invariant measure.

Examples.
- $G = \mathbb{R}$; $d\mu(x) = dx$, the Lebesgue measure.
- $G = \mathbb{R}^*$; $d\mu(x) = \frac{dx}{|x|}$.
- $G = \mathbb{T}$; $\mu(f) = \frac{1}{2\pi}\int_0^{2\pi} f(\mathrm{e}^{i\theta})\, d\theta$ $[= \int_0^1 f(\mathrm{e}^{2\pi i\theta})\, d\theta]$.

- If G is a discrete group, e.g. $G = \mathbb{Z}$ or G is a finite group, then $C_c(G)$ consists of functions on G with finite support. The measure μ defined by

$$\mu(f) = \sum_{x \in G} f(x),$$

called the "counting measure", is an invariant measure on G.

All the above measures are bi-invariant (i.e. both left- and right-invariant).

- Let now G be the group of matrices $\left(\begin{smallmatrix} \lambda & \mu \\ 0 & 1 \end{smallmatrix}\right)$ with $\lambda > 0$, $\mu \in \mathbb{R}$. Then a left-invariant measure is given by $\frac{d\lambda d\mu}{\lambda^2}$, a right-invariant measure is $\frac{d\lambda d\mu}{\lambda}$.

Later on we will give more examples. The following theorem holds:

Theorem 4.2.1 (Haar 1933, existence; von Neumann 1934, uniqueness). *Every locally compact group admits a left-invariant positive measure, which is non-trivial and unique up to a positive constant.*

Notation and terminology. We call such a measure from now on a left *Haar measure* and denote it also by dx. So for a left Haar measure we have

$$\int_G f(a^{-1}x)dx = \int_G f(x)dx \qquad (f \in C_c(X)).$$

Alfred Haar (Budapest 11 October 1885–Szeged 16 March 1933). Hungarian mathematician, in 1912 professor at the universities of Cluj and Szeged. He has publications on many subjects in analysis and algebra, in particular group theory.

John von Neumann (Budapest 28 December 1903–Washington D. C. 8 February 1957). Has been assistant of Hilbert at Göttingen and professor in Princeton, USA. He went to the USA in 1937. He is the founder of modern computer science and was one of the greatest mathematicians of his time. See his Collected Papers.

Properties of the Haar measure

(i) *Let μ be a left Haar measure on G, $f \in C_c(G)$, $f \geq 0$. If f is not identically zero, then $\mu(f) > 0$.*

Suppose there exists $f_0 \in C_c(X)$, $f_0 \geq 0$ with $f_0(x_0) > 0$ for some $x_0 \in G$ and $\mu(f_0) = 0$. Let $f \in C_c(G)$, $f \geq 0$ be arbitrary. There are $a_1, \ldots, a_n$ in G such that $x \mapsto \sum_{i=1}^n f_0(a_i^{-1}x)$ is strictly positive on Supp f. Then there is a scalar $\lambda > 0$ with

$$\lambda \sum_{i=1}^n f_0(a_i^{-1}x) \geq f(x)$$

for all $x \in G$ and hence $\mu(f) \leq 0$. So $\mu(f) = 0$ for all $f \in C_c(G)$, $f \geq 0$, and therefore $\mu(f) = 0$ for all $f \in C_c(G)$. This is a contradiction.

A straightforward consequence is: $\operatorname{Supp}\mu = G$, and every compact neighbourhood of e has strictly positive Haar measure.

(ii) The *proof of existence* of the Haar measure is long and ingenious, but you don't learn more from it than that the Haar measure exists. Proofs can be found in [5, Chapter 8], [58], [29], [34], a.o. The *uniqueness* is easy to show. We shall show:

Proposition 4.2.2. *If μ_1 is a positive and μ_2 a complex left-invariant measure on G, then there is a complex constant c such that $\mu_2 = c\mu_1$.*

We apply Fubini's theorem: if μ_1 and μ_2 are measures on X and Y respectively (X and Y locally compact spaces), then one has for any $k \in C_c(X \times Y)$

$$\int_Y \int_X k(x,y)\,d\mu_1(x)d\mu_2(y) = \int_X \int_Y k(x,y)\,d\mu_2(y)d\mu_1(x).$$

Let μ_1 be a positive and μ_2 an arbitrary complex left-invariant measure on G. Given $f, g \in C_c(G)$ the function $(x,y) \mapsto f(x)g(x^{-1}y)$ is continuous on $G \times G$ and has compact support. So

$$\iint f(x)\,g(x^{-1}y)\,d\mu_1(x)d\mu_2(y) = \int f(x)\,d\mu_1(x) \cdot \int g(y)\,d\mu_2(y) = \mu_1(f)\,\mu_2(g),$$

by Fubini's theorem. This is also equal to

$$\begin{aligned}\int \Big\{ \int f(x)g(x^{-1}y)d\mu_1(x) \Big\} d\mu_2(y) &= \int \Big\{ \int f(yx)g(x^{-1})\,d\mu_1(x) \Big\} d\mu_2(y) \\ &= \int g(x^{-1}) \Big\{ \int f(yx)\,d\mu_2(y) \Big\} d\mu_1(x).\end{aligned}$$

Now fix $f \in C_c(X)$, $f \geq 0$ with $\mu_1(f) \neq 0$. Define $F(x) = \int f(yx^{-1})\,d\mu_2(y)$. Then we have

$$\mu_1(f) \int g(x)\,d\mu_2(x) = \int g(x)F(x)\,d\mu_1(x^{-1})$$

for all g. Consequently $d\mu_2(x) = \theta(x)d\mu_1(x^{-1})$ with $\theta(x) = \frac{F(x)}{\mu_1(f)}$.

Since $\operatorname{Supp}\mu_1 = G$, there is only one continuous function θ determined by the relation $d\mu_2(x) = \theta(x)\,d\mu_1(x^{-1})$. So θ is independent of the choice of the function f, hence

$$\theta(e) = \frac{F(e)}{\mu_1(f)} = \frac{\mu_2(f)}{\mu_1(f)}$$

is independent of f, so $\mu_2(f) = c \cdot \mu_1(f)$ for all f with $f \geq 0$, $\mu_1(f) \neq 0$. By (i) this is true for all f, so $\mu_2 = c\,\mu_1$ for some constant $c \in \mathbb{C}$.

(iii) For many locally compact groups a left Haar measure is also a right Haar measure. A group with this property is said to be *unimodular*. This is certainly the case for abelian groups. In other situations there is a *modulus*, which we are now going to introduce.

Let dx be a left Haar measure on G. For any $a \in G$,

$$f \mapsto \int f(xa^{-1})\, dx \qquad (f \in C_c(G))$$

is again a left-invariant and positive measure. Hence there is a constant $\Delta(a) > 0$ such that

$$\int_G f(xa^{-1})\, dx = \Delta(a) \int_G f(x)\, dx \tag{4.2.1}$$

for all $f \in C_c(G)$ and $a \in G$.

The mapping Δ is a continuous homomorphism from G into $\mathbb{R}_+^*$ and is independent of the choice of dx. It is called the *Haar modulus* of G.

If G is the group of matrices $\left(\begin{smallmatrix} \lambda & \mu \\ 0 & 1 \end{smallmatrix}\right)$ with $\lambda > 0$, $\mu \in \mathbb{R}$, then its Haar modulus is equal to $\Delta : \left(\begin{smallmatrix} \lambda & \mu \\ 0 & 1 \end{smallmatrix}\right) \mapsto \frac{1}{\lambda}$.

Furthermore,

$$f \mapsto \int_G f(x)\, \Delta(x^{-1})\, dx$$

is a right Haar measure. Hence $f \mapsto \int_G f(x^{-1})\Delta(x^{-1})\, dx$ is again a left Haar measure, so

$$\int_G f(x^{-1})\Delta(x^{-1})\, dx = c \int_G f(x)\, dx \qquad (f \in C_c(G))$$

for some $c > 0$. Replacing $f(x)$ by $f(x^{-1})\Delta(x^{-1})$, which is again in $C_c(G)$, implies $c^2 = 1$, hence $c = 1$. Therefore

$$\int_G f(x^{-1})\Delta(x^{-1})\, dx = \int_G f(x)\, dx. \tag{4.2.2}$$

It also follows

$$\int_G f(x^{-1})\, dx = \int_G f(x)\Delta(x^{-1})\, dx, \tag{4.2.3}$$

or $dx^{-1} = \Delta(x^{-1})dx$.

Remark 4.2.3. The null sets, the local null sets and the measurable sets for the left and right Haar measure coincide. In any non-unimodular group there are open sets which are integrable for the left Haar measure, but not for the right Haar measure.

Compact groups are unimodular: $\Delta(G)$ is compact in $\mathbb{R}_+^*$, hence equal to $\{1\}$.

(iv) *The group G has finite Haar measure if and only if G is compact.*

Take $f \neq 0$, $0 \leq f \leq 1$, $f \in C_c(G)$. Let $C = \operatorname{Supp} f$. If G is not compact, there is a sequence $(a_n)_{n \geq 1}$ in G such that the subsets $a_n C$ are mutually disjoint for $n \geq 1$. Indeed, choose a_1 arbitrary and if $a_1, \ldots, a_n$ have been selected, take a_{n+1} outside $(\bigcup_{m=1}^{n} a_m C)\, C^{-1}$. Then we have

$$\int_G \sum_{1 \leq n \leq N} f(a_n^{-1}x)\, dx = N \int_G f(x)\, dx$$

and $\sum_{1 \leq n \leq N} f(a_n^{-1}x) \leq 1$ for all x and all N. This would imply that the Haar measure is unbounded, leading to a contradiction.

(v) Define for any function f on G and any element $a \in G$ the function $L_a f$ by $L_a f(x) = f(a^{-1}x)$ $(x \in G)$. In practice the following situation often occurs. In some open neighbourhood V of e in G a positive measure $d_V(x) \neq 0$ is given which is *locally left-invariant*: for all $f \in C_c(G)$ with $\operatorname{Supp} f \subset V$ one has

$$\int_V f(a^{-1}x)\, d_V(x) = \int_V f(x)\, d_V(x)$$

for all $a \in G$ with $\operatorname{Supp} L_a f \subset V$.

Then $d_V(x)$ is the restriction to V of a unique left Haar measure dx on G, and dx can be defined explicitly in terms of $d_V(x)$.

See [38, Chapter 3, §3]. The idea of the proof is clear. Select a symmetric open neighbourhood U of e with $U^2 \subset V$. For all $f \in C_c(G)$ with $\operatorname{Supp} f \subset U$ we have by assumption

$$\int_V f(y^{-1}x)\, d_V(x) = \int_V f(x)\, d_V(x) \tag{4.2.4}$$

for all $y \in G$ with $\operatorname{Supp} L_y f \subset U$.

Let now $f \in C_c(G)$ be such that $\operatorname{Supp} L_s f \subset U$ for some $s \in G$. We shall say that f has 'small support'. We define for such f (with respect to V)

$$\int_G f(x)\, dx = \int_V (L_s f)(x)\, d_V(x).$$

This definition depends only on f, not on $s \in G$, because of property (4.2.4). In this way we get a left-invariant integral for $f \in C_c(G)$ with small support. We have to show (exercise):

(a) Every $f \in C_c(G)$ can be written as a finite sum $f = \sum f_n$ with $f_n \in C_c(G)$ and with small support; if $f \geq 0$ then we can take $f_n \geq 0$ for all n.

(b) If $\sum_n f_n = \sum_j g_j$, where $f_n, g_j \in C_c(G)$ and all f_n, g_j have small support, then $\sum_n \int f_n(x)\, dx = \sum_j \int g_j(x)\, dx$.

If (a) and (b) are shown to be true, the definition of $\int_G f(x)\, dx$ for $f \in C_c(G)$ arbitrary, is obvious.

4.3 Weil's formula

(i) Let H be a *closed subgroup* of the locally compact group G. The group H is itself locally compact and thus has a left Haar measure, say $d\xi$. For $f \in C_c(G)$ we consider

$$x \mapsto \int_H f(x\xi)\,d\xi \qquad (x \in G).$$

This is, in fact, a function on G/H, which we shall denote by $T_H f$. So

$$(T_H f)(\dot{x}) = \int_H f(x\xi)\,d\xi \qquad (\dot{x} = \pi_H(x)). \tag{4.3.1}$$

The function $T_H f$ vanishes outside $\pi_H(\operatorname{Supp} f)$, so it has compact support. It is also a continuous function on G/H. We shall show a little more.

Lemma 4.3.1. *Let $f \in C_c(G)$ and $\varepsilon > 0$. There is a neighbourhood U_ε of e such that for all $x \in G$*

$$\int_H |f(y^{-1}x\xi) - f(x\xi)|\,d\xi < \varepsilon$$

for all $y \in U_\varepsilon$.

Let V be a compact symmetric neighbourhood of e and take $g \in C_c(G)$, $g \geq 0$, such that $g(x) = 1$ on $V \cdot \operatorname{Supp} f$. Then

$$|f(y^{-1}x) - f(x)| \leq |f(y^{-1}x) - f(x)|\,g(x) \qquad (x \in G, y \in V).$$

Set $M_x = 1 + \int_H g(x\xi)\,d\xi$. Given $\varepsilon > 0$ there is a neighbourhood $U(\varepsilon, M_x)$ of e, contained in V, such that

$$|f(y^{-1}x) - f(x)| < \varepsilon/M_x$$

for all $y \in U(\varepsilon, M_x)$ and $x \in G$. Hence

$$|f(y^{-1}x\xi) - f(x\xi)| \leq \varepsilon/M_x \cdot g(x\xi)$$

for all $y \in U(\varepsilon, M_x)$, $x \in G$, $\xi \in H$.

This implies that $T_H f$ is continuous, so $T_H f \in C_c(G/H)$ for all $f \in C_c(G)$. In particular, $T_H g$ is continuous and has compact support. Now replace M_x by $M = 1 + \sup T_H g(\dot{x})$ and $U(\varepsilon, M_x)$ by U_ε, applying the right uniform continuity of f. This gives the lemma.

We now have: T_H is a linear mapping from $C_c(G)$ into $C_c(G/H)$. We shall show that T_H *is surjective*.

Set $\dot{K} = \operatorname{Supp} \dot{f}$ for $\dot{f} \in C_c(G/H)$. There is a compact subset $K \subset G$ with $\pi_H(K) = \dot{K}$. Take $g \in C_c(G)$, $g \geq 0$, $g(x) > 0$ for all $x \in K$. Then $T_H g(\dot{x}) > 0$ for all $\dot{x} \in \dot{K}$. Now define

$$f(x) = \begin{cases} \dot{f} \circ \pi_H(x) \frac{g(x)}{T_H g(\pi_H(x))} & \text{if } x \in \pi_H^{-1}(\dot{K}), \\ 0 & \text{otherwise.} \end{cases}$$

Then f is continuous, in $C_c(G)$, and $T_H f = \dot{f}$.

(ii) Let now H be a *closed normal subgroup* of G. Then the quotient space $\dot{G} = G/H$ is a locally compact group, see 3.4. Denote its own Haar measure by $d\dot{x}$. Consider

$$f \mapsto \int_{G/H} T_H f(\dot{x})\, d\dot{x} \qquad (f \in C_c(G)).$$

This is a positive linear functional on $C_c(G)$, hence a measure. It is not identically zero and left-invariant. Therefore there is a positive constant c such that

$$\int_{G/H} T_H f(\dot{x})\, d\dot{x} = c \int_G f(x)\, dx$$

for all $f \in C_c(G)$.

Suppose now that two out of the three Haar measures on G, H and G/H are given. Then we normalize the third one such that $c = 1$. We then have

$$\int_{G/H} \left\{ \int_H f(x\xi)\, d\xi \right\} d\dot{x} = \int_G f(x)\, dx \qquad (f \in C_c(G)),$$

which is often denoted in a formal way by $dx = d\xi d\dot{x}$.

This formula is called *Weil's formula*, after the French mathematician André Weil (1906–1998).

(iii) Let $G = G_1 \times G_2$ and let dx_1 and dx_2 be left Haar measures on G_1 and G_2 respectively. Then $dx = dx_1 dx_2$ is a Haar measure on G. Weil's formula is actually a generalization of this result.

(iv) Let H be a *closed normal* subgroup of G. Denote its Haar modulus by Δ_H. Let Δ be the Haar modulus of G. *Then* $\Delta(\xi) = \Delta_H(\xi)$ *for all* $\xi \in H$.

Fix $\eta \in H$ and set $f_\eta(x) = f(x\eta^{-1})$ for $x \in G$, $f \in C_c(G)$. Then we have by Weil's formula

$$\int_{G/H} \left\{ \int_H f_\eta(x\xi)\, d\xi \right\} d\dot{x} = \int_G f_\eta(x)\, dx = \int_G f(x\eta^{-1})\, dx.$$

Hence

$$\Delta_H(\eta) \int_{G/H} \{f(x\xi)\, d\xi\}\, d\dot{x} = \Delta(\eta) \int_G f(x)\, dx,$$

so $\Delta_H(\eta) = \Delta(\eta)$ for $\eta \in H$.

4.4 Haar measures for specific groups

In this section we consider Haar measures for specific groups. Proofs are omitted and left to the reader.

(a) Let $G = \mathrm{GL}(n, \mathbb{R})$ denote the group of $n \times n$ real matrices $x = (x_{ij})_{1 \leq i,j \leq n}$. For every $x = (x_{ij})$ the matrix entries give a global coordinate system on G. In these coordinates $dx = \prod dx_{ij}/|\det x|^n$ is both a left and a right Haar measure. The group G is thus unimodular. The same holds for its subgroup $\mathrm{SL}(n, \mathbb{R})$ of $n \times n$ real matrices with determinant 1, since it is a closed normal subgroup of G.

(b) Let $\mathrm{ST}_1(n, \mathbb{R})$ be the group of $n \times n$ real matrices $x = (x_{ij})$ with $x_{ii} = 1$ $(1 \leq i \leq n)$, $x_{ij} = 0$ $(1 \leq j < i \leq n)$. This group is unimodular. The Haar measure is given by $\prod_{i<j} dx_{ij}$ if we choose x_{ij} $(i < j)$ as global coordinates.

(c) The group $G = \mathrm{GL}(n, \mathbb{C})$ it topologically an open subset of $\mathbb{C}^{n^2} = \mathbb{R}^{2n^2}$. The Haar measure is given by

$$\prod_{i,j} \frac{dz_{ij}\, d\overline{z_{ij}}}{|\det z|^{2n}} \qquad (z = (z_{ij}) \in G).$$

The group G is unimodular.

For those readers that are familiar with Lie groups, we remark that the following groups are *unimodular*:

(i) *Lie groups G with the property that* $\mathrm{Ad}(G)$ *is compact,*

(ii) *connected semisimple and reductive Lie groups,*

(iii) *connected nilpotent Lie groups.*

4.5 Quasi-invariant measures on quotient spaces

Literature: [38, Chapter 8].

Let H be a closed subgroup of a locally compact group G and consider the quotient space G/H. If H is normal then G/H has a Haar measure and Weil's formula holds. We shall now consider a generalization of Weil's formula for arbitrary closed subgroups H. This important generalization is due to Mackey and Bruhat.

Recall that T_H, previously defined, is a surjective linear mapping from $C_c(G)$ to $C_c(G/H)$.

(i) Let μ and $\dot{\mu}$ be complex measures on G and G/H respectively, such that

$$\int_{G/H} \left\{ \int_H f(x\xi)\, d\xi \right\} d\dot{\mu}(\dot{x}) = \int_G f(x)\, d\mu(x) \qquad (f \in C_c(G)). \tag{4.5.1}$$

Compare this formula with Weil's formula.

Replace in (4.5.1) the function f by f_η defined by

$$f_\eta(x) = f(x\eta^{-1}) \qquad (\eta \in H,\ x \in G).$$

Then μ clearly satisfies

$$\Delta_H(\eta) \int_G f(x)\, d\mu(x) = \int_G f(x\eta^{-1})\, d\mu(x) \qquad (\eta \in H,\ f \in C_c(G)). \quad (4.5.2)$$

Conversely, let μ satisfy (4.5.2). *Then there exists a unique measure* $\dot{\mu}$ *on* G/H *such that* (4.5.1) *holds.*

Fix $f \in C_c(G)$. For all $g \in C_c(G)$ we have

$$\begin{aligned}
&\int_G f(x)\Big\{\int_H g(x\xi)\, d\xi\Big\} d\mu(x) \\
&\quad = \int_H \Big\{\int_G f(x)g(x\xi)\, d\mu(x)\Big\} d\xi \\
&\quad = \int_H \Delta_H(\xi^{-1})\Big\{\int_G f(x\xi^{-1})g(x)\, d\mu(x)\Big\} d\xi \qquad \text{(by (4.5.2))} \\
&\quad = \int_G g(x)\Big\{\int_H f(x\xi)\, d\xi\Big\}\, d\mu(x).
\end{aligned}$$

We can choose g such that $\int_H g(x\xi)\, d\xi = 1$ for all $x \in \operatorname{Supp} f$. Hence, if $\int_H f(x\xi)\, d\xi = 0$ for all $x \in G$, then $\int_G f(x)\, d\mu(x) = 0$. Now set $\dot{\mu}(\dot{f}) = \int_G f(x)\, d\mu(x)$ for $\dot{f} \in C_c(G/H)$, where f is such that $f \in C_c(G)$, $T_H f = \dot{f}$. The functional $\dot{\mu}$ is well-defined and is a measure on G/H. Indeed, if $\dot{K} \subset G/H$ is compact, take then $f_1 \in C_c(G)$ such that $T_H f_1 = 1$ on $\dot{K}$. For all $\dot{f} \in C_c(G/H)$ with $\operatorname{Supp} \dot{f} \subset \dot{K}$ one can now take $f = (\dot{f} \circ \pi_H) f_1$ and therefore

$$|\dot{\mu}(\dot{f})| = \left|\int_G f(x)\, d\mu(x)\right| \le C_{\dot{K}}\, \|\dot{f}\|_\infty,$$

where $C_{\dot{K}} = \int_G |f_1(x)|\, d|\mu(x)|$, which is independent of $\dot{f}$. The measure $\dot{\mu}$ satisfies formula (4.5.1).

Summarizing:

Let μ *be a measure on* G. *Then condition* (4.5.2) *is necessary and sufficient for the existence of a measure* $\dot{\mu}$ *on* G/H *which satisfies* (4.5.1)

(ii) Now take a special kind of measure: $d\mu(x) = q(x)\, dx$, q a continuous function on G, dx a left Haar measure. The relation (4.5.2) becomes

$$\int_G f(x\eta^{-1})\, q(x)\, dx = \Delta_H(\eta) \int_G f(x)\, q(x)\, dx \qquad (\eta \in H).$$

Hence (4.5.2) is equivalent to

$$\int_G f(x)\,q(x\eta)\,\Delta(\eta)\,dx = \Delta_H(\eta)\int_G f(x)\,q(x)\,dx,$$

hence with

$$q(x\eta) = q(x)\frac{\Delta_H(\eta)}{\Delta(\eta)} \qquad (x \in G, \eta \in H). \tag{4.5.3}$$

One can show that there always exists (i.e. for any H) a continuous and strictly positive function q on G that satisfies (4.5.3) (cf. [38, Chapter 8, §1.7–1.9]).

Take such a function q. Then there exists a positive measure $d_q\dot{x}$ on G/H such that

$$\int_{G/H}\left\{\int_H f(x\xi)\,d\xi\right\}d_q\dot{x} = \int_G f(x)q(x)\,dx \qquad (f \in C_c(G)). \tag{4.5.4}$$

If H is a *normal* subgroup, then we can take $q = 1$ and we get Weil's formula, with $d_q\dot{x} = d\dot{x}$.

(iii) Let again q be a continuous, strictly positive, function on G satisfying (4.5.3). Set

$$\lambda_y(\dot{x}) = \frac{q(yx)}{q(x)} \qquad (\dot{x} = \pi_H(x)). \tag{4.5.5}$$

Notice that indeed λ_y is a function on G/H. The group G acts on G/H by

$$\Lambda_y(xH) = y^{-1}xH \qquad (y \in G).$$

These are homeomorphisms of G/H. The measure $d_q\dot{x}$ has the property: for all $\dot{f} \in C_c(G/H)$ one has

$$\int_{G/H}\dot{f}(\Lambda_y\dot{x})\,d_q\dot{x} = \int_{G/H}\dot{f}(\dot{x})\,\lambda_y(\dot{x})\,d_q\dot{x} \qquad (y \in G). \tag{4.5.6}$$

Choose f such that $T_H f = \dot{f}$ and apply (4.5.4).

We call $d_q(\dot{x})$ a *quasi-invariant measure* on G/H. If λ_y depends only on y (not on $\dot{x}$), then $d_q\dot{x}$ is said to be *relatively invariant*. If $\lambda_y = 1$ then $d_q\dot{x}$ is a *G-invariant measure* on G/H.

(iv) Suppose there is a continuous and strictly positive function r on G such that

$$\begin{cases} \text{(a)} & r(xy) = r(x)\,r(y) \quad (x, y \in G), \\ \text{(b)} & r(\xi) = \frac{\Delta_H(\xi)}{\Delta(\xi)} \qquad (\xi \in H). \end{cases} \tag{4.5.7}$$

If we then set $q(x) = r(x)$, then (4.5.3) holds. Furthermore we obtain $\lambda_y(\dot{x}) = r(y)$, so that $d_q\dot{x}$ is relatively invariant. Thus there exists a relatively invariant measure $d_r\dot{x}$ on G/H such that for all $\dot{f} \in C_c(G/H)$ one has

$$\int_{G/H} \dot{f}(\Lambda_y\dot{x})\, d_r\dot{x} = r(y) \int_{G/H} \dot{f}(\dot{x})\, d_r\dot{x} \tag{4.5.8}$$

and (4.5.4) holds with $q = r$. In particular: if

$$\Delta_H(\xi) = \Delta(\xi) \qquad (\xi \in H), \tag{4.5.9}$$

then there exists an *invariant* positive measure on G/H.

Conversely, if G/H admits a relatively invariant positive measure, then one might take $r(y)$ as in (4.5.8). One concludes that (4.5.7) (a) and (b) are satisfied and that r is continuous and strictly positive on G. In the case of an invariant measure, so $r = 1$, the property (4.5.9) follows from (4.5.7) (b).

Example. Let H be a closed *unimodular* subgroup of G, so $\Delta_H = 1$. Take $r(x) = \Delta(x)^{-1}$. Then G/H admits a relatively invariant measure. If H is compact, then G/H admits an invariant measure. The same holds if H is any closed *normal* subgroup.

Let both G and H be unimodular, then G/H admits an invariant measure. If G is unimodular and G/H admits an invariant measure, then H is unimodular.

(v) We shall now consider a *practical example*, where q can be determined in a simple way.

Assume $G = G_1 \cdot H$, where G_1 and H are closed subgroups with $G_1 \cap H = \{e\}$. So every $x \in G$ can be written uniquely in the form $x = g_1 h$ $(g_1 \in G_1,\ h \in H)$. Let us assume in addition that g_1 and h depend continuously on x. Then $x \mapsto (g_1, h)$ is a homeomorphism from G onto $G_1 \times H$.

Example. The Iwasawa decomposition of a semisimple Lie group, $G = K \cdot H$, $H = AN$. For $\mathrm{SL}(2, \mathbb{R})$ we have for example

$$K = \left\{ \begin{pmatrix} \cos\theta & \sin\theta \\ -\sin\theta & \cos\theta \end{pmatrix} : 0 \le \theta < 2\pi \right\},$$
$$A = \left\{ \begin{pmatrix} e^t & 0 \\ 0 & e^{-t} \end{pmatrix} : t \in \mathbb{R} \right\}, \quad N = \left\{ \begin{pmatrix} 1 & x \\ 0 & 1 \end{pmatrix} : x \in \mathbb{R} \right\}.$$

Define q as follows. For $h \in H$ let $q(h) = \frac{\Delta_H(h)}{\Delta(h)}$ and for general $x \in G$ set

$$q(x) = q(h) \text{ if } x = g_1 h \qquad (g_1 \in G_1,\ h \in H).$$

Then q is continuous, strictly positive, and satisfies (4.5.3). Since $q(g_1x) = q(x)$ $(g_1 \in G_1, x \in G)$, the measure $d_q\dot{x}$ is invariant under G_1. Since G/H is topologically isomorphic to G_1, the measure $d_q\dot{x}$ is a left Haar measure on G_1. So formula (4.5.4) becomes here

$$\int_{G_1}\left\{\int_H f(g_1h)\,dh\right\}dg_1 = \int_G f(x)\,q(x)\,dx \qquad (f \in C_c(G))$$

or, if we replace f by f/q,

$$\int_G f(x)dx = \int_{G_1}\left\{\int_H f(g_1h)\frac{\Delta(h)}{\Delta_H(h)}\,dh\right\}dg_1 \qquad (f \in C_c(G)). \quad (4.5.10)$$

Special cases of (4.5.10):

(a) G is *unimodular*. Then

$$\int_G f(x)dx = \int_{G_1}\left\{\int_H f(g_1h)\,\Delta_H(h^{-1})\,dh\right\}dg_1 \qquad (f \in C_c(G)).$$

Notice that $\Delta_H(h^{-1})dh$ is simply a right Haar measure on H.

(b) H is *normal*. Then

$$\int_G f(x)dx = \int_{G_1}\left\{\int_H f(g_1h)\,dh\right\}dg_1 \qquad (f \in C_c(G)).$$

This is also a direct consequence of Weil's formula.

We leave it to the reader to show that several of the formulae derived above also hold for functions $f \in L^1(G)$ etc. In most cases an exact formulation obvious.

4.6 The convolution product on G. Properties of $L^1(G)$

Literature: [29, §35], [34, §28].

As before, G will be a locally compact group with left Haar measure dx. We denote by $L^1(G)$ the set of (equivalence classes of) complex-valued integrable functions on G.

(i) Similar to the case $G = \mathbb{R}$ (see Section 2.1), we can define the *convolution product* of $f, g \in L^1(G)$ by

$$(f * g)(x) = \int_G f(y)\,g(y^{-1}x)\,dy \qquad (x \in G). \qquad (4.6.1)$$

One has $f * g \in L^1(G)$ and $\|f * g\|_1 \leq \|f\|_1 \|g\|_1$. With this operation as multiplication, $L^1(G)$ is an associative algebra and $f \mapsto \|f\|_1$ is a norm on $L^1(G)$. Even $L^1(G)$ is a Banach algebra. Moreover there is an *involution* on $L^1(G)$, $f \mapsto f^*$, given by

$$f^*(x) = \overline{f(x^{-1})}\,\Delta(x^{-1}) \qquad (x \in G).$$

One clearly has $\|f^*\|_1 = \|f\|_1$ and $(f * g)^* = g^* * f^*$. So $L^1(G)$ is a Banach algebra with involution.

(ii) Similar to the case $G = \mathbb{R}$ we have:

Theorem 4.6.1. *If $f \in L^1(G)$ and $g \in L^p(G)$ $(p \geq 1)$, then $f * g \in L^p(G)$ and $\|f * g\|_p \leq \|f\|_1 \|g\|_p$.*

For any function f on G set $\widetilde{f}(x) = \overline{f(x^{-1})}$ $(x \in G)$.

Theorem 4.6.2. (a) *If $f \in L^p(G)$, $g \in L^q(G)$ $(\frac{1}{p} + \frac{1}{q} = 1,\ p > 1)$, then $f * \widetilde{g}$ is a continuous function vanishing at infinity. In addition, $\|f * \widetilde{g}\|_\infty \leq \|f\|_p \|g\|_q$.*

(b) *If $f \in L^1(G)$ and $g \in L^\infty(G)$, then $f * \widetilde{g}$ is a bounded continuous function. Furthermore, $\|f * \widetilde{g}\|_\infty \leq \|f\|_1 \|g\|_\infty$.*

(iii) Fix a real number $p \geq 1$. For any function f on G set

$$L_y f(x) = f(y^{-1}x) \quad \text{and} \quad R_y f(x) = f(xy)\,\Delta(y)^{1/p} \qquad (x, y \in G).$$

The operators L_y and R_y are *isometries* in $L^p(G)$ for all $p \geq 1$. Furthermore the mappings $y \mapsto L_y f$ and $y \mapsto R_y f$ are *continuous* from G to $L^p(G)$ for each f in $L^p(G)$. This follows immediately using Section 3.6 (i).

(iv) *The Banach space $L^p(G)$ $(1 \leq p < \infty)$ is separable (i.e. has a countable dense set of elements) if G satisfies the second axiom of countability.*

(v) *The convolution algebra $L^1(G)$ is commutative if and only if G is an abelian group.*

If G is abelian, then G is unimodular, so we have for $f, g \in L^1(G)$,

$$\begin{aligned} f * g(x) &= \int_G f(y)\,g(y^{-1}x)\,dy = \int_G f(xy^{-1})g(y)\,dy \\ &= \int_G g(y)\,f(y^{-1}x)\,dy = g * f(x). \end{aligned}$$

Let, conversely, $L^1(G)$ be commutative. Take $f, g \in C_c(G)$. Then

$$0 = f * g(x) - g * f(x) = \int_G [f(y)g(y^{-1}x) - g(xy)f(y^{-1})]\, dy$$
$$= \int_G f(y)\,[g(y^{-1}x) - g(xy^{-1})\Delta(y^{-1})]\, dy.$$

Since this holds for all $f \in C_c(G)$, we get $g(y^{-1}x) = g(xy^{-1})\Delta(y^{-1})$ for all $x, y \in G$ and all $g \in C_c(G)$. Taking $x = e$ we get $\Delta = 1$, so $g(xy) = g(yx)$ for all $x, y \in G$ and all $g \in C_c(G)$. Since $C_c(G)$ separates the points of G, we obtain $xy = yx$ for all x and y. So G is an abelian group.

(vi) *The algebra $L^1(G)$ has a unit element if and only if G is discrete.*

Let G be discrete and let us assume that every point in G has mass equal to 1 with respect to the Haar measure. Let δ be the function that is equal to 1 at e and zero elsewhere. Then δ is the unit element of $L^1(G)$, i.e.

$$f * \delta(x) = \sum_{y \in G} f(y)\,\delta(y^{-1}x) = f(x),$$

similarly $\delta * f = f$.

Conversely, let $L^1(G)$ have a unit element, called δ. We assert that there is a positive lower bound for the masses of the non-empty open subsets. Indeed, otherwise there would be for any $\varepsilon > 0$ an open neighbourhood V of e with mass $V < \varepsilon$ and therefore one with $\int_V |\delta(x)|dx < \varepsilon$. Choose a symmetric neighbourhood U with $U^2 \subset V$ and let ξ be the characteristic function of U. Then we have for $x \in U$

$$1 = \xi(x) = \xi * \delta(x) = \int_G \delta(y)\xi(y^{-1}x)dx = \int_{xU} \delta(y)dy \le \int_V |\delta(y)|dy < \varepsilon.$$

We get a contradiction. So there is $\alpha > 0$ such that the mass of all non-empty open sets is greater than or equal to α. Select now an open subset V with compact closure. Then mass $V \ge n\alpha$ for $n = 1, 2, \ldots$ as soon as V contains at least n points. So any open set with compact closure is finite, so any point is open, hence G is discrete.

(vii) *The algebra $L^1(G)$ has an 'approximate unit'; even more: given $f \in L^p(G)$ $(1 \le p < \infty)$ and $\varepsilon > 0$, there exists a neighbourhood V of e such that $\|f * u - f\|_p < \varepsilon$ and $\|u * f - f\|_p < \varepsilon$ for all $u \in C_c(G)$ with $u \ge 0$,* Supp u *$\subset V$ and $\int_G u(x)\,dx = 1$.*

Take $h \in L^q(G)$ $(\frac{1}{p} + \frac{1}{q} = 1)$. Then one has

$$|(u * f - f, h)| = \left| \int_G \int_G u(y)\,\{f(y^{-1}x) - f(x)\}\overline{h(x)}\,dydx \right|$$
$$\le \|h\|_q \int_G \|L_y f - f\|_p u(y)dy,$$

by Hölder's inequality. So $\|u * f - f\|_p \leq \int_G \|L_y f - f\|_p u(y)dy$. Now choose V such that for all $y \in V$ one has $\|L_y f - f\|_p < \varepsilon$ (see (iii)). Then $\|u * f - f\|_p < \varepsilon$.

For $f * u$ the proof is similar, but a little more complicated because the group G need not be unimodular.

Otherwise formulated: select for any neighbourhood V of e a function $u = u_V$ as above. Then $\lim_{V \to \{e\}} u_V * f = f$ and $\lim_{V \to \{e\}} f * u_V = f$ for $f \in L^p(G)$ (convergence in $L^p(G)$). The family $\{u_V\}$ is called an *approximate unit.*

(viii) By Theorem 4.6.1, the space $L^p(G)$ $(p \geq 1)$ is a left $L^1(G)$-module. One has:

In the space $L^p(G)$ the closed left $L^1(G)$-submodules coincide with the closed left-invariant subspaces.

Let M be a closed left $L^1(G)$-submodule of $L^p(G)$. Choose an approximate unit $\{u_V\}$. If $f \in M$, then also $(L_x u_V) * f \in M$ for all $x \in G$. But $(L_x u_V) * f = L_x (u_V * f) \to L_x f$ if $V \to \{e\}$ in $L^p(G)$. Hence M is left-invariant.

Let now M be a closed left-invariant subspace of $L^p(G)$. Set $\mathrm{Ann}(M) = \{g \in L^q(G) : (f, g) = 0 \text{ for all } f \in M\}$. Then $f \in M$ if and only if $f \in \mathrm{Ann}(\mathrm{Ann}(M))$. Let now $h \in L^1(G)$, $f \in M$, $g \in \mathrm{Ann}(M)$. Then

$$(h * f, g) = \iint h(y) f(y^{-1}x)\overline{g(x)}\, dy dx = \int h(y) \Big[\int f(y^{-1}x)\overline{g(x)}\, dx \Big] dy = 0.$$

So $h * f \in M$.

As a corollary we have:

In $L^1(G)$ the closed left-(right-)ideals coincide with the closed left-(right-)invariant subspaces.

(ix) One can extend the convolution product to $M^1(G)$, the space of bounded measures on G. For $\mu_1, \mu_2 \in M^1(G)$ define

$$\mu_1 * \mu_2\,(f) = \int_G \int_G f(xy)\, d\mu_1(x) d\mu_2(y) \qquad (f \in C_c(G)).$$

The product is well-defined (use Section 3.6), $M^1(G)$ is thus a convolution algebra and $L^1(G)$ can be regarded as a subalgebra of $M^1(G)$, by associating to $f \in L^1(G)$ the measure $f(x)\, dx$.

Chapter 5
Harmonic Analysis on Locally Compact Abelian Groups

Literature: [7].

In this chapter G will be a locally compact (later on, abelian) group, satisfying *the second axiom of countability*, with left Haar measure dx.

5.1 Positive-definite functions and unitary representations

(i) Unitary representations

Let $\mathcal{H}$ be a complex Hilbert space with scalar product denoted by $\langle\,,\rangle$. We set $\mathrm{End}(\mathcal{H})$ for the algebra of continuous linear operators on $\mathcal{H}$ and denote by $\mathrm{GL}(\mathcal{H})$ the group of invertible elements in $\mathrm{End}(\mathcal{H})$. An operator U is said to be *unitary* if $UU^* = U^*U = I$.

A *representation* of G is a pair $(\pi, \mathcal{H})$ of a Hilbert space $\mathcal{H}$ and a homomorphism $\pi : G \to \mathrm{GL}(\mathcal{H})$ such that the mapping $(x, v) \mapsto \pi(x)v$ from $G \times \mathcal{H}$ to $\mathcal{H}$ is continuous.

One calls $(\pi, \mathcal{H})$ a *unitary representation* if $\pi(x)$ is a unitary operator for all $x \in G$.

Lemma 5.1.1. *Let π be a homomorphism of G into $\mathrm{GL}(\mathcal{H})$ such that $\pi(x)$ is unitary for all $x \in G$. Then the following statements are equivalent:*

(a) *$(x, v) \mapsto \pi(x)v$ is continuous from $G \times \mathcal{H}$ to $\mathcal{H}$,*

(b) *$x \mapsto \pi(x)v$ is continuous for every $v \in \mathcal{H}$,*

(c) *$x \mapsto \langle \pi(x)v, w\rangle$ is a continuous function on G for each pair of vectors $v, w \in \mathcal{H}$.*

Remark 5.1.2. Lemma 5.1.1 is also true without the assumption of unitarity of $\pi(x)$. However, the proof becomes more difficult, see [57, Prop. 4.2.2.1]. The proof of Lemma 5.1.1 (for unitary π) is straightforward.

A linear subspace $\mathcal{H}_1 \subset \mathcal{H}$ is called *invariant* if $\pi(x)\mathcal{H}_1 \subset \mathcal{H}_1$ for all $x \in G$.

If $(\pi, \mathcal{H})$ and (σ, W) are representations of G, we denote by $\mathrm{Hom}_G(\pi, \sigma)$ the space of all bounded operators $A : \mathcal{H} \to W$ satisfying $A\,\pi(x) = \pi(x)\,A$ for all $x \in G$. Such operators A are called *intertwining operators*. The two representations are said to be *equivalent* if there is an $A \in \mathrm{Hom}_G(\pi, \sigma)$ which is bijective. If $(\pi, \mathcal{H})$ and (σ, W) are both unitary representations, then we say that π and σ are *unitarily equivalent* if there is an isometric bijection in $\mathrm{Hom}_G(\pi, \sigma)$.

It is known that equivalence for unitary representations implies unitary equivalence. This can be seen by applying polar decomposition of continuous linear operators, see for example [8, 2.2.2].

A representation $(\pi, \mathcal{H})$ of G is called *irreducible* (also: topologically irreducible) if the only *closed* invariant subspaces of $\mathcal{H}$ are $\{0\}$ and $\mathcal{H}$ itself.

It is customary to denote by $\hat{G}$ the set of unitary equivalence classes of irreducible unitary representations of G.

Lemma 5.1.3. *Let $(G, \mathcal{H})$ be a unitary representation.*

(a) *If $\mathcal{H}_1 \subset \mathcal{H}$ is an invariant subspace, then so is $\mathcal{H}^{\perp}$.*

(b) *If $\mathcal{H}_1 \subset \mathcal{H}$ is a closed subspace and P is the orthogonal projection of $\mathcal{H}$ onto $\mathcal{H}_1$, then $\mathcal{H}_1$ is an invariant subspace if and only if P commutes with $\pi(x)$ for all $x \in G$.*

For *unitary* representations we have the following *criteria* for irreducibility.

Theorem 5.1.4. *Let $(\pi, \mathcal{H})$ be a unitary representation. The following criteria are equivalent:*

(a) *The only orthogonal projections commuting with all $\pi(x)$ $(x \in G)$ are the zero operator and the identity operator I.*

(b) $\mathrm{Hom}_G(\pi, \pi) = \mathbb{C} \cdot I$ *(Schur's lemma).*

(c) *Every vector $v \neq 0$ in $\mathcal{H}$ is cyclic, i.e. the closure of* $\mathrm{span}(\pi(G)v)$ *is $\mathcal{H}$.*

(d) *$(\pi, \mathcal{H})$ is irreducible.*

Criterion (a) implies (b). This is commonly proved with the spectral decomposition of bounded self-adjoint operators. Let A be in $\mathrm{Hom}_G(\pi, \pi)$. Then $A^* \in \mathrm{Hom}_G(\pi, \pi)$. So for the proof we may assume A to be self-adjoint. Write $A = \int_{-\infty}^{\infty} \lambda dE_\lambda$ (spectral decomposition of A). Since $\pi(x)A\pi(x^{-1}) = A$ for all $x \in G$, the same holds for all E_λ: $\pi(x)E_\lambda\pi(x^{-1}) = E_\lambda$ for all $x \in G$. By (a) we get $E_\lambda = O$ or I. So $A = c\,I$ for some constant c. So (a) and (b) are equivalent. The criteria (a) and (c) are easily shown to imply irreducibility and to be equivalent, applying Lemma 5.1.3 (b). Furthermore, (a) and (d) are equivalent, again by this lemma.

Let $(\pi, \mathcal{H})$ be unitary and $f \in L^1(G)$. For any pair of vectors $v, w \in \mathcal{H}$ the integral

$$\int_G f(x) \langle \pi(x)v, w \rangle \, dx$$

exists and one has

$$\left| \int_G f(x) \langle \pi(x)v, w \rangle \, dx \right| \leq \|f\|_1 \, \|v\| \, \|w\|.$$

Therefore $A \in \mathrm{End}(\mathcal{H})$ exists with $\|A\| \leq \|f\|_1$ and

$$\langle Av, w \rangle = \int_G f(x) \langle \pi(x)v, w \rangle \, dx \qquad (v, w \in \mathcal{H}).$$

We shall write $A = \pi(f)$ (also: $\hat{f}(\pi)$), so in particular $\|\pi(f)\| \leq \|f\|_1$.

Examples. 1. Let G be as before. For $s \in G$ let L_s be the operator on $L^2(G)$ defined by

$$(L_s f)(x) = f(s^{-1}x) \qquad (f \in L^2(G), x \in G).$$

We shall denote the mapping $s \mapsto L_s$ by L. One verifies that $(L, L^2(G))$ is a unitary representation of G, the *left regular representation.* Let Δ be the Haar modulus of G. For $s \in G$ let R_s be the operator on $L^2(G)$ defined by

$$(R_s f)(x) = \Delta^{1/2}(s) \, f(xs) \qquad (f \in L^2(G), \, x \in G).$$

Denoting the mapping $s \mapsto R_s$ by R, one easily checks that $(R, L^2(G))$ is a unitary representation of G as well, the *right regular representation.*

For $f \in L^2(G)$ define $f' \in L^2(G)$ by

$$f'(x) = \Delta^{-1/2}(x) \, f(x^{-1}).$$

Set $Af = f'$. Then A is a unitary operator from $L^2(G)$ onto $L^2(G)$ and, in addition, $A \in \mathrm{Hom}_G(L, R)$. *Therefore L and R are unitarily equivalent.*

2. Any irreducible unitary representation of a locally compact *abelian* group G is one-dimensional. This follows at once from Schur's lemma. If $(\pi, \mathcal{H})$ is such a representation, then one can write $\pi(x) = \chi(x)I$ for all $x \in G$, where χ is a (continuous) homomorphism of G into $\mathbb{T}$. We call χ a (unitary) *character* of G. We leave it to the reader to show that in the case $G = \mathbb{R}$ the characters are given by $\chi_y(x) = \mathrm{e}^{2\pi i x y}$ $(x \in \mathbb{R})$, where y runs over $\mathbb{R}$.

The following results are standard, and will not be proved here. See [53].

3. *Any finite-dimensional representation of a finite group is equivalent to a unitary representation.*

4. *Any representation of a compact group on a Hilbert space is equivalent to a unitary representation.*

5. *The group* $G = \mathrm{SL}(2,\mathbb{R})$ *admits no non-trivial finite-dimensional unitary representations.*

(ii) Positive-definite functions

We begin with an unusual definition.

Definition 5.1.5. A complex-valued locally integrable function φ is said to be positive-definite if

$$\int_G \int_G \varphi(x^{-1}y)\, f(x)\, \overline{f(y)}\, dx dy \geq 0$$

for all $f \in C_c(G)$.

Such a function φ defines in a natural way a Hilbert space $\mathcal{H}_\varphi$ as follows. For $f, g \in C_c(G)$ define

$$(f, g)_\varphi = \int_G \int_G \varphi(x^{-1}y)\, f(x)\, \overline{f(y)}\, dx dy. \tag{5.1.1}$$

This gives a positive sesqui-linear Hermitean form on $C_c(G)$. Let $K_\varphi = \{f : (f, f)_\varphi = 0\}$. From the Cauchy–Schwarz inequality (which holds for such forms) it follows that K_φ is a linear subspace of $C_c(G)$. In a natural way we thus have on $C_c(G)/K_\varphi$ a genuine scalar product. Let $\mathcal{H}_\varphi$ be the Hilbert space completion of this space.

For $s \in G$ let L_s be defined on $C_c(G)$ by $L_s f(x) = f(s^{-1}x)$, $x \in G$. It is clear that $(L_s f, L_s g)_\varphi = (f, g)_\varphi$ for $f, g \in C_c(G)$. Thus L_s can be uniquely extended to a unitary operator on $\mathcal{H}_\varphi$. We even obtain a *unitary representation* of G on $\mathcal{H}_\varphi$: (L, H_φ). To show the continuity of this representation, it suffices to show that

$$s \mapsto (L_s f, g)_\varphi$$

is continuous for any pair $f, g \in C_c(G)$. It even suffice to show the continuity at $s = e$. One has

$$\begin{aligned}
|(L_s f - f, g)_\varphi| &= \left| \int\!\!\int \varphi(x^{-1}y)\, \{L_s f(x) - f(x)\}\, \overline{g(y)}\, dx dy \right| \\
&= \left| \int\!\!\int \varphi(y)\, \{L_s f(x) - f(x)\}\, \overline{g(xy)}\, dx dy \right| \\
&\leq \|L_s f - f\|_1 \cdot \|g\|_\infty \cdot \int_K |\varphi(y)|\, dy,
\end{aligned}$$

where K is a compact subset of G, depending on Supp f and Supp g. Now apply Section 4.6 (iii).

Theorem 5.1.6. *Any bounded positive-definite function φ (i.e. $\varphi \in L^\infty$) coincides almost everywhere with a continuous positive-definite function.*

Let V be a neighbourhood of e and choose $\{u_V\}$ as in Section 4.6 (vii). For $f \in C_c(G)$ one has

$$(f, u_V)_\varphi \to \int_G f(x)\,\overline{\varphi(x)}\,dx \quad \text{as} \quad V \to \{e\}. \tag{5.1.2}$$

Indeed,

$$\begin{aligned}(f, u_V)_\varphi &= \int_G \int_G f(x)\,\overline{u_V(y)}\,\varphi(x^{-1}y)\,dx dy\\ &= \int_G \int_G \overline{u_V(x)}\,f(y)\,\overline{\varphi(x^{-1}y)}\,dx dy = \int_G \overline{u_V(x)}\,f * \widetilde{\varphi}(x)\,dx\\ &\to f * \widetilde{\varphi}(e) \quad \text{as} \quad V \to \{e\},\end{aligned}$$

because $f * \widetilde{\varphi}$ is continuous. Furthermore we have for $f \in C_c(G)$

$$\left| \int_G f(x)\,\overline{\varphi(x)}\,dx \right| \le \|\varphi\|_\infty^{1/2}\,\|f\|_\varphi. \tag{5.1.3}$$

Indeed,

$$|(f, u_V)_\varphi|^2 \le (f, f)_\varphi\,(u_V, u_V)_\varphi \le \|\varphi\|_\infty \cdot \|f\|_\varphi^2$$

for all V. So we can extend $f \mapsto \int_G f(x)\,\overline{\varphi(x)}\,dx$ to a bounded linear form on $\mathcal{H}_\varphi$. Consequently $\varepsilon \in \mathcal{H}_\varphi$ exists such that

$$(f, \varepsilon)_\varphi = \int_G f(x)\,\overline{\varphi(x)}\,dx \quad \text{for} \quad f \in C_c(G).$$

Then also

$$(L_{s^{-1}} f, \varepsilon)_\varphi = \int_G f(s^{-1}x)\,\overline{\varphi(x)}\,dx$$

and

$$(f, L_s \varepsilon)_\varphi = \int_G f(x)\,\overline{\varphi(s^{-1}x)}\,dx$$

for $f \in C_c(G)$. We now have

$$(f, g)_\varphi = \int_G (f, L_s\,\varepsilon)_\varphi\,\overline{g(s)}\,ds \tag{5.1.4}$$

for all $f, g \in C_c(G)$.

Both sides of (5.1.4) are continuous in f, hence (5.1.4) holds for all $f \in \mathcal{H}_\varphi$ and $g \in C_c(G)$. In particular

$$(\varepsilon, g)_\varphi = \int_G (\varepsilon, L_s\,\varepsilon)_\varphi\,\overline{g(s)}\,ds = \overline{(g,\varepsilon)_\varphi}$$
$$= \int_G \varphi(s)\,\overline{g(s)}\,ds$$

for all $g \in C_c(G)$. We may conclude that $\varphi(s) = (\varepsilon, L_s\,\varepsilon)_\varphi$ almost everywhere. The right-hand side is continuous (and positive-definite).

Remark 5.1.7. (1) Apparently any bounded *continuous* positive-definite function φ is of the form $\varphi(x) = \langle \varepsilon, \pi(x)\varepsilon\rangle$ $(x \in G)$, where π is a unitary representation of G on a Hilbert space $\mathcal{H}$ and $\varepsilon \in \mathcal{H}$.

(2) The vector $\varepsilon \in \mathcal{H}_\varphi$ is *cyclic*: is $(f, L_s\,\varepsilon)_\varphi = 0$ for all $s \in G$ and for some $f \in \mathcal{H}_\varphi$, then $(f, g)_\varphi = 0$ for all $g \in C_c(G)$ (see (5.1.4)), hence $f = 0$.

(3) Given a unitary representation $(\pi, \mathcal{H})$ of G and $\varepsilon \in \mathcal{H}$, then the function $\varphi(x) = \langle \varepsilon, \pi(x)\,\varepsilon\rangle$ is continuous and positive-definite. Indeed we have

$$\int_G\int_G f(x)\,\overline{f(y)}\,\langle \varepsilon, \pi(x^{-1}y)\,\varepsilon\rangle\,dx dy$$
$$= \int_G\int_G f(x)\,\overline{f(y)}\,\langle \pi(x)\varepsilon, \pi(y)\,\varepsilon\rangle\,dx dy = \|\pi(f)\varepsilon\|^2 \geq 0$$

for all $f \in C_c(G)$.

(4) One can show (exercise): *if $(\pi, \mathcal{H})$ is a unitary representation of G with cyclic vector v, and $\varphi(x) = \langle v, \pi(x)v\rangle$ $(x \in G)$, then $(\pi, \mathcal{H})$ is unitarily equivalent to $(L, \mathcal{H}_\varphi)$ by means of a unitary intertwining operator $A : \mathcal{H} \to \mathcal{H}_\varphi$ which maps v to ε.*

(5) *Any continuous positive-definite function φ is also positive-definite in the following sense: for any n, any n-tuples $x_1, \dots, x_n \in G$ and $\alpha_1, \dots, \alpha_n \in \mathbb{C}$ one has*

$$\sum_{i,j=1}^{n} \alpha_i\overline{\alpha_j}\varphi(x_i^{-1}x_j) \geq 0.$$

The converse is also true (cf. [8, 13.4.4]).

Similar to Section 2.7 one can show:

Lemma 5.1.8. *Let φ be a continuous positive-definite function. Then*

(a) $\varphi(e) \geq 0$,

(b) $\varphi = \widetilde{\varphi}$,

(c) $|\varphi(x)| \leq \varphi(e)$ *for all x.*

In particular, any continuous positive-definite function is bounded.

We conclude this section with two exercises:

(1) If φ_1 and φ_2 are continuous positive-definite functions, then the pointwise product $\varphi_1 \cdot \varphi_2$ is also a continuous positive-definite function.

(2) Let φ be a continuous positive-definite function with $\varphi(e) = 1$. The set of all $x \in G$ for which $\varphi(x) = 1$ is a closed subgroup of G.

5.2 Some functional analysis

(i) Weak topologies

Let X be a normed complex linear space with dual space X^*. The weak topology on X, denoted by $\sigma(X, X^*)$, is the weakest topology such that all linear forms $x \mapsto (x, x^*)$ $(x^* \in X^*)$ are continuous.

A *weak neighbourhood* basis of x_0 in X is given by the sets of the form

$$O(x_0; x_1^*, \dots, x_n^*; \varepsilon) = \{x \in X : |(x - x_0, x_i^*)| < \varepsilon \ (i = 1, \dots, n)\},$$

where $x_1^*, \dots, x_n^* \in X^*$ and $\varepsilon > 0$.

Provided with this topology X is a topological vector space.

A sequence $\{x_n\}$ in X *converges weakly* to $x_0 \in X$ if $\lim_{n\to\infty}(x_n, x^*) = (x_0, x^*)$ for every $x^* \in X^*$.

The norm topology is stronger than the weak topology, hence a weakly closed linear subspace of X is certainly strongly closed.

Theorem 5.2.1. *A strongly closed linear subspace L of X is also weakly closed.*

If $x_0 \notin L$, then $x_0^* \in X^*$ exists with $x_0^*(L) = \{0\}$ and $(x_0, x_0^*) = 1$ (Hahn–Banach's theorem). So $O(x_0; x_0^*; 1) \cap L = \emptyset$. Hence the complement of L is weakly open.

One also provides X^* with a weak topology, denoted by $\sigma(X^*, X)$. It is the weakest topology making all forms $x^* \mapsto (x, x^*)$ $(x^* \in X^*; x \in X)$ continuous. A neighbourhood basis of $x^* \in X^*$ is then determined by finitely many elements $x_1, \dots, x_n$ in X and $\varepsilon > 0$:

$$O(x_0^*; x_1, \dots, x_n; \varepsilon) = \{x^* \in X^* : |(x^* - x_0^*, x_i)| < \varepsilon \ (i = 1, \dots, n)\}.$$

Notice the difference between the way of defining $\sigma(X, X^*)$ and $\sigma(X^*, X)$. If X is *reflexive* both definitions are similar.

Theorem 5.2.2 (Alaoglu's theorem). *For any normed linear space X the unit ball $S^* = \{x^* : \|x^*\| \leq 1\}$ in X^* is compact in the weak topology $\sigma(X^*, X)$.*

Assign to every $x \in X$ the compact disk $C_x = \{\lambda \in \mathbb{C} : |\lambda| \leq \|x\|\}$. The Cartesian product $P = \prod\{C_x : x \in X\}$ consists of all (uncountable) "sequences" $\{\lambda_x\}$ with $\lambda_x \in C_x$ for all $x \in X$. According to Tychonov's theorem P is compact in the product topology. Any $x^* \in X^*$ is determined by its values (x, x^*) $(x \in X)$. Assign to $x^* \in S^*$ the "sequence" $\{\lambda_x\} = \{(x, x^*)\}$. Since $\|x^*\| \leq 1$ we have $|\lambda_x| \leq \|x\|$, so $\lambda_x \in C_x$ and $\{\lambda_x\} \in P$. By $x^* \mapsto \{(x, x^*)\}$ we get a mapping from S^* onto a set $T \subset P$. From the definitions of the topology in P and S^* it follows that this mapping is topological. It is therefore sufficient to show that T is closed in P.

Take a point $\{\mu_x\}$ in $\overline{T}$, the closure of T. Every neighbourhood of this point contains an element of T, thus a continuous linear form on X; we shall show that $\{\mu_x\}$ shares this property with its neighbours. For $x_0, y_0 \in X$ let z stand for x_0, y_0, cx_0 or x_0+y_0 $(c \in \mathbb{C})$, while $\varepsilon > 0$. The set of all elements $\{\lambda_x\} \in P$ with $|\lambda_z - \mu_z| < \varepsilon$ is a neighbourhood of $\{\mu_x\}$ and contains a point $\{(x, x^*)\}$ of T. This yields four inequalities, e.g. $|(x_0 + y_0, x^*) - \mu_{x_0+y_0}| < \varepsilon$ for $z = x_0 + y_0$; we get

$$|\mu_{x_0+y_0} - \mu_{x_0} - \mu_{y_0}| < 3\varepsilon, \; |c\mu_{x_0} - \mu_{cx_0}| < (|c| + 1)\,\varepsilon \quad \text{for all } \varepsilon > 0.$$

Consequently $\{\mu_x\} \in T$, so $T = \overline{T}$.

(ii) The adjoint of an operator

Let X and Y be complex normed linear spaces and T a continuous linear operator from X into Y, so $T \in \mathrm{Hom}\,(X, Y)$. We shall write $\mathrm{End}(X)$ for $\mathrm{Hom}(X, X)$. For $y^* \in Y^*$ the mapping $x \mapsto (Tx, y^*)$ is a continuous linear form on X, because

$$|(Tx, y^*)| \leq \|y^*\|\,\|T\|\,\|x\|,$$

so in X^*. Call this form T^*y^*. We thus have

$$(Tx, y^*) = (x, T^*y^*) \qquad (x \in X,\ y^* \in Y^*).$$

We immediately see that $T^* \in \mathrm{Hom}\,(Y^*, X^*)$. The operator T^* is called the *adjoint* of T (this definition is due to Banach). We observe that $\|T^*\| \leq \|T\|$.

Lemma 5.2.3. *Let $T \in \mathrm{Hom}(X, Y)$. Then $\|T^*\| = \|T\|$.*

From $\|Tx\| = \sup\{|(Tx, y^*)| : \|y^*\| \leq 1\}$ (by Hahn–Banach's theorem) it follows

$$\begin{aligned}\|T^*\| &= \sup_{\|y^*\|\leq 1} \|T^*y^*\| = \sup_{\|y^*\|\leq 1}\sup_{\|x\|\leq 1} |(Tx, y^*)| \\ &= \sup_{\|x\|\leq 1} \|Tx\| = \|T\|.\end{aligned}$$

Observe that there is a slight difference with the definition of the (Hilbert-) adjoint of $A \in \mathrm{End}(\mathcal{H})$, when $\mathcal{H}$ is a Hilbert space.

The proof of the next theorem is left to the reader.

Theorem 5.2.4. *Let X and Y be complex normed linear spaces.*

(a) *If $T \in \mathrm{Hom}(X, Y)$, then $T^* : Y^* \to X^*$ is continuous with respect to the weak topologies of Y^* and X^*.*

(b) *If $T \in \mathrm{Hom}(X, Y)$ and $S \in \mathrm{Hom}(Y, Z)$, then $(ST)^* = T^*S^*$.*

(c) *The adjoint of the identity operator on X is the identity operator on X^*.*

The mapping $T \mapsto T^*$ is therefore an anti-isomorphism from $\mathrm{End}(X)$ into $\mathrm{End}(X^*)$.

(iii) Krein–Milman's theorem

For the first part of this paragraph we refer to [34, I §3] and [4].

Let X be a real or complex vector space. One calls X a *topological linear space* or topological vector space if X is provided with a Hausdorff topology such that addition and scalar multiplication are continuous operations.

A topological linear space X is said to be *locally convex* if zero has a neighbourhood basis consisting of symmetric convex sets (i.e. sets S satisfying: if $x \in S$ then also $\alpha x \in S$ for all scalars α with $|\alpha| \leq 1$).

Let K be a *convex* set in a locally convex space X, which contains zero as an interior point. The set K thus contains a symmetric convex neighbourhood U of zero. Let $x \in X$. There exists $\delta > 0$ such that $\alpha x \in U$ for $|\alpha| < \delta$ and thus $\alpha x \in K$ for $|\alpha| < \delta$. So $x \in \frac{1}{\alpha} K$ for $|\alpha| < \delta$.

Define $p(x) = p_K(x) = \inf\{\beta : \beta > 0,\ x \in \beta K\}$ $(x \in X)$.

Lemma 5.2.5. *The function p is a convex functional: $p(x) \geq 0$; $p(x + y) \leq p(x) + p(y)$; $p(\alpha x) = \alpha p(x)$ for $\alpha > 0$.*

Clearly $p(x) \geq 0$. Furthermore it easily follows from the definition of p that $p(\alpha x) = \alpha p(x)$ for $\alpha \geq 0$. For any $\varepsilon > 0$ we have $x \in \{p(x) + \varepsilon\} K$. So $\frac{1}{p(x)+\varepsilon} x \in K$ and (similarly) $\frac{1}{p(y)+\varepsilon} y \in K$. Since K is convex, we also have $(1-t)\frac{1}{p(x)+\varepsilon} x + t\frac{1}{p(y)+\varepsilon} y \in K$ for $0 \leq t \leq 1$. Solving t from the equation $\frac{1-t}{p(x)+\varepsilon} = \frac{t}{p(y)+\varepsilon}$ gives $t = \frac{p(y)+\varepsilon}{p(x)+p(y)+\varepsilon}$. Substituting this value of t yields: $\frac{x+y}{p(x)+p(y)+\varepsilon} \in K$, so $x+y \in (p(x)+p(y)+2\varepsilon) K$ for all $\varepsilon > 0$. This gives $p(x + y) \leq p(x) + p(y) + 2\varepsilon$ and hence $p(x + y) \leq p(x) + p(y)$.

Clearly $p = p_K$ is symmetric, i.e. $p(\alpha x) = |\alpha|\, p(x)$ (α a scalar, $x \in X$), as soon as K is a symmetric (convex) set.

For any $x \in K$ we have $p(x) \leq 1$. On the other hand, if $p(x) \leq 1$ then $x \in K$. So $p(x) \geq 1$ for all $x \notin K$.

Notice that p is *continuous* on X. Therefore the set $\dot{K}$ of interior points of K is equal to $\dot{K} = \{x : p(x) < 1\}$. Furthermore the *boundary* of K is equal to $\overline{K} \backslash \dot{K} = \{x : p(x) = 1\}$.

Examples of locally convex linear spaces are: normed linear spaces, normed linear spaces X provided with the topology $\sigma(X, X^*)$ and X^* provided with $\sigma(X^*, X)$.

The following lemma is interesting, but it is not used in this section. Nevertheless, we like to mention it.

Lemma 5.2.6. *A locally convex linear space admits sufficiently many continuous linear forms to separate its points.*

Let $x_0 \neq 0$ in X. According to the above theory there is a symmetric convex functional p with $p(x_0) > 0$. Set $f(\alpha x_0) = \alpha p(x_0)$. The function f is a linear form on $\mathbb{C}x_0$, such that $|f(x)| \leq p(x)$ for all $x \in \mathbb{C}x_0$ and $f(x_0) = p(x_0)$. According to Hahn–Banach's theorem, f can be extended to a linear form on X such that $|f(x)| \leq p(x)$ for all $x \in X$. Since p is continuous, f is continuous as well.

After these introductory remarks, we can formulate Krein–Milman's theorem.

Let K be a set in a locally convex linear space X. A point $x_0 \in K$ is said to be an *extremal point* of K if x_0 can not be written in the form $ta + (1-t)b$ for some $a, b \in K$ and $0 < t < 1$. We shall denote the set of extremal points of K by $\mathrm{ext}(K)$. Example: $X = \mathbb{R}^n$, $K = \{x : \|x| \leq 1\}$; $\mathrm{ext}(K) = \{x : \|x\| = 1\}$. By $\overline{\mathrm{co}}\,(K)$ we denote the *closed convex hull* of K, the smallest closed convex set containing K.

Theorem 5.2.7 (Krein–Milman). *If K is a compact subset of a locally convex linear space X and E the set of its extremal points, then $\overline{\mathrm{co}}(E) \supset K$. So $\overline{\mathrm{co}}(E) = \overline{\mathrm{co}}(K)$ and $\overline{\mathrm{co}}\,(E) = K$ if K is convex.*

We shall prove this theorem in a number of steps. The proof is similar to the one given in [10, V §8].

First of all we take X *real* and we shall write X_r to emphasize this fact.

Let S_1 and S_2 be non-trivial subsets of X_r. We shall say that S_1 *can be separated by* S_2 if there exists a continuous linear form $u \neq 0$ on X_r and a constant c such that $u(S_1) \leq c$ and $u(S_2) \geq c$. We shall write in that case S_1 *sep* S_2.

The next proposition is evident.

Proposition 5.2.8. *Let S_1 and S_2 be as above. Then one has:*

(a) *S_1 sep S_2 if and only if S_2 sep S_1 (symmetry).*

(b) *S_1 sep S_2 if and only if $S_1 - S_2$ sep $\{0\}$ (transitivity).*

(c) *S_1 sep S_2 if and only if $S_1 - x$ sep $S_2 - x$ $(x \in X)$.*

Lemma 5.2.9 (Separation theorem). *Let M and N be non-trivial, disjoint, convex sets. Assume that M has an interior point m. Then M sep N.*

According to Proposition 5.2.8 (c) we can take $m = 0$. Fix $p \in N$. Then $-p \in M - N$ and $-p$ is an interior point of $M - N$. Hence zero is an interior point of $Q = M - N + p$. Since $M \cap N = \emptyset$, we have $0 \notin M - N$, so $p \notin Q$. The set Q is a convex neighbourhood of zero. Then p_Q is defined and define $u_p(\alpha\, p) = \alpha\, p_Q(p)$ $(\alpha \in \mathbb{R})$. We have $p_Q(p) \geq 1$. Extend (using Hahn–Banach's theorem) the linear form u_p on $\mathbb{R}\, p$ to all of X_r, satisfying $|u_p(x)| \leq p_Q(x)$ for all x. Then $u_p(Q) \leq 1$, $u_p(p) \geq 1$. Hence Q *sep* $\{p\}$, so $M - N$ *sep* $\{0\}$, so M *sep* N.

Lemma 5.2.10. *Let C be a closed convex subset of X_r and $p \notin C$. Then there is $u \in X_r^*$ and $\varepsilon > 0$ such that*

$$u(C) \leq c - \varepsilon < c = u(p).$$

The set $p - C$ is closed and convex, $0 \notin p - C$. There is a convex symmetric neighbourhood U_0 of 0 with $U_0 \cap (p - C) = \emptyset$. By the separation theorem there exists $u \in X_r^*$ and a constant d with $u(p - C) \geq d$ and $u(U_0) \leq d$, $u \neq 0$. We may assume $d > 0$, since $u \neq 0$. Hence $u(p) \geq u(C) + d$. In addition $u(p) \geq u(p) - d \geq u(C)$. Take $c = u(p)$ and $\varepsilon = d$.

We shall now consider the *complex* case. So let X_c be a complex locally convex linear space. Of course, X_c can also be seen as a real locally convex linear space, which we call then X_r. If $f \in X_c^*$, then $\operatorname{Re} f \in X_r^*$. Conversely, every $g \in X_r^*$ is the real part of some $f \in X_c^*$; indeed take

$$f(x) = g(x) + i g(ix).$$

The correspondence $f \mapsto \operatorname{Re} f$ from X_c^* to X_r^* is one-to-one (and onto). The following lemma is obvious.

Lemma 5.2.11. *Let C be a closed, convex subset of X_c and let $p \notin C$. Then there is $u \in X_c^*$ and $\varepsilon > 0$ such that*

$$\operatorname{Re} u(C) \leq c - \varepsilon < c = \operatorname{Re} u(p).$$

Let K be a subset of a real or complex linear space. A non-empty subset A of K is called an *extremal subset* of K if the following property is satisfied: if a strictly convex combination $\{tk_1 + (1 - t)k_2 | 0 < t < 1\}$, of two points $k_1, k_2 \in K$ is contained in A, then both k_1 and k_2 belong to A. An extremal subset of K consisting of a single point is an *extremal point* of K.

Lemma 5.2.12. *Any non-empty compact subset of a locally convex linear space has extremal points.*

Let $K \subset X$ be compact, X locally convex, K non-empty. Denote by $\mathcal{A}$ the (non-empty) family of closed non-empty extremal subsets of K. We order $\mathcal{A}$ by inclusion. One easily sees that if $\mathcal{A}_1$ is a totally ordered subfamily of $\mathcal{A}$, then $\bigcap_{A \in \mathcal{A}_1} A$ is a non-empty, closed, extremal subset of K, which is a lower bound for $\mathcal{A}_1$. By Zorn's lemma there exists a minimal element A_0 in $\mathcal{A}$. Suppose that A_0 contains at least two different points p and q. Then we can find $x^* \in X^*$ with $\operatorname{Re} x^*(p) \neq \operatorname{Re} x^*(q)$. Hence $A_1 = \{x : x \in A_0, \operatorname{Re} x^*(x) = \inf_{y \in A_0} \operatorname{Re} x^*(y)\}$ is a proper, non-empty, subset of A_0. On the other hand, if $k_1, k_2 \in K$ and $t k_1 + (1-t) k_2 \in A_1$ for some $0 < t < 1$, then $k_1, k_2 \in A_0$. According to the definition of A_1 we even have $k_1, k_2 \in A_1$. So A_1 is a proper, closed, extremal subset of A_0, which yields a contradiction. Hence A_0 contains only one point, an extremal point of K.

Lemma 5.2.13. *Let K be a subset of a linear space, A_1 an extremal subset of K and A_2 an extremal subset of A_1. Then A_2 is an extremal subset of K.*

The proof of this lemma is easy and left to the reader.

We are now ready to prove Krein–Milman's theorem.

Let there exist a $p \in K$ with $p \notin \overline{\text{co}}(E)$. By Lemma 5.2.11 there is a $x^* \in X^*$, a constant c and $\varepsilon > 0$ with $\operatorname{Re} x^*(p) \le c$, $\operatorname{Re} x^*(\overline{\text{co}}\,(E)) \ge c + \varepsilon$. Set

$$K_1 = \{x : x \in K,\ \operatorname{Re} x^*(x) = \inf_{y \in K} \operatorname{Re} x^*(y)\}.$$

The set K_1 is a closed extremal subset of K and $K_1 \cap E = \emptyset$. By Lemma 5.2.13 , K_1 has no extremal points. But this contradicts Lemma 5.2.12. So $\overline{\text{co}}\,(E) \supset K$. The remaining statements are clear.

5.3 Elementary positive-definite functions

Let $\mathcal{P}_0$ be the set of all continuous positive-definite functions φ on G satisfying $\varphi(e) \le 1$. We identify $\mathcal{P}_0$ with the subset of $L^\infty(G)$ consisting of the positive-definite functions φ with $\|\varphi\|_\infty \le 1$. Clearly $\mathcal{P}_0$ is weakly, i.e. $\sigma(L^\infty,\ L^1)$-, *closed*, so by Alaoglu's theorem it is weakly compact. It is therefore a compact convex subset of the space $L^\infty(G)$, which is clearly a locally convex linear space with respect to the topology $\sigma(L^\infty, L^1)$. By Krein–Milman's theorem we conclude that $\mathcal{P}_0$ is the closed convex hull of its extremal points.

Lemma 5.3.1. *The extremal points of $\mathcal{P}_0$ are given by*

(a) *the zero function,*

(b) *the functions $\varphi \in \mathcal{P}_0$ with $\varphi(e) = 1$ (φ continuous), which satisfy the condition: if $\varphi = \varphi_1 + \varphi_2$ ($\varphi_1, \varphi_2 \in \mathcal{P}_0$) then there is $\lambda \ge 0$ such that $\varphi_1 = \lambda\varphi$, $\varphi_2 = (1-\lambda)\varphi$.*

The functions of type (b) are called *elementary positive-definite functions.*

Theorem 5.3.2. *Let φ be a continuous positive-definite function on G with $\varphi(e) = 1$. Then φ is elementary if and only if the associated unitary representation of G in $\mathcal{H}_\varphi$ is irreducible.*

Let φ be an elementary positive-definite function with $\varphi(e) = 1$ and let $(L, \mathcal{H}_\varphi)$ be as usual. Then $\varphi(x) = (\varepsilon, L_x\varepsilon)_\varphi$ $(x \in G)$ where ε a cyclic vector with $\|\varepsilon\|_\varphi = 1$. We shall show that $(L, \mathcal{H}_\varphi)$ is irreducible. Let P be an orthogonal projection onto a closed linear subspace of $\mathcal{H}_\varphi$, which commutes with all L_x $(x \in G)$. We may then write

$$\begin{aligned}\varphi(x) = (\varepsilon, L_x\varepsilon)_\varphi &= (P\varepsilon, L_x\varepsilon)_\varphi + (\varepsilon - P\varepsilon, L_x\varepsilon)_\varphi \\ &= (P\varepsilon, L_x P\varepsilon)_\varphi + (\varepsilon - P\varepsilon, L_x(\varepsilon - P\varepsilon))_\varphi \qquad (x \in G).\end{aligned}$$

Hence $(P\varepsilon, L_x\varepsilon)_\varphi = \lambda\,(\varepsilon, L_x\varepsilon)_\varphi$ for some $\lambda \geq 0$ and all $x \in G$, so $P\varepsilon = \lambda\,\varepsilon$, and since ε is cyclic and L_x commutes with P, $P = \lambda\, I$, so $P = 0$ or $P = I$. This proves the irreducibility.

Let now φ be continuous, positive-definite, $\varphi(e) = 1$, and let $(L, \mathcal{H}_\varphi)$ be irreducible. Suppose $\varphi = \varphi_1 + \varphi_2$ for some $\varphi_1, \varphi_2 \in \mathcal{P}_0$. It is clear that $(f, f)_{\varphi_1} \leq (f, f)_\varphi$ for all $f \in C_c(G)$.

Since $|(f, g)_{\varphi_1}|^2 \leq (f, f)_\varphi\,(g, g)_\varphi$ $(f, g \in C_c(G))$, it follows that $(f, g)_{\varphi_1}$ defines a bounded sesqui-linear form on H_φ. Therefore there exists a self-adjoint $A \in \mathrm{End}(\mathcal{H}_\varphi)$ with

$$(f, g)_{\varphi_1} = (Af, g)_\varphi$$

for all $f, g \in C_c(G)$. Furthermore we have, for all $f, g \in C_c(G)$,

$$(AL_s f, g)_\varphi = (L_s f, g)_{\varphi_1} = (f, L_{s^{-1}} g)_{\varphi_1} = (Af, L_{s^{-1}} g)_\varphi = (L_s Af, g)_\varphi$$

for all $s \in G$, hence $AL_s = L_s A$ for all $s \in G$. By Schur's lemma we get $A = \lambda\, I$ for some $\lambda \geq 0$. Consequently, $(f, g)_{\varphi_1} = \lambda\,(f, g)_\varphi$ for all $f, g \in C_c(G)$, so $\varphi_1 = \lambda\varphi$. This completes the proof of the theorem.

We now proceed with the special case that G is abelian.

Theorem 5.3.3. *Let G be an abelian locally compact group. The elementary positive-definite functions are precisely the unitary characters of G.*

Let χ be a unitary character of G. Then clearly $\chi(e) = 1$ and

$$\int_G \int_G \chi(x^{-1}y)\, f(x)\, \overline{f(y)}\, dx dy = \left| \int_G \overline{\chi(x)}\, f(x)\, dx \right|^2 \geq 0 \qquad (f \in C_c(G)),$$

so χ is positive-definite, and it easily follows that $\dim \mathcal{H}_\chi = 1$. Hence $(L, \mathcal{H}_\chi)$ is irreducible and thus χ elementary.

Conversely, let φ be an elementary positive-definite function. Then the space $\mathcal{H}_\varphi$ is one-dimensional, $L_x = \chi(x)\, I$, where χ is a unitary character of G. Moreover $\varphi(x) = (\varepsilon, L_x\varepsilon)_\varphi = \overline{\chi(x)}$ for all $x \in G$, since $\|\varepsilon\|_\varphi = 1$.

5.4 Fourier transform, Riemann–Lebesgue lemma and Bochner's theorem

From now on G will be *abelian* in this chapter.

(i) The dual group

The unitary characters of G and the zero function together form the set of extremal points G' of $\mathcal{P}_0$. The subset of all unitary characters will be denoted by $\widehat{G}$. Clearly, $\widehat{G}$ is an *abelian group* with respect to pointwise multiplication.

Theorem 5.4.1. (a) *The set G' is closed in the topology $\sigma(L^\infty, L^1)$, hence G' is weakly compact and thus $\widehat{G}$ is locally compact.*

(b) *On $\widehat{G}$ the topology $\sigma(L^\infty, L^1)$ coincides with the topology of uniform convergence on compact subsets of G.*

Remark 5.4.2. (1) The topology of uniform convergence on compact subsets has the following basis (and neighbourhood basis): fixing $\chi_0 \in \widehat{G}$, $K \subset G$ compact and $\varepsilon > 0$, we consider

$$\widehat{U}(\chi_0, K, \varepsilon) = \{\chi \in \widehat{G} : |\chi(x) - \chi_0(x)| < \varepsilon \text{ for all } x \in K\}.$$

(2) Pontryagin defined the above topology on $\widehat{G}$ in case $\widehat{G}$ is countable, Van Kampen considered, independently, the more general situation where the group G satisfies the second axiom of countability.

(3) Provided with the topology of uniform convergence on compact subsets, $\widehat{G}$ is a *topological group*.

We shall now give the proof of Theorem 5.4.1.

(a) Let $\varphi_n \in \widehat{G}$ converge weakly to $\varphi_0 \in \mathcal{P}_0$, $\varphi_0 \neq 0$. (If G does not satisfy the second axiom of countability, we have to consider filters instead of sequences.)

Take $f_1, f_2 \in L^1(G)$. From the relation

$$\int_G (f_1 * f_2)(x)\,\overline{\varphi_n(x)}\,dx = \int_G \int_G f_1(x)\, f_2(y)\,\overline{\varphi_n(x)}\,\overline{\varphi_n(y)}\,dx dy$$

we get the same relation for φ_0. Writing

$$\int_G (f_1 * f_2)(x)\,\overline{\varphi_0(x)}\,dx = \int_G \int_G f_1(x)\, f_2(y)\,\overline{\varphi_0(xy)}\,dx dy,$$

we obtain $\varphi_0(xy) = \varphi_0(x)\varphi_0(y)$ for all $x, y \in G$. We know $\varphi_0(e) \neq 0$ and thus, in particular, we have $\varphi_0(e) = 1$. Since φ_0 is bounded, we finally get $|\varphi_0(x)| = 1$ for all $x \in G$, hence $\varphi_0 \in \widehat{G}$. Therefore G' is weakly closed.

(b) Let $\varphi \in \widehat{G}$. A weak neighbourhood $U(\varphi; f_1, \dots, f_n; \varepsilon) \cap \widehat{G}$ of φ, $(f_1, \dots, f_n \in L^1; \varepsilon > 0)$ always contains a neighbourhood of the form $\widehat{U}(\varphi; K'; \varepsilon')$. Indeed, for sufficiently large K' and sufficiently small ε', the relation $|\varphi(x) - \chi(x)| < \varepsilon'$ for all $x \in K'$ implies

$$\left| \int_G f_i(x)\,\overline{\varphi(x)}\,dx - \int_G f_i(x)\,\overline{\chi(x)}\,dx \right| < \varepsilon$$

for $i = 1, \dots, n$.

Conversely, let $\widehat{U}(\varphi, K, \varepsilon)$ be given. We shall show that this set contains a set of the form $U(\varphi; f_1, \dots, f_n; \varepsilon') \cap \widehat{G}$ for some $f_1, \dots, f_n \in L^1$ and $\varepsilon' > 0$. This is more difficult to show. The proof is in two steps.

(1) Let $f \in L^1(G)$ be given. For any K and $\varepsilon > 0$ there is a weak neighbourhood V_φ of φ such that

$$\left| \int_G f(xy)\,\overline{\varphi(y)}\,dy - \int_G f(xy)\,\overline{\chi(y)}\,dy \right| < \varepsilon$$

for all $x \in K$ and $\chi \in \widehat{G} \cap V_\varphi$.

(2) Select f such that $\int_G f(x)\,\overline{\varphi(x)}\,dx \neq 0$. There is a weak neighbourhood U_φ of φ such that

$$\left| \frac{\int_G f(xy)\,\overline{\varphi(y)}\,dy}{\int_G f(y)\,\overline{\varphi(y)}\,dy} - \frac{\int_G f(xy)\,\overline{\chi(y)}\,dy}{\int_G f(y)\,\overline{\chi(y)}\,dy} \right| < \varepsilon$$

for all $x \in K$ and $\chi \in U_\varphi \cap \widehat{G}$.

It now easily follows from (2) that $|\varphi(x) - \chi(x)| < \varepsilon$ for all $x \in K$ and $\chi \in U_\varphi \cap \widehat{G}$, hence $U_\varphi \cap \widehat{G} \subset \widehat{U}(\varphi, K, \varepsilon)$.

Let us first show (1). There exists a neighbourhood W of e in G such that $\int_G |f(xy) - f(y)|\,dy < \varepsilon/3$ for all $x \in W$. The translates Wz $(z \in K)$ of W cover K. Hence there are $z_1, \dots, z_N$ such that $(Wz_n)_{1 \le n \le N}$ already cover K. Set $f_n(y) = f(z_n y)$. Choose V_φ such that

$$\left| \int_G f_n(y)\,\overline{\varphi(y)}\,dy - \int_G f_n(y)\,\overline{\chi(y)}\,dy \right| < \varepsilon/3$$

for $1 \le n \le N$ and $\chi \in V_\varphi \cap \widehat{G}$. This means

$$\left| \int_G f(z_n y)\,\overline{\varphi(y)}\,dy - \int_G f(z_n y)\,\overline{\chi(y)}\,dy \right| < \varepsilon/3.$$

Choose $z \in K$, say $z \in W z_n$. Then we have: if $\chi \in V_\varphi \cap \widehat{G}$, then

$$\begin{aligned}\Big| \int_G f(zy)\, \overline{\varphi(y)}\, dy - \int_G f(zy)\, \overline{\chi(y)}\, dy \Big| &\le \Big| \int_G \{f(zy) - f(z_n y)\}\, \overline{\varphi(y)}\, dy \Big| + \Big| \int_G \{f(zy) - f(z_n y)\}\, \overline{\chi(y)}\, dy \Big| \\ &\quad + \Big| \int_G f(z_n y)\, \{\overline{\varphi(y)} - \overline{\chi(y)}\}\, dy \Big| \\ &< \varepsilon.\end{aligned}$$

Statement (2) follows easily by choosing U_φ smaller than V_φ, namely such that for sufficiently small $\varepsilon' > 0$ in addition holds

$$\Big| \int_G f(y)\, \overline{\varphi(y)}\, dy - \int_G f(y)\, \overline{\chi(y)}\, dy \Big| < \varepsilon'.$$

The group $\widehat{G}$ is thus a locally compact group, the *dual group* of G. The elements of G' we denote by $\hat{x}$, $\hat{y}$, $\hat{z}$ etc. and we shall write $(x, \hat{x})$ for $\hat{x}(x)$. Observe that $\hat{x}$ may be the zero function.

(ii) The Fourier transform of functions in $L^1(G)$

The Fourier transform $\hat{f}$ of a function $f \in L^1(G)$ is defined by

$$\hat{f}(\hat{x}) = \int_G f(x)\, \overline{(x, \hat{x})}\, dx \qquad (\hat{x} \in G').$$

It is immediately seen that $\hat{f}$ is continuous on G' in the weak topology of L^∞. Moreover $\hat{f}(0) = 0$.

We shall identify now the point zero of G' with the point "∞" (infinity) of $\widehat{G}$. Adding the point ∞ makes $\widehat{G}$ compact, the one-point compactification of $\widehat{G}$. We thus get the *Riemann–Lebesgue lemma*: $\hat{f}$, restricted to $\widehat{G}$, vanishes at infinity.

Here are some *properties* of the Fourier transform:

(i) $\hat{f}$ is continuous on $\widehat{G}$; $|\hat{f}(\hat{x})| \le \|f\|_1$ for all $\hat{x} \in \widehat{G}$, $f \in L^1(G)$,

(ii) $(\lambda f + \mu g)\widehat{\ } = \lambda \hat{f} + \mu \hat{g} \quad (\lambda, \mu \in \mathbb{C};\ f, g \in L^1(G))$,

(iii) $(f * g)\widehat{\ } = \hat{f}\,\hat{g} \quad (f, g \in L^1(G))$,

(iv) $(L_a f)\widehat{\ }(\hat{x}) = \overline{(a, \hat{x})}\, \hat{f}(\hat{x}) \quad (a \in G;\ f \in L^1(G))$,

(v) $(L_{\hat{a}} \hat{f})(\hat{x}) = (\hat{a}\, f)\widehat{\ }(\hat{x}) \quad (\hat{a} \in \widehat{G};\ f \in L^1(G))$,

(vi) $\widehat{\overline{f}} = \overline{\hat{f}} \qquad (f \in L^1(G))$.

We conclude that the set of all $\hat{f}$ with $f \in L^1(G)$ is an algebra under the pointwise multiplication, called $\mathcal{F}^1(G')$, with the following additional properties:

(1) any $F \in \mathcal{F}^1(G')$ is continuous on the compact space G';

(2) if $F \in \mathcal{F}^1(G')$, then $\overline{F} \in \mathcal{F}^1(G')$;

(3) $F(0) = 0$ for all $F \in \mathcal{F}^1(G')$. If $\hat{x} \neq \hat{y}$ then there exists $F \in \mathcal{F}^1(G')$ with $F(\hat{x}) \neq F(\hat{y})$.

By Stone–Weierstrass' theorem, every complex-valued continuous function on G' which vanishes at $\hat{x} = 0$, can be uniformly approximated by functions from $\mathcal{F}^1(G')$. So if μ is a bounded measure on $\widehat{G}$ satisfying $\int_{\widehat{G}} \hat{f}(\hat{x})\, d\mu(\hat{x}) = 0$ for all $f \in L^1(G)$, then $\mu = 0$. This remains true if $\int_{\widehat{G}} \hat{f}(\hat{x})\, d\mu(\hat{x}) = 0$ for f in a dense subset of $L^1(G)$.

(iii) Inverse Fourier transform of a measure on $\widehat{G}$. Bochner's theorem

Let μ be a bounded measure on $\widehat{G}$. Set

$$(T^*\mu)(x) = \int_{\widehat{G}} \overline{(x, \hat{x})}\, d\mu(\hat{x}).$$

By approximating μ by measures with compact support, and observing that $(x, \hat{x})$ is continuous on $G \times \widehat{G}$ (use the topology of uniform convergence on compact sets), it follows that $T^*\mu$ is a bounded continuous function on G. Furthermore we have, if $f \in L^1(G)$,

$$\begin{aligned} \int_{\widehat{G}} \hat{f}(\hat{x})\, d\mu(\hat{x}) &= \int_{\widehat{G}} \Big\{ \int_G f(x) \overline{(x, \hat{x})}\, dx \Big\}\, d\mu(\hat{x}) \\ &= \int_G f(x) \Big\{ \int_{\widehat{G}} \overline{(x, \hat{x})}\, d\mu(\hat{x}) \Big\}\, dx = \int_G f(x)\, (T^*\mu)(x)\, dx. \end{aligned}$$

Let $T : L^1(G) \to \mathcal{C}_0(\widehat{G})$ be given by $Tf = \hat{f}$. Then clearly T^* is the adjoint of T.

Because the image of T is dense in $\mathcal{C}_0(\widehat{G})$, T^* is one-to-one.

Theorem 5.4.3 (Bochner's theorem). *Let $\mathcal{M}_0 = \{\mu \in M^1(\widehat{G}) : \mu \geq 0,\ \|\mu\| \leq 1\}$. Then $\mathcal{P}_0 = T^*(\mathcal{M}_0)$.*

(1) $\mathcal{M}_0$ is closed in the topology $\sigma(M^1(\widehat{G}), \mathcal{C}_0(\widehat{G}))$, so by Alaoglu's theorem compact.

(2) T^* is continuous with respect to the weak topologies on $M^1(\widehat{G})$ and $L^\infty(G)$.

(3) $T^*(\mathcal{M}_0)$ is compact in the topology $\sigma(L^\infty(G), L^1(G))$.

(4) $T^*(\mathcal{M}_0) \subset \mathcal{P}_0$.

(5) $T^*(\mathcal{M}_0)$ is convex and contains G'.

(6) By Krein–Milman's theorem we now have $T^*(\mathcal{M}_0) = \mathcal{P}_0$.

Notice that $T^* : \mathcal{M}_0 \to \mathcal{P}_0$ is a topological bijection with respect to the weak topologies considered above.

5.5 The inversion theorem

(i) Formulation of the theorem

Let again G be a locally compact abelian group, $\widehat{G}$ its dual group. The dual group is a locally compact group with respect to the topology of uniform convergence on compact subsets. Let $d\hat{x}$ be a Haar measure on $\widehat{G}$. This measure is determined up to a positive constant, which we now shall fix.

Let $\mathcal{V}(G)$ be the set of all complex linear combinations of continuous positive-definite functions on G (the so-called *Fourier–Stieltjes algebra* of G). As usual, denote by $M^1(\widehat{G})$ the set of bounded measures on $\widehat{G}$. The operator T^*, defined in Section 5.4, maps $M^1(\widehat{G})$ one-to-one onto $\mathcal{V}(G)$.

Set $\mathcal{V}^1(G) = \mathcal{V}(G) \cap L^1(G)$, $\mathcal{V}^2(G) = \mathcal{V}(G) \cap L^2(G)$. For any $f \in \mathcal{V}(G)$ there is a unique measure $\mu_f \in M^1(\widehat{G})$ such that

$$f(x) = \int_{\widehat{G}} (x, \hat{x})\, d\mu_f(\hat{x}) \qquad (x \in G).$$

Theorem 5.5.1 (Inversion theorem). *The Haar measure $d\hat{x}$ on $\widehat{G}$ can be normalized in such a way that for any $f \in \mathcal{V}^1(G)$ one has $d\mu_f(\hat{x}) = \hat{f}(\hat{x})\, d\hat{x}$. More explicitly: if $f \in L^1(G)$ is a complex linear combination of continuous positive-definite functions, then $\hat{f} \in L^1(\widehat{G})$ and*

$$f(x) = \int_{\widehat{G}} (x, \hat{x})\, \hat{f}(\hat{x})\, d\hat{x}$$

for a suitable normalization of $d\hat{x}$, independent of f.

Before we proceed with the proof, we give a few remarks. If $f \in C_c(G)$ then $f * \widetilde{f} \in C_c(G)$ is positive-definite, hence $f * \widetilde{f} \in \mathcal{V}^1(G)$. So $f * g \in \mathcal{V}^1(G)$ for all $f, g \in C_c(G)$. Consequently, $\mathcal{V}^1(G)$ is dense in $L^1(G)$. In a similar way $\mathcal{V}^2(G)$ is dense in $L^2(G)$. Furthermore $\mathcal{V}^1(G) \subset \mathcal{V}^2(G)$ since the functions in $\mathcal{V}(G)$ are bounded.

(ii) Proof of the theorem

We shall present a proof which is a combination of the one by Cartan and Godement [7] and Rudin [39].

(a) *If* $f, g \in \mathcal{V}^1(G)$, *then* $\hat{g}\, d\mu_f = \hat{f}\, d\mu_g$.

Set $T'(\mu) = \overline{T^*(\overline{\mu})}$ for all bounded measures μ on $\widehat{G}$. Then we have

$$\begin{aligned} T'(\hat{g}\, d\mu_f) &= \int_{\widehat{G}} (x, \hat{x})\, \hat{g}(\hat{x})\, d\mu_f(\hat{x}) = \int_{\widehat{G}} \int_G (x, \hat{x})\, \overline{(y, \hat{x})}\, g(y)\, dy\, d\mu_f(\hat{x}) \\ &= \int_G \Big\{ \int_{\widehat{G}} (xy^{-1}, \hat{x})\, d\mu_f(\hat{x}) \Big\}\, g(y)\, dy = \int_G f(xy^{-1})\, g(y)\, dy = g * f\,(x). \end{aligned}$$

In a similar way $T'(\hat{f}\, d\mu_g) = f * g\,(x)$. Since T' is one-to-one, statement (a) follows.

(b) *Define (see* [39]*) the functional* ν *on* $C_c(\widehat{G})$ *as follows:*

$$\nu(k) = \int_{\widehat{G}} \frac{k(\hat{x})}{\hat{g}(\hat{x})}\, d\mu_g(\hat{x}) \qquad (k \in C_c(\widehat{G})),$$

where g *is a function in* $\mathcal{V}^1(G)$ *such that* $\hat{g}$ *is strictly positive on* $\operatorname{Supp} k$. *This functional is well-defined.*

(1) Given $k \in C_c(\widehat{G})$, at least one function $g \in \mathcal{V}^1(G)$ exists with the above property. Indeed, for every $\hat{x} \in \operatorname{Supp} k$ there is a $u \in C_c(G)$ with $\hat{u}(\hat{x}) \neq 0$. Hence also $(u * \widetilde{u})\hat{\ }(\hat{x}) = |\hat{u}(\hat{x})|^2 > 0$, the function $u * \widetilde{u}$ being in $\mathcal{V}^1(G)$. There is a neighbourhood of $\hat{x}$ where $(u * \widetilde{u})\hat{\ }(\hat{y}) > 0$ for $\hat{y}$ in that neighbourhood. Applying the compactness of $\operatorname{Supp} k$, we can take for g a function of the form $g = u_1 * \widetilde{u}_1 + \cdots + u_N * \widetilde{u}_N$ with functions $u_i \in C_c(G)$.

(2) The definition of $\nu(k)$ does not depend on the special choice of g. Let $f \in \mathcal{V}^1(G)$ and $\hat{f}$ be strictly positive on $\operatorname{Supp} k$, then we have by (a)

$$\begin{aligned} \int_{\widehat{G}} \frac{k(\hat{x})}{\hat{f}(\hat{x})}\, d\mu_f(\hat{x}) &= \int_{\widehat{G}} \frac{k(\hat{x})}{\hat{f}(\hat{x})\hat{g}(\hat{x})} \hat{g}(\hat{x})\, d\mu_f(\hat{x}) = \int_{\widehat{G}} \frac{k(\hat{x})}{\hat{f}(\hat{x})\hat{g}(\hat{x})} \hat{f}(\hat{x})\, d\mu_g(\hat{x}) \\ &= \int_{\widehat{G}} \frac{k(\hat{x})}{\hat{g}(\hat{x})}\, d\mu_g(\hat{x}). \end{aligned}$$

(c) *The functional* ν *is a non-trivial positive linear form on* $C_c(\widehat{G})$, *hence* ν *is a non-trivial positive measure on* $\widehat{G}$.

- ν is *linear*: $\nu(\alpha_1 k_1 + \alpha_2 k_2) = \alpha_1 \nu(k_1) + \alpha_2 \nu(k_2)$. To see this, select therefore $g \in \mathcal{V}^1(G)$ such that $\hat{g}$ is strictly positive on $\operatorname{Supp} k_1 \cup \operatorname{Supp} k_2$.
- ν is *positive*: take g of the form $g = u_1 * \widetilde{u}_1 + \cdots + u_N * \widetilde{u}_N$ with $u_i \in C_c(G)$. Then μ_g is a positive measure, by Bochner's theorem.

- ν is *non-trivial*: take $f \in \mathcal{V}^1(G)$, $f \neq 0$. Then $d\mu_f \neq 0$. So there exists $k \in C_c(\widehat{G})$ with $\int_{\widehat{G}} k\, d\mu_f \neq 0$. Consider now $\nu(k\hat{f})$. Observe that $k\hat{f} \in C_c(\widehat{G})$. We have $\nu(k\hat{f}) = \int_{\widehat{G}} \frac{k\hat{f}}{\hat{g}}\, d\mu_g$ for $g \in \mathcal{V}^1(G)$, $\hat{g}$ strictly positive on $\operatorname{Supp} k$. Furthermore we have

$$\nu(k\hat{f}) = \int_{\widehat{G}} \frac{k}{\hat{g}} \hat{f}\, d\mu_g = \int_{\widehat{G}} \frac{k}{\hat{g}} \hat{g}\, d\mu_f = \int_{\widehat{G}} k\, d\mu_f \neq 0.$$

(d) *The measure ν is a Haar measure on $\widehat{G}$.*

For $\hat{y}_0 \in \widehat{G}$ let $(L_{\hat{y}_0}k)(\hat{x}) = k(\hat{y}_0^{-1}\hat{x})$ $(k \in C_c(\widehat{G}))$. We have to show that $\nu(L_{\hat{y}_0}k) = \nu(k)$ for all $k \in C_c(\widehat{G})$. Observe first of all that if $g \in \mathcal{V}^1(G)$ and $\hat{y}_0 \in \widehat{G}$ then $g_0 = \hat{y}_0 \cdot g \in \mathcal{V}^1(G)$ and $\hat{g}_0 = L_{\hat{y}_0}\hat{g}$, $\mu_{g_0} = L_{\hat{y}_0}\mu_g$. The latter property follows from the identity $g(x) = \int_G (x, \hat{x})\, d\mu_g(\hat{x})$, hence

$$g_0(x) = \int_G (x, \hat{y}_0\hat{x})\, d\mu_g(\hat{x}) = \int_G (x, \hat{x})\, d\mu_g(\hat{y}_0^{-1}\hat{x}) = \int_G (x, \hat{x})\, d(L_{\hat{y}_0}\mu_g)(\hat{x})$$

(by definition).

We now arrive at the invariance of ν.

Let $k \in C_c(\widehat{G})$ and $\hat{y}_0 \in \widehat{G}$. Take $g \in \mathcal{V}^1(G)$ such that $\hat{g}$ is strictly positive on $\hat{y}_0^{-1} \cdot \operatorname{Supp} k$. Then we have

$$\begin{aligned}\nu(L_{\hat{y}_0^{-1}}k) &= \int_{\widehat{G}} \frac{k(y_0\hat{x})}{\hat{g}(\hat{x})}\, d\mu_g(\hat{x}) = \int_{\widehat{G}} \frac{k(\hat{x})}{\hat{g}(\hat{y}_0^{-1}\hat{x})}\, d\mu_g(\hat{y}_0^{-1}\hat{x}) = \int_{\widehat{G}} \frac{k(\hat{x})}{\hat{g}_0(\hat{x})}\, d\mu_{g_0}(\hat{x}) \\ &= \nu(k).\end{aligned}$$

We shall write for ν also $d\hat{x}$. As we already saw under (c), for every $f \in \mathcal{V}^1(G)$ one has

$$\nu(k\hat{f}) = \int_{\widehat{G}} k\, d\mu_f \qquad (k \in C_c(\widehat{G})).$$

So $\int_{\widehat{G}} k\hat{f}\, d\hat{x} = \int_{\widehat{G}} k\, d\mu_f$ for all $k \in C_c(\widehat{G})$ or $\hat{f}(\hat{x})\, d\hat{x} = d\mu_f(\hat{x})$ for $f \in \mathcal{V}^1(G)$. Since μ_f is a bounded measure, this implies that $\hat{f} \in L^1(\widehat{G})$. In addition we have

$$f(x) = \int_{\widehat{G}} (x, \hat{x})\, d\mu_f(\hat{x}) = \int_{\widehat{G}} (x, \hat{x})\, \hat{f}(\hat{x})\, d\hat{x}$$

for all $f \in \mathcal{V}^1(G)$ and all $x \in G$. This completes the proof of the inversion theorem.

(iii) Corollaries of the inversion theorem

Corollary 5.5.2. *Let $f \in L^1(G)$ be continuous and positive-definite. Then $\hat{f}$ is a positive function on $\widehat{G}$.*

By the inversion theorem $d\mu_f(\hat{x}) = \hat{f}(\hat{x})\, d\hat{x}$, hence $\hat{f}(\hat{x}) \geq 0$ for all $\hat{x}$ because μ_f is a positive measure and $\hat{f}$ is continuous.

In particular we have $\hat{f}(\hat{e}) = \int_{\widehat{G}} f(x)\,dx \geq 0$.

Corollary 5.5.3. *Let* $F \in L^1(\widehat{G})$ *and set* $f(x) = \int_{\widehat{G}}(x,\hat{x})\,F(\hat{x})\,d\hat{x}$. *If* $f \in L^1(G)$, *then* $F(\hat{x}) = \hat{f}(\hat{x})$ *almost everywhere.*

Observe that $f \in \mathcal{V}^1(G)$ since $F(\hat{x})\,d\hat{x} \in M^1(\widehat{G})$. Hence by the inversion theorem, $f(x) = \int_{\widehat{G}}(x,\hat{x})\,\hat{f}(\hat{x})\,d\hat{x}$. Since T' is one-to-one, we get $F(\hat{x})\,d\hat{x} = \hat{f}(\hat{x})\,d\hat{x}$, so $F = \hat{f}$ almost everywhere.

Remark 5.5.4. If $f \in \mathcal{V}^1(G)$, then $\hat{f} \in \mathcal{V}^1(\widehat{G})$.

By the inversion theorem $\hat{f} \in L^1(\widehat{G})$. If we write f in the form $f = f_1 - f_2 + i(f_3 - f_4)$ with $f_i \geq 0$, $f_i \in L^1(G)$, then the functions $\widehat{f_i}$ are clearly continuous and positive-definite on $\widehat{G}$. Hence $\widehat{f_i} \in \mathcal{V}^1(\widehat{G})$.

Later on we shall see, applying *Pontryagin's duality theorem*, see Section 5.7, that the converse of Remark 5.5.4 is also true.

5.6 Plancherel's theorem

Let $f \in L^1(G) \cap L^2(G)$. The function $g = f * \widetilde{f}$ is then in $\mathcal{V}^1(G)$ and $\hat{g} = |\hat{f}|^2$. By the inversion theorem $\hat{g} \in L^1(\widehat{G})$, hence $\hat{f} \in L^2(\widehat{G})$. Moreover, $g(x) = \int_{\widehat{G}}(x,\hat{x})\,|\hat{f}(\hat{x})|^2\,d\hat{x}$, in particular

$$g(e) = \int_{\widehat{G}} |\hat{f}(\hat{x})|^2\,d\hat{x} = \int_G |f(x)|^2\,dx.$$

We have thus shown:

Lemma 5.6.1. *If* $f \in L^1(G) \cap L^2(G)$, *then* $\hat{f} \in L^2(\widehat{G})$ *and* $\|f\|_2 = \|\hat{f}\|_2$.

The Fourier transform T is therefore an isometric linear mapping from the dense linear subspace $L^1(G) \cap L^2(G)$ of $L^2(G)$ to $L^2(\widehat{G})$. We can extend T *isometrically* to $L^2(G)$.

Theorem 5.6.2 (Plancherel). *T is an isometry from $L^2(G)$ onto $L^2(\widehat{G})$.*

We start with some preparations for the proof. Let $F \in L^1(\widehat{G}) \cap L^2(\widehat{G})$ and set, as usual,

$$(T'F)(x) = \int_{\widehat{G}} (x,\hat{x})\,F(\hat{x})\,d\hat{x}.$$

Then $T'F$ certainly is a bounded continuous function. For any $f \in L^1(G) \cap L^2(G)$ we have

$$\int_G f(x)\,\overline{(T'F)(x)}\,dx = \int_{\widehat{G}} \hat{f}(\hat{x})\,\overline{F(\hat{x})}\,d\hat{x}.$$

Notice that $\hat{f} \in L^2(\widehat{G})$, hence

$$\Big|\int_G f(x)\,\overline{(T'F)(x)}\,dx\Big| \le \|\hat{f}\|_2\|F\|_2 = \|f\|_2\|F\|_2.$$

Thus $f \mapsto \langle f, T'F\rangle = \int_G f(x)\,\overline{(T'F)(x)}\,dx$ can be continued to a continuous linear form on $L^2(G)$, hence $T'F \in L^2(G)$ and $\|T'F\|_2 \le \|F\|_2$.

Lemma 5.6.3. *Every function $F * G$ ($F, G \in C_c(\widehat{G})$) is the Fourier transform of a function in $L^1(G) \cap L^2(G)$.*

One easily verifies that $T'(F * G) = T'(F) \cdot T'(G)$. Because both $T'F$ and $T'G$ are in $L^2(G)$, $T'(F * G)$ is in $L^1(G)$, hence by Corollary 5.5.3, $F * G$ is the Fourier transform of $T'(F * G)$. We also have of course $T'(F * G) \in L^2(G)$.

Proof of Plancherel's theorem

The space $T(L^2(G))$ is closed in $L^2(\widehat{G})$ (a Banach space in a Banach space is closed). From Lemma 5.6.3 it follows that $T(L^2(G))$ is dense in $L^2(\widehat{G})$. Hence T is onto. This completes the proof of Theorem 5.6.2.

Corollary 5.6.4 (Uniqueness theorem). *The Fourier transform is one-to-one on $L^1(G)$.*

Let $f \in L^1(G)$ and $\hat{f} = 0$. For all $u \in C_c(G)$ we have $f * u \in L^1(G) \cap L^2(G)$ and $\widehat{(f * u)} = 0$, therefore by Lemma 5.6.1 $f * u = 0$. Taking an approximate unit for u, it follows that $f = 0$.

5.7 Pontryagin's duality theorem

To any $x \in G$ we assign the character $\alpha(x) \in \widehat{\widehat{G}}$ defined by

$$\alpha(x)(\hat{x}) = \overline{(x, \hat{x})} \qquad (\hat{x} \in \widehat{G}).$$

Notice that $\alpha(x)$ is continuous.

Theorem 5.7.1. *The mapping $x \mapsto \alpha(x)$ is an algebraic and topological isomorphism of G onto $\widehat{\widehat{G}}$.*

Pontryagin [35] has proved this theorem for compact and discrete groups, while Van Kampen [54] gave a proof for locally compact abelian groups satisfying the second axiom of countability. Compare this with Remark 5.4.2 (2).

The proof is in four steps:

(1) α is an algebraic isomorphism from G onto $\alpha(G)$,

(2) α is a topological isomorphism from G onto $\alpha(G)$,

(3) $\alpha(G)$ is closed in $\widehat{\widehat{G}}$,

(4) $\alpha(G)$ is dense in $\widehat{\widehat{G}}$.

Lemma 5.7.2. *The functions of the form $u * v$ with $u, v \in C_c(G)$ separate the points of G.*

Let $a \in G$ and let U be a neighbourhood of the unit element e. We shall show that there exists a function f of the required type with $f(a) \neq 0$, $\operatorname{Supp} f \subset aU$. Let V be a symmetric neighbourhood of e with $V^2 \subset U$ and choose $k \in C_c(G)$ with the properties: $k \geq 0$, $k(x) = k(x^{-1})$ $(x \in G)$, $\operatorname{Supp} k \subset V$ and $\int_G k^2(x)\,dx = 1$. Then $f = L_a(k * k) = L_a k * k$ satisfies the above conditions.

We shall now prove Theorem 5.7.1 step by step

(1) The mapping α clearly is an algebraic homomorphism from G to $\widehat{\widehat{G}}$. Suppose that for certain $x, y \in G$ we have $(x, \hat{x}) = (y, \hat{x})$ for all $\hat{x} \in \widehat{G}$. If $f \in \mathcal{V}^1(G)$ we then get, by the inversion theorem, $f(x) = \int_{\widehat{G}} (x, \hat{x})\, \hat{f}(\hat{x})\, d\hat{x} = f(y)$. Since all functions of the form $u * v$ $(u, v \in C_c(G))$ are in $\mathcal{V}^1(G)$, it follows from Lemma 5.7.2 that $x = y$. So α is one-to-one.

We have shown: a locally compact abelian group has sufficiently many characters. In a more general setting one can show: a locally compact group admits sufficiently many irreducible unitary representations (proved by I. M. Gelfand and D. Raikov, for compact groups by H. Weyl).

(2) The following result from general topology is very useful here.

Lemma 5.7.3. *Let X be a locally compact topological space and $\mathcal{F}$ a family of complex-valued continuous functions on X satisfying*

(a) *all functions in $\mathcal{F}$ vanish at infinity,*

(b) *$\mathcal{F}$ separates the points of X,*

(c) *there is no point in X where all functions in $\mathcal{F}$ vanish.*

Then the weak topology on X defined by $\mathcal{F}$ (i.e. the weakest topology on X such that all functions in $\mathcal{F}$ are continuous) coincides with the topology on X.

Let X_∞ be the compactification of X by ∞. The functions in $\mathcal{F}$ can be viewed as continuous functions on X_∞ vanishing at ∞. Let $\tau_{\mathcal{F}}$ be the topology on X_∞ defined by $\mathcal{F}$. This topology is Hausdorff by (b) and (c). Clearly $\tau_{\mathcal{F}}$ is weaker than the existing topology on X_∞. Since X_∞ is compact, the two topologies coincide on X_∞, so on X (use that any compact subset in a compact Hausdorff space is closed).

Let $\mathcal{B}$ be the complex linear space spanned by functions of the form $u * v$ with $u, v \in C_c(G)$. The space $\mathcal{B}$ is dense in $L^1(G)$. Notice that $\mathcal{B} \subset \mathcal{V}^1(G)$.

Lemma 5.7.4. *$T(\mathcal{B})$ is dense in $L^1(\widehat{G})$.*

Let $f \in C_c(\widehat{G})$, $f \geq 0$. Set $g = \sqrt{f}$, so $f = g^2$, $g \in C_c(\widehat{G})$. By Plancherel's theorem there is $u \in C_c(G)$ with

$$\|g - \hat{u}\|_2 < \varepsilon \qquad (\varepsilon > 0,\ \text{given}).$$

Then we now have

$$\begin{aligned}\|f - \widehat{(u * u)}\|_1 &= \|g \cdot g - \hat{u} \cdot \hat{u}\|_1 \\ &\leq \|g(g - \hat{u})\|_1 + \|\hat{u}(g - \hat{u})\|_1 \leq \|g\|_2 \|g - \hat{u}\|_2 + \|\hat{u}\|_2 \|g - \hat{u}\|_2 \\ &\leq \|g|_2\, \varepsilon + (\|g\|_2 + \varepsilon)\, \varepsilon = \varepsilon\, (2\|g\|_2 + \varepsilon).\end{aligned}$$

We now apply Lemma 5.7.3 to $\mathcal{F} = T(\mathcal{B})$. Any open neighbourhood of $x \in G$ is given by

$$\begin{aligned}&U(x; f_1, \ldots, f_n; \varepsilon) \\ &= \left\{y \in G : \left| \int_{\widehat{G}} \hat{f_i}(\hat{x})\, (x, \hat{x})\, d\hat{x} - \int_{\widehat{G}} \hat{f_i}(\hat{x})\, (y, \hat{x})\, d\hat{x} \right| < \varepsilon \text{ for } 1 \leq i \leq n\right\},\end{aligned}$$

where $f_1, \ldots, f_n \in \mathcal{B}$. Transfer this neighbourhood to $\alpha(G)$:

$$\begin{aligned}&U(\alpha(x); f_1, \ldots, f_n; \varepsilon) \\ &= \left\{\alpha(y) : \left| \int_{\widehat{G}} \hat{f_i}(\hat{x})\, \overline{\alpha(x)(\hat{x})}\, d\hat{x} - \int_{\widehat{G}} \hat{f_i}(\hat{x})\, \overline{\alpha(y)(\hat{x})}\, d\hat{x} \right| < \varepsilon \text{ for } 1 \leq i \leq n\right\}.\end{aligned}$$

Since all elements of $\widehat{\widehat{G}}$ have norm equal to 1 in $L^\infty(\widehat{G})$, it follows now from Lemma 5.7.4 that the transferred topology on $\alpha(G)$ coincides with the restriction of the topology $\sigma(L^\infty(\widehat{G}), L^1(\widehat{G}))$ on $\widehat{\widehat{G}}$ to $\alpha(G)$. Hence α is a topological isomorphism from G onto $\alpha(G)$.

(3) *A subgroup of a locally compact group, which itself is locally compact in the induced topology, is closed.*

From this we immediately conclude that $\alpha(G)$ is closed in $\widehat{\widehat{G}}$.

Since G satisfies the second axiom of countability, G is metrizable and complete by Section 3.3 (vii). If $H \subset G$ is locally compact, the same holds. A complete subspace of a complete space is certainly closed.

(4) Suppose $\alpha(G)$ is not dense in $\widehat{\widehat{G}}$. Then there is $a \in \widehat{\widehat{G}}$ and a neighbourhood U_a of a such that $U_a \cap \alpha(G) = \emptyset$. There exists a function F of the form $F = F_1 * F_2$ $(F_1, F_2 \in C_c(\widehat{\widehat{G}}))$ with $F(a) \neq 0$ and $\operatorname{Supp} F \subset U_a$ (see the proof of Lemma 5.7.2). By Lemma 5.6.3 there is $H \in L^1(\widehat{G})$ such that

$$F(\hat{\hat{x}}) = \int_{\widehat{G}} \overline{(\hat{x}, \hat{\hat{x}})}\, H(\hat{x})\, d\hat{x} \qquad (\hat{\hat{x}} \in \widehat{\widehat{G}}).$$

Hence $\int_{\widehat{G}} (x, \hat{x})\, H(\hat{x})\, d\hat{x} = 0$ for all $x \in G$. Since T' is one-to-one, $H = 0$ a.e., so $F = 0$, which gives a contradiction.

The theorem now follows from (1), (2), (3) and (4).

5.8 Subgroups and quotient groups

Given a closed subgroup H of G, let $H^\perp = \{\hat{x} \in \widehat{G} : (x, \hat{x}) = 1 \text{ for all } x \in H\}$.

Theorem 5.8.1. *Let H be a closed subgroup of G. Every element x of G with $(x, \hat{x}) = 1$ for all $\hat{x} \in H^\perp$ belongs to H.*

The proof is based on the following lemma.

Lemma 5.8.2. *In order that $f \in \mathcal{V}(G)$ is right-invariant under H, it is necessary and sufficient that* $\operatorname{Supp} \mu_f \subset H^\perp$.

The condition is clearly sufficient. Now assume $f(xh) = f(x)$ for all $x \in G$ and $h \in H$. Then we get

$$\int_{\widehat{G}} (x, \hat{x})\, (h, \hat{x})\, d\mu_f(\hat{x}) = \int_{\widehat{G}} (x, \hat{x})\, d\mu_f(\hat{x})$$

for all $x \in G$ and $h \in H$. Therefore $(h, \hat{x})\, d\mu_f(\hat{x}) = d\mu_f(\hat{x})$ for all $h \in H$, so $(h, \hat{x}) = 1$ for $\hat{x} \in \operatorname{Supp} \mu_f$.

Proof of the theorem

Clearly $H \subset (H^\perp)^\perp$. From Lemma 5.8.2 we see that any H-invariant $f \in \mathcal{V}(G)$ is also $(H^\perp)^\perp$-invariant. It is thus sufficient to construct, for every $x \notin H$, a function $f \in \mathcal{V}(G)$, invariant under H, with $f(x) \neq f(e)$. This can easily be achieved by working on G/H and applying Lemma 5.7.2. This completes the proof.

Corollary 5.8.3. *Let H be a closed subgroup of G. Then $(G/H)\widehat{\ } \simeq H^{\perp}$ and $\widehat{H} \simeq \widehat{G}/H^{\perp}$.*

The first statement is easy: the natural isomorphism is clearly an algebraic isomorphism. It is also a topological isomorphism since any compact set in G/H is the image under the projection π_H of a compact set in G, see Section 3.5 (i). The second statement follows by duality and Theorem 5.8.1.

5.9 Compact and discrete abelian groups

Theorem 5.9.1. *If G is discrete, then $\widehat{G}$ is compact.*

It suffices to show that $0 \in G'$ is an isolated point (notations as in Section 5.4). Let us consider the function ε on G satisfying $\varepsilon(e) = 1$, $\varepsilon(x) = 0$ for $x \neq e$. Clearly $\varepsilon \in L^1(G)$. Let us take the usual counting measure on G as Haar measure. So every point of G has mass equal to one. The Fourier transform $\widehat{\varepsilon}$ is the function that is equal to 1 on $\widehat{G}$ and satisfies $\widehat{\varepsilon}(0) = 0$. Since $\widehat{\varepsilon}$ is continuous on G', the theorem follows. Moreover, by the inversion theorem, the dual Haar measure on $\widehat{G}$ has total mass 1.

Theorem 5.9.2. *If G is compact, then $\widehat{G}$ is discrete.*

Indeed, let us take the Haar measure on G with total mass 1. Let now $\varepsilon(x) = 1$ for all $x \in G$. One clearly has $\widehat{\varepsilon}(\hat{e}) = 1$. One also has $(y, \hat{x})\widehat{\varepsilon}(\hat{x}) = \widehat{\varepsilon}(\hat{x})$ for all $\hat{x} \in \widehat{G}$, $y \in G$. So, if $\widehat{\varepsilon}(\hat{x}) \neq 0$, then $(y, \hat{x}) = 1$ for all $y \in G$, so $\hat{x} = \hat{e}$. Hence $\widehat{\varepsilon}(\hat{x}) = 0$ for all $\hat{x} \neq \hat{e}$. Hence $\widehat{G}$ is discrete and, by the inversion theorem, its dual Haar measure is clearly seen to be the counting measure on $\widehat{G}$.

Chapter 6
Classical Theory of Gelfand Pairs

Literature: [15], [18], [21], [51].

6.1 Gelfand pairs and spherical functions

(i) Definition of a Gelfand pair

Let G be a locally compact group with left Haar measure dx and let $C_c(G)$ be, as usual, the convolution algebra of continuous, complex-valued functions on G with compact support. Let K be a compact subgroup of G with normalized Haar measure dk (i.e. $\text{vol}(K) = \int_K dk = 1$) and denote by $C_c^\#(G)$ the space of functions in $C_c(G)$ which are bi-invariant with respect to K, i.e. functions f which satisfy

$$f(kxk') = f(x) \qquad (x \in G; k, k' \in K).$$

The space $C_c^\#(G)$ is a subalgebra of the convolution algebra $C_c(G)$. Given $f \in C_c(G)$, notice that the projection

$$f^\#(x) = \int_K \int_K f(kxk')\, dk\, dk' \tag{6.1.1}$$

is in $C_c^\#(G)$.

Definition 6.1.1. The pair (G, K) is said to be a *Gelfand pair* if the convolution algebra $C_c^\#(G)$ is commutative.

The first example of a Gelfand pair is that of an abelian locally compact group G with K reduced to the unit element, $K = \{e\}$. We shall meet other examples in the next chapter.

Proposition 6.1.2 (C. Berg). *Let (G, K) be a Gelfand pair. Then G is unimodular.*

Let Δ denote the Haar modulus of the group G. One has

$$\int_G f(x)\, dx = \int_G f(x^{-1})\, \Delta(x^{-1})\, dx$$

for any function $f \in C_c(G)$. So it is sufficient to show, applying the projection (6.1.1) to f, that for any $f \in C_c^\#(G)$ one has

$$\int_G f(x)\,dx = \int_G f(x^{-1})\,dx,$$

because Δ is bi-K-invariant.

Let g be a function in $C_c^\#(G)$ equal to 1 on the compact set $\operatorname{Supp} f \cup (\operatorname{Supp} f)^{-1}$. We get

$$\int_G f(x)\,dx = f * g(e) = g * f(e) = \int_G f(x^{-1})\,dx.$$

Proposition 6.1.3. *Let G be a locally compact group and K a compact subgroup of G. Assume there exists a continuous involutive automorphism θ of G such that*

$$\theta(x) \in Kx^{-1}K$$

for all $x \in G$. Then (G, K) is a Gelfand pair.

For $f \in C_c(G)$ set $f^\theta(x) = f(\theta(x))$, $x \in G$. The mapping $f \mapsto \int_G f^\theta(x)\,dx$ is a left-invariant positive measure on G, hence there exists a constant $c > 0$ such that

$$\int_G f^\theta(x)\,dx = c \int_G f(x)\,dx.$$

Since $\theta^2 = 1$, we get $c^2 = 1$, so $c = 1$. This implies $(f * g)^\theta = f^\theta * g^\theta$ for all $f, g \in C_c(G)$. On the other hand we have, for $f \in C_c(G)$,

$$(f * g)\check{} = \check{g} * \check{f}$$

where $\check{f}(x) = f(x^{-1})$ $(x \in G)$.

For f, bi-invariant with respect to K, one has by assumption $\check{f} = f^\theta$, so, if $f, g \in C_c^\#(G)$,

$$(f * g)^\theta = (f * g)\check{} = \check{g} * \check{f} = g^\theta * f^\theta = (g * f)^\theta,$$

hence $f * g = g * f$.

(ii) Spherical functions

Let (G, K) be a Gelfand pair.

Definition 6.1.4. Let φ be a continuous, bi-K-invariant function on G. The function φ is called a *spherical function* if the functional χ defined by

$$\chi(f) = \int_G f(x)\,\varphi(x^{-1})\,dx$$

is a non-trivial character of the convolution algebra $C_c^{\#}(G)$, i.e.

$$\chi(f * g) = \chi(f) \cdot \chi(g)$$

for all $f, g \in C_c^{\#}(G)$.

If $G = \mathbb{R}$ and $K = \{0\}$, the spherical functions are the exponential functions

$$\varphi(x) = \mathrm{e}^{\lambda x} \qquad (\lambda \in \mathbb{C}).$$

The spherical functions play the role of the exponential functions for the Gelfand pairs.

Proposition 6.1.5. *Let φ be a continuous function on G, bi-invariant under K, $\varphi \neq 0$. Then φ is a spherical function if and only if for all $x, y \in G$*

$$\int_K \varphi(xky)\, dk = \varphi(x)\, \varphi(y),$$

where dk denotes the normalized Haar measure on K. In particular, $\varphi(e) = 1$.

This relation replaces the functional relation of the exponentials.

We apply the projection mapping (6.1.1) again. Moreover observe that $f \mapsto \check{f}$ is an automorphism of $C_c^{\#}(G)$. Set

$$\Phi(f) = \int_G f(x)\, \varphi(x)\, dx \qquad (f \in C_c(G)).$$

For any $f, g \in C_c(G)$ we have

$$\begin{aligned} \Phi(f^{\#} * g^{\#}) - \Phi(f^{\#})\, \Phi(g^{\#}) &= \int_G \int_G \{\varphi(xy) - \varphi(x)\varphi(y)\}\, f^{\#}(x)\, g^{\#}(y)\, dxdy \\ &= \int_G \int_G \Big\{ \int_K \varphi(xky)\, dk - \varphi(x)\varphi(y) \Big\}\, f(x)\, g(y)\, dxdy. \end{aligned}$$

From this relation the proposition follows easily.

Proposition 6.1.6. *Let φ be a bi-K-invariant, continuous function on G. The function φ is spherical if and only if*

(a) $\varphi(e) = 1$,

(b) *for every $f \in C_c^{\#}(G)$ there is a complex number $\chi(f)$ such that $f * \varphi = \chi(f)\, \varphi$.*

Let φ be a spherical function and $f \in C_c^\#(G)$. Then by the previous proposition

$$\begin{aligned} f * \varphi(x) &= \int_G f(y)\,\varphi(y^{-1}x)\,dy \\ &= \int_G f(y)\left\{\int_K \varphi(y^{-1}kx)\,dk\right\}dy \\ &= \varphi(x)\int_G f(y)\,\varphi(y^{-1})\,dy, \end{aligned}$$

because of the left K-invariance of f.

If, conversely, φ satisfies (a) and (b), then the mapping

$$f \mapsto \int_G f(y)\,\varphi(y^{-1})\,dy = \chi(f)$$

clearly is a non-trivial character of $C_c^\#(G)$.

(iii) Bounded, symmetric and positive-definite spherical functions

Denote by $L^1(G)^\#$ the space of integrable functions on G which are bi-invariant under K. This is a convolution algebra too. Suppose now that (G, K) is a Gelfand pair. Then $L^1(G)^\#$ is a *commutative Banach algebra*, even with an *involution* given by

$$f^*(x) = \widetilde{f}(x) = \overline{f(x^{-1})} \qquad (x \in G;\ f \in L^1(G)^\#).$$

We are interested in the Gelfand theory of this algebra; cf. [29, §23].

Theorem 6.1.7. *Let φ be a bounded spherical function. The mapping*

$$f \mapsto \chi(f) = \int_G f(x)\,\varphi(x^{-1})\,dx$$

is a character of $L^1(G)^\#$, and each non-trivial character of $L^1(G)^\#$ is of this form.

If φ is a bounded spherical function, the result follows from the fact that $C_c^\#(G)$ is dense in $L^1(G)^\#$. Let us prove the converse. Let χ be a non-trivial character of $L^1(G)^\#$. By [29, §35A], any such χ is continuous with norm equal to 1. Therefore there exists a function φ in $L^\infty(G)$, bi-invariant under K, with $\|\varphi\|_\infty = 1$ and such that

$$\chi(f) = \int_G f(x)\,\varphi(x^{-1})\,dx.$$

The relation $\chi(f * g) = \chi(f) \cdot \chi(g)$ gives

$$\int_G (f * \varphi)(x)\,g(x^{-1})\,dx = \chi(f)\int_G \varphi(x)\,g(x^{-1})\,dx.$$

This being true for all $f, g \in C_c^{\#}(G)$, we get

$$(f * \varphi)(x) = \chi(f) \cdot \varphi(x) \quad \text{a.e.}$$

Hence, by Proposition 6.1.6, the function φ coincides a.e. with a spherical function.

We conclude that the spectrum of $L^1(G)^{\#}$ (the space of its maximal ideals) can be identified with the *bounded* spherical functions. Actually, the supremum-norm of a bounded spherical function equals one. Among these spherical functions there are some that correspond to *symmetric* maximal ideals, namely all φ that satisfy the relation $\varphi = \widetilde{\varphi}$. Since we shall discuss harmonic analysis on $L^1(G)^{\#}$, the *positive-definite* spherical functions take our attention. Let us recall that such functions are automatically bounded and symmetric, see Lemma 5.1.8.

6.2 Positive-definite spherical functions and unitary representations

The general theory of positive-definite functions and their relation to unitary representations was discussed in Section 5.1. We will use the notations and results of that section in relation with this theory.

Let π be a unitary representation of G. For every function $f \in L^1(G)$ the operator $\pi(f)$ was defined in Section 5.1. In a completely similar way, we define $\pi(\mu)$ for any bounded measure μ on G. By abuse of notation we shall write

$$\pi(\mu) = \int_G \pi(x)\, d\mu(x).$$

Let ν be a measure on K; ν may also be considered as a measure on G with compact support, hence as a bounded measure, hence $\pi(\nu)$ is defined. In particular $\pi(h)$ is defined for every function $h \in L^1(K, dk)$.

Let us denote the space of bounded measures on G by $M^1(G)$ (as before, see Section 3.7). Clearly $M^1(G)$ is a Banach algebra under convolution. An involution is given by $\mu^* = \widetilde{\mu}$, where $\widetilde{\mu}$ is defined by $\widetilde{\mu}(f) = \overline{\mu(\widetilde{f})}$ ($f \in C_c(G)$). The subspace of bi-K-invariant bounded measures will be denoted by $M^1(G)^{\#}$; it is a $*$-subalgebra under convolution and the above involution.

Let e denote the characteristic function of K. Consider again a unitary representation π of G on a complex Hilbert space $\mathcal{H}$. The operator $\pi(e) = \int_K \pi(k)\, dk$, defined on $\mathcal{H}$, is an orthogonal projection, call its image $\mathcal{H}_e$; it is the space of vectors in $\mathcal{H}$ fixed under the K-action. We have

$$\pi(\mu)\, \pi(e) = \pi(e)\, \pi(\mu) = \pi(\mu)$$

for all bounded bi-K-invariant measures on G. Therefore

$$\pi(\mu)\mathcal{H}_e \subset \mathcal{H}_e, \ \pi(\mu)\mathcal{H}_e^\perp = \{0\}$$

for all $\mu \in M^1(G)^\#$. Here $\mathcal{H}_e^\perp$ denotes the ortho-complement of $\mathcal{H}_e$ in $\mathcal{H}$. So by restricting the $*$-representation $\mu \mapsto \pi(\mu)$ of the algebra $M^1(G)$ to $M^1(G)^\#$, we actually get a $*$-representation of $M^1(G)^\#$ on the closed subspace $\mathcal{H}_e$. We call this representation π_e.

A $*$-representation π of a $*$-algebra A on a Hilbert space $\mathcal{H}$ is a homomorphism π of A into $\mathrm{End}(\mathcal{H})$ such that $\pi(a^*) = \pi(a)^*$ for all $a \in A$.

Such a representation π is called *non-degenerate* if the closed subspace of $\mathcal{H}$ generated by the vectors $\pi(a)\,\xi$ $(a \in A; \xi \in \mathcal{H})$ equals $\mathcal{H}$. It is called irreducible if there are no non-trivial closed invariant subspaces.

There is a one-to-one correspondence between unitary representations of G and non-degenerate $*$-representations of $M^1(G)$. See [8, 13.3]. If π is an irreducible unitary representation of G, then the associated $*$-representation of $M^1(G)$ is irreducible and conversely. Indeed, apply that $\pi(\delta_x) = \pi(x)$ for $x \in G$, where δ_x is the point mass concentrated at x.

Lemma 6.2.1. *Let π be an irreducible unitary representation of G on a Hilbert space $\mathcal{H}$. Then either $\mathcal{H}_e = \{0\}$ (and hence $\pi_e = 0$) or π_e is a (topologically) irreducible $*$-representation of $M^1(G)^\#$ on $\mathcal{H}_e$.*

Assume that $\mathcal{H}_e$ is non-trivial. The $*$-representation $\mu \mapsto \pi(\mu)$ of $M^1(G)$ on $\mathcal{H}$ is irreducible, since π is. Let ξ be a vector in $\mathcal{H}_e$, $\xi \neq 0$. Since ξ is cyclic for π, the vectors $\pi(\mu)\xi$ where μ runs through $M^1(G)$, span a dense subspace of $\mathcal{H}$. Hence the vectors $\pi(e)\pi(\mu)\xi = \pi(e)\pi(\mu)\pi(e)\xi = \pi(e * \mu * e)\xi$ span a dense subspace of $\mathcal{H}_e$.

To any continuous positive-definite function φ on G we have associated a unitary representation π of G on some Hilbert space $\mathcal{H}$, such that $\varphi(x) = \langle \varepsilon, \pi(x)\,\varepsilon \rangle$ $(x \in G)$, where ε is a cyclic vector for π in $\mathcal{H}$ (cf. Section 5.1). Let us assume that $\varphi(e) = 1$. The representation π is irreducible if and only if φ is elementary (Theorem 5.3.2). Again $\mathcal{H}_e$ can be defined.

Lemma 6.2.2. *The cyclic vector ε belongs to $\mathcal{H}_e$ if and only if φ is bi-K-invariant.*

Suppose $\varepsilon \in \mathcal{H}_e$. Then clearly φ is bi-K-invariant. Assume now that φ is bi-K-invariant. Then we have for all $x \in G$ and $k \in K$

$$\langle \varepsilon, \pi(x)\,\varepsilon \rangle = \varphi(x) = \varphi(k^{-1}x) = \langle \varepsilon, \pi(k^{-1})\pi(x)\,\varepsilon \rangle = \langle \pi(k)\,\varepsilon, \pi(x)\,\varepsilon \rangle.$$

Since ε is a cyclic vector, we get $\pi(k)\,\varepsilon = \varepsilon$ for all $k \in K$, so $\varepsilon \in \mathcal{H}_e$.

Lemma 6.2.3. *Let π be a unitary representation of G on $\mathcal{H}$ admitting a K-invariant cyclic vector. If* $\dim \mathcal{H}_e = 1$*, then the representation π is irreducible.*

Let T be a continuous linear operator on $\mathcal{H}$ commuting with all $\pi(x)$, $x \in G$. In particular we then have $T\,\pi(k)\,\varepsilon = \pi(k)\,T\,\varepsilon$ for all $k \in K$, hence $\pi(e)\,T = T\,\pi(e)$. Therefore $T(\mathcal{H}_e) \subset \mathcal{H}_e$ and hence $T\,\varepsilon = \lambda\,\varepsilon$ for some complex number λ. For arbitrary $x \in G$ we obtain $T\,\pi(x)\,\varepsilon = \pi(x)\,T\,\varepsilon = \lambda\,\pi(x)\,\varepsilon$. So $T = \lambda\,I$ where I is the identity operator on $\mathcal{H}$, because ε is a cyclic vector. Consequently π is irreducible (Theorem 5.1.4).

The preceding lemmas were independent of the commutativity of the algebra $C_c^\#(G)$. From now on we renew for this section the basic assumption that $C_c^\#(G)$ is *commutative*.

Lemma 6.2.4. *Let A be a commutative $*$-Banach algebra and π a non-trivial $*$-representation of A on a Hilbert space $\mathcal{H}$. If π is irreducible then the dimension of $\mathcal{H}$ is equal to one.*

The proof is similar to that of Theorem 5.1.4 and is left to the reader. In other words: Schur's lemma holds in the context of $*$-representations of $*$-Banach algebras.

Theorem 6.2.5. *Let φ be a continuous, positive-definite function, that is bi-K-invariant, and satisfies $\varphi(e) = 1$. Then φ is a (positive-definite) spherical function if and only if φ is elementary.*

Let φ be elementary, whence the representation π on $\mathcal{H}$ associated with φ is irreducible. Set $\varphi(x) = \langle \varepsilon, \pi(x)\,\varepsilon \rangle$ $(x \in G)$, ε a cyclic vector in $\mathcal{H}$. Since $\varepsilon \in \mathcal{H}_e$ by Lemma 6.2.2, π_e is an irreducible $*$-representation of $M^1(G)^\#$ on $\mathcal{H}_e$, by Lemma 6.2.1. By assumption $C_c^\#(G)$ is commutative and hence so is $M^1(G)^\#$, because $C_c^\#(G)$ is dense in $M^1(G)^\#$. Therefore, by Lemma 6.2.3, $\dim \mathcal{H}_e = 1$. For all $f \in C_c^\#(G)$ we now have

$$\begin{aligned} f * \varphi &= \int_G f(y)\,\langle \varepsilon, \pi(y^{-1}x)\,\varepsilon \rangle \, dy \\ &= \langle \pi_e(f)\,\varepsilon, \pi(x)\,\varepsilon \rangle = \chi(f)\,\langle \varepsilon, \pi(x)\,\varepsilon \rangle = \chi(f)\,\varphi(x) \qquad (x \in G), \end{aligned}$$

where $\chi(f)$ is a constant depending on f. From Proposition 6.1.6 it now follows that φ is a spherical function.

Conversely, let φ be a spherical function. Then $f * \varphi = \chi(f)\,\varphi$ for $f \in C_c^\#(G)$, where $\chi(f)$ is a constant depending on f. If we write this down more explicitly, we obtain

$$\langle \pi_e(f)\,\varepsilon, \pi(x)\,\varepsilon \rangle = f * \varphi(x) = \chi(f)\,\varphi(x) = \langle \chi(f)\,\varepsilon, \pi(x)\,\varepsilon \rangle \qquad (x \in G).$$

Hence $\pi_e(f)\,\varepsilon = \chi(f)\,\varepsilon$, since ε is a cyclic vector. Now ε is also a cyclic vector for π_e in $\mathcal{H}_e$, so we have $\dim \mathcal{H}_e = 1$. By Lemma 6.2.3 π is irreducible and hence φ is elementary. This completes the proof.

Corollary 6.2.6. *Two positive-definite spherical functions are equal if and only if the irreducible unitary representations associated with them are (unitarily) equivalent.*

Let φ_1, φ_2 be two positive-definite spherical functions. After obvious identifications we may restrict ourselves to the following situation: there exists an irreducible unitary representation π of G on some Hilbert space $\mathcal{H}$ admitting two cyclic vectors ε_1 and ε_2 such that $\varphi_1(x) = \langle \varepsilon_1, \pi(x)\,\varepsilon_1 \rangle, \varphi_2(x) = \langle \varepsilon_2, \pi(x)\,\varepsilon_2 \rangle$, $x \in G$. By Lemma 6.2.2 we have $\varepsilon_1 \in \mathcal{H}_e$, $\varepsilon_2 \in \mathcal{H}_e$. Since $\dim \mathcal{H}_e = 1$, we have $\varepsilon_1 = \lambda\,\varepsilon_2$ for some complex λ with $|\lambda| = 1$. Hence $\varphi_1(x) = \varphi_2(x)$ for all $x \in G$ and the corollary follows.

6.3 Representations of class one

In this section we present a criterion for the commutativity of the convolution algebra $C_c^{\#}(G)$ in terms of certain properties of the irreducible unitary representations of G. Also we determine the representations which are associated with positive-definite spherical functions.

Proposition 6.3.1. *The algebra $C_c^{\#}(G)$ is commutative if and only if for every irreducible unitary representation π of G on a Hilbert space $\mathcal{H}$ the subspace $\mathcal{H}_e$ of K-fixed vectors is at most one-dimensional.*

First assume that $C_c^{\#}(G)$ is commutative. Let π be an irreducible unitary representation of G on $\mathcal{H}$. Suppose $\mathcal{H}_e \neq \{0\}$. By Lemma 6.2.1, π_e is an irreducible $*$-representation of $M^1(G)^{\#}$ on $\mathcal{H}_e$. Hence $\dim \mathcal{H}_e = 1$ because $M^1(G)^{\#}$ is also commutative.

The converse is due to the fact that $C_c^{\#}(G)$ possesses sufficiently many one-dimensional $*$-representations (assuming that $\dim \mathcal{H}_e \leq 1$ for any irreducible unitary representation of G on a Hilbert space $\mathcal{H}$). Indeed, take a function $f \in C_c^{\#}(G)$, $f \neq 0$. By well-known results, due to Gelfand and Raikov (cf. [8, 13.5 and 13.8]), we can find a continuous elementary positive-definite function φ on G satisfying $\int_G f(x)\,\overline{\varphi(x)}\,dx \neq 0$. Denote the irreducible unitary representation of G associated with φ by π. Let $\varphi(x) = \langle \varepsilon, \pi(x)\,\varepsilon \rangle$ $(x \in G)$, ε a cyclic vector in $\mathcal{H}$. Then we have

$$\int_G f(x)\,\overline{\varphi(x)}\,dx = \int_G f(x)\,\overline{\langle \varepsilon, \pi(x)\,\varepsilon \rangle}\,dx = \langle \pi(f)\,\varepsilon, \pi(e)\,\varepsilon \rangle \neq 0.$$

Therefore $\pi(e)\varepsilon \neq 0$ and hence $\mathcal{H}_e \neq \{0\}$, so, by assumption, $\dim \mathcal{H}_e = 1$. Thus π_e is a one-dimensional $*$-representation of $C_c^{\#}(G)$ on $\mathcal{H}_e$ and $\pi_e(f) \neq 0$. We have shown that $C_c^{\#}(G)$ possesses sufficiently many one-dimensional $*$-representations. To complete the proof, we observe that for any two functions $f_1, f_2 \in C_c^{\#}(G)$ and any one-dimensional $*$-representation ρ of $C_c^{\#}(G)$ we have $\rho\,(f_1 * f_2 - f_2 * f_1) = 0$. Hence by the considerations made above $f_1 * f_2 = f_2 * f_1$. So we have shown that $C_c^{\#}(G)$ is commutative and the proposition follows.

Definition 6.3.2. An irreducible unitary representation π of G on a Hilbert space $\mathcal{H}$ is said to be of class one if the subspace of K-fixed vectors $\mathcal{H}_e$ is non-trivial.

Corollary 6.3.3. *Let (G, K) be a Gelfand pair. The positive-definite spherical functions on G correspond one-to-one to the equivalence classes of irreducible unitary representations of G of class one.*

This follows easily from Theorem 6.3.1 and Corollary 6.2.6.

6.4 Harmonic analysis on Gelfand pairs

In this section we assume that (G, K) is a Gelfand pair. We shall present a sketch of the most essential part of the abstract harmonic analysis for (G, K). The well-known tools of harmonic analysis on abelian groups will appear frequently (see Chapter 5). In almost all cases we shall omit proofs.

(i) The dual space Z

Let $\mathcal{P}_0^\#$ be the set of continuous, bi-K-invariant, positive-definite functions φ on G with $\varphi(e) \leq 1$. We may consider $\mathcal{P}_0^\#$ as a convex subset of $L^\infty(G)$. Clearly $\mathcal{P}_0^\#$ is weakly closed, i.e. $\sigma(L^\infty, L^1)$-closed, so by Alaoglu's theorem it is weakly compact.

Proposition 6.4.1. *The extremal points of $\mathcal{P}_0^\#$ consist of the set of positive-definite spherical functions and the zero function.*

Compare this proposition with Lemma 5.3.1. Denote by Z the set of positive-definite spherical functions, provided with the topology $\sigma(L^\infty, L^1)$.

Proposition 6.4.2. (a) *The space Z is locally compact.*

(b) *The topology on Z coincides with the topology of uniform convergence on compact subsets of G on Z.*

The proof is similar to that of Theorem 5.4.1.

The space Z is called the *dual space* of the pair (G, K).

(ii) The Fourier transform on $L^1(G)^\#$

Definition 6.4.3. The Fourier transform $\hat{f}$ of a function $f \in L^1(G)^\#$ is the function defined on Z by

$$\hat{f}(\varphi) = \int_G f(x)\, \varphi(x^{-1})\, dx \qquad (\varphi \in Z).$$

The following properties hold:

(a) $\hat{f}$ is a continuous function on Z, vanishing at 'infinity'; furthermore, we have $|f(\varphi)| \le \|f\|_1$ for all $\varphi \in Z$,

(b) the mapping $f \mapsto \hat{f}$ is a linear transformation,

(c) $\widehat{(f * g)} = \hat{f} \cdot \hat{g}$ for all $f, g \in L^1(G)^{\#}$,

(d) $\widehat{(\tilde{f})} = \overline{\hat{f}}$ for all $f \in L^1(G)^{\#}$.

Furthermore, one has the usual property: any continuous function on Z that vanishes at infinity can be approximated uniformly on Z by functions of the form $\hat{f}$ $(f \in C_c^{\#}(G))$. We refer to Section 5.4 (ii).

(iii) The analog of Bochner's theorem

Let $\mathcal{C}_0(Z)$ be the space of continuous complex-valued functions on Z that vanish at infinity, provided with the supremum norm. Denote by $M^1(Z)$ the space of bounded complex measures on Z and by $\mathcal{V}(G)^{\#}$ the space of all complex linear combinations of continuous positive-definite functions on G that are bi-K-invariant.

Define the mapping $T^* : M^1(Z) \to L^\infty(G)$ by

$$(T^*\mu)(x) = \int_Z \varphi(x)\, d\mu(\varphi).$$

One easily checks that T^* is the adjoint of T, defined by $Tf = \hat{f}$ $(f \in L^1(G)^{\#})$. Clearly image$(T^*) \subset \mathcal{V}(G)^{\#}$. Since image$(T)$ is dense in $\mathcal{C}_0(Z)$, T^* is injective. Observe that $T^*\mu \in \mathcal{P}_0^{\#}$ if μ is a positive measure.

Set $\mathcal{M}_0(Z) = \{\mu \in M^1(Z) : \mu \text{ is positive}, \|\mu\| \le 1\}$.

Theorem 6.4.4. *T^* maps $\mathcal{M}_0(Z)$ onto $\mathcal{P}_0^{\#}$.*

Again Krein–Milman's theorem is crucial in the proof of Theorem 6.4.4; compare with Theorem 5.4.3. One can even show that T^* is a *homeomorphism* from $\mathcal{M}_0(Z)$ provided with the topology induced by $\sigma(M^1(Z), \mathcal{C}_0(Z))$ onto $\mathcal{P}_0^{\#}$ provided with the topology induced by $\sigma(L^\infty, L^1)$.

(iv) Inversion theorem

Set $\mathcal{V}^1(G)^{\#} = \mathcal{V}(G)^{\#} \cap L^1(G)$.

To any $f \in \mathcal{V}^1(G)^{\#}$ there corresponds, by Theorem 6.4.4, a unique measure $\mu_f \in M^1(Z)$ such that

$$f(x) = \int_Z \varphi(x)\, d\mu_f(\varphi) \qquad (x \in G).$$

Theorem 6.4.5. *There exists a unique positive measure* ν *on* Z *such that* $d\mu_f(\varphi) = \hat{f}(\varphi)\, d\mu_f(\varphi)$ *for all* $f \in \mathcal{V}^1(G)^\#$. *More precisely: a unique positive measure* ν *on* Z *exists such that for all* $f \in \mathcal{V}^1(G)^\#$ *one has*

(a) $\hat{f} \in L^1(Z, \nu)$,

(b) $f(x) = \int_Z \varphi(x)\, \hat{f}(\varphi)\, d\nu(\varphi) \quad (x \in G)$.

The proof is similar to that of Theorem 5.5.1 and depends heavily on the commutativity of the algebra $L^1(G)^\#$.

(v) Plancherel's theorem

Let ν be the positive measure on Z obtained in (iv).

Theorem 6.4.6. *For every* $f \in C_c^\#(G)$ *one has*

(a) $\hat{f} \in L^2(Z, \nu)$,

(b) $\int_G |f(x)|^2\, dx = \int_Z |\hat{f}(\varphi)|^2\, d\nu(\varphi)$.

If one extends the mapping $f \mapsto \hat{f}$ *from* $C_c^\#(G)$ *to* $L^2(G)^\#$, *the closure of* $C_c^\#(G)$ *in* $L^2(G)^\#$, *one obtains an isometric isomorphism from* $L^2(G)^\#$ *onto* $L^2(Z, \nu)$.

For $f \in C_c^\#(G)$, set $g = f * \widetilde{f}$. Then $g \in \mathcal{V}^1(G)^\#$ and hence $\hat{g} \in L^1(Z, \nu)$. Moreover

$$g(x) = \int_Z \hat{g}(\varphi)\, \varphi(x)\, d\nu(\varphi)$$

for all $x \in G$, by (iv). Since $\hat{g}(\varphi) = |\hat{f}(\varphi)|^2$ $(\varphi \in Z)$ and $g(e) = \int_G |f(x)|^2\, dx$, we obtain

$$\int_G |f(x)|^2\, dx = \int_Z |\hat{f}(\varphi)|^2\, d\nu(\varphi).$$

It follows that $f \mapsto \hat{f}$ can be extended isometrically to $L^2(G)^\#$ with values in $L^2(Z, \nu)$. We will show that this mapping is surjective. The arguments have to be different from those in 5.6. Let F be a continuous function on Z with compact support. Let $\varepsilon > 0$ be given. Functions F' and u exist ($F' \in C_c(Z)$, $u \in C_c^\#(G)$) such that (α) $\hat{u}$ is strictly positive on the support of F, (β) $F = F' \cdot \hat{u}$.

By (ii) F' can be approximated uniformly on Z by a function of the form $\hat{h}$ ($h \in C_c^\#(G)$), so we have

$$|F'(\varphi) - \hat{h}(\varphi)| < \varepsilon$$

for all $\varphi \in Z$. We obtain, using (α) and (β),

$$|F(\varphi) - (u * h)\hat{}(\varphi)| = |F(\varphi) - \hat{u}(\varphi)\hat{h}(\varphi)| = |\hat{u}(\varphi)||F'(\varphi) - \hat{h}(\varphi)| \le \varepsilon\, |\hat{u}(\varphi)|$$

for all $\varphi \in Z$. Therefore

$$\int_Z |F(\varphi) - (u * h)\hat{}(\varphi)|^2\, d\nu(\varphi) \le \varepsilon \int_Z |\hat{u}(\varphi)|^2\, d\nu(\varphi).$$

Since ε was arbitrary, we conclude that we have embedded $T(L^2(G)^\#)$ as a dense subspace into $L^2(Z, \nu)$. The theorem now follows easily.

Remark 6.4.7. (1) If G is abelian and $K = \{e\}$, the set Z has a structure of an abelian group, it is the dual group $\widehat{G}$. Furthermore, ν is a Haar measure on $\widehat{G}$.

(2) We clearly have the inclusion

$$\mathrm{Supp}(\nu) \subset Z.$$

In certain cases this inclusion is strict, as we shall see later on from the examples.

(3) The measure ν is commonly called the *Plancherel measure* for the Gelfand pair (G, K).

6.5 Compact Gelfand pairs

We recall some properties of representations of *compact* groups G.

(a) *Any representation of G is equivalent to a unitary representation.*

(b) *Any irreducible unitary representation is finite-dimensional.*

(c) *Any unitary representation is the direct sum of irreducible unitary representations.*

For the above, and more, we refer to [53].

Let G be a compact group and let K be a closed subgroup of G. We assume that (G, K) is a Gelfand pair and we denote by dx and dk the normalized Haar measures on G and K respectively.

Theorem 6.5.1. *Every spherical function is positive-definite. Let $(\pi, \mathcal{H})$ be the unitary representation associated with a spherical function φ. The space $\mathcal{H}$ is finite-dimensional. Let χ be the character of π, i.e.*

$$\chi(x) = \operatorname{tr} \pi(x) \qquad (\text{trace of } \pi(x)).$$

Then

$$\varphi(x) = \int_K \chi(x^{-1}k)\, dk$$

and

$$\int_G |\varphi(x)|^2\, dx = \frac{1}{d},$$

where d is the dimension of $\mathcal{H}$.

(a) Let $(\pi, \mathcal{H})$ be an irreducible unitary representation of G with a vector ε, fixed under K and with norm equal to 1. The function φ defined by

$$\varphi(x) = \langle \varepsilon, \pi(x)\,\varepsilon \rangle \qquad (x \in G)$$

is a positive-definite spherical function. Since G is compact, $\mathcal{H}$ is finite-dimensional. The subspace $\mathcal{H}_e$ of K-fixed vectors in $\mathcal{H}$ is one-dimensional and spanned by ε, by Proposition 6.3.1. The projection on $\mathcal{H}_e$ is given by $\pi(e) = \int_K \pi(k)\,dk$. Consider an orthonormal basis of $\mathcal{H}$, say $\varepsilon_1, \varepsilon_2, \ldots, \varepsilon_d$, such that $\varepsilon_1 = \varepsilon$. We have $\pi(e)\,\varepsilon_i = 0$ for $i \geq 2$ and

$$\operatorname{tr} \pi(e)\pi(x)\pi(e) = \varphi(x^{-1}) = \operatorname{tr} \pi(x)\pi(e) = \int_K \chi(xk)\,dk,$$

hence $\varphi(x) = \int_K \chi(x^{-1}k)\,dk$.

(b) Let φ be a spherical function and set

$$V_\varphi = \{f = \mu * \varphi \mid \mu \in M_0(G)\}$$

where $M_0(G)$ denotes the space of measures of the form

$$\mu = \sum_{i=1}^{N} a_i\,\delta_{x_i} \qquad (x_1, \ldots, x_N \in G).$$

Let us first show that V_φ is finite-dimensional. For $h \in C^\#(G)$ set

$$T_h f = f * h \qquad (f \in V_\varphi).$$

Then T_h leaves $C(G)$ invariant and is a compact operator. If f is a function in V_φ, then

$$\begin{aligned} T_h f &= \mu * \varphi * h \\ &= \mu * \chi(h)\varphi = \chi(h) f \end{aligned}$$

where

$$\chi(h) = \int_G \varphi(x^{-1})\,h(x)\,dx.$$

So V_φ is an eigenspace of T_h with eigenvalue $\chi(h)$. We can obviously choose h such that $\chi(h) \neq 0$, hence V_φ is finite-dimensional.

Consider V_φ as a subspace of $L^2(G)$ and define the unitary representation π of G on V_φ by

$$(\pi(x)f)(y) = f(x^{-1}y).$$

Let $\pi(e)$ be the orthogonal projection on the subspace $(V_\varphi)_e$ of the K-invariant functions in V_φ. We have

$$\pi(e)(\mu * \varphi) = \chi(\mu)\,\varphi$$

where

$$\chi(\mu) = \int_G \varphi(x^{-1})\,d\mu(x),$$

hence $\dim(V_\varphi)_e = 1$ and $(V_\varphi)_e$ is spanned by φ. Moreover, (π, V_φ) is irreducible by Lemma 6.2.3, since φ is also a cyclic vector in V_φ.

Observe that for each function f in V_φ one has

$$f * \varphi = \chi(\varphi)\, f. \tag{1}$$

Consider an orthonormal basis $f_1, \ldots, f_d$ of V_φ and set

$$H(x, y) = \sum_{i=1}^{d} f_i(x)\,\overline{f_i(y)}.$$

The function H is called the *reproducing kernel* of the space V_φ: for all $f \in V_\varphi$ one has

$$f(x) = \int_G H(x, y)\, f(y)\, dy. \tag{2}$$

Comparing the relations (1) and (2) yields

$$\chi(\varphi)\, H(x, y) = \varphi(y^{-1}x).$$

Taking $x = y$ and integrating over G, we obtain

$$\chi(\varphi)\, d = \varphi(e) = 1,$$

hence

$$\sum_{i=1}^{d} f_i(x)\overline{f_i(y)} = d\, \varphi(y^{-1}x),$$

so φ is positive-definite. Moreover

$$\int_G |\varphi(x)|^2\, dx = \chi(\varphi) = \frac{1}{d}$$

and

$$\langle \varphi, \pi(x)\,\varphi \rangle = \frac{1}{d}\, \varphi(x).$$

This completes the proof of the theorem.

Observe that a fine example of a compact Gelfand pair is given by $G = G' \times G'$ where G' is a compact group and $K = \operatorname{diag} G = \{(x, x) : x \in G'\}$. One shows that this example yields the 'classical' analysis on compact groups (Peter–Weyl theorem, etc.). For more examples we refer to the next chapter.

Chapter 7

Examples of Gelfand Pairs

Literature: [15], [21].

7.1 Euclidean motion groups

(i) Introduction

Consider on $\mathbb{R}^n$ the usual scalar product

$$(x, y) = x_1 y_1 + \cdots + x_n y_n,$$

where $x = (x_1, \ldots, x_n)$, $y = (y_1, \ldots, y_n)$. The group of Euclidean motions of $\mathbb{R}^n$ is the semi-direct product

$$G = K \ltimes \mathbb{R}^n$$

with $K = \mathrm{SO}(n, \mathbb{R}^n)$. Elements of G are written as pairs $g = (k, a)$ with $k \in K, a \in \mathbb{R}^n$. Such a pair has to be viewed as the product of the rotation k and the translation over a, considered as operating on $\mathbb{R}^n$:

$$g \cdot x = k \cdot x + a \qquad (x \in \mathbb{R}^n).$$

Hence the product in G is given by

$$(k, a)\,(k', a') = (kk', k' \cdot a + a').$$

Clearly $(k, a) = (k, 0)\,(1, a)$. Define the mapping θ by $\theta\,(k, a) = (k, -a)$. Then θ is a continuous involutive automorphism of G and

$$\theta(k, a) = \theta((k, 0)\,(1, a)) = (k, 0)\,(1, -a) = (k, 0)\,(k, a)^{-1}\,(k, 0).$$

So $\theta(g) \in K g^{-1} K$ for all $g \in G$, hence (G, K) is a Gelfand pair, by Proposition 6.1.3.

This fact can also be seen in another way. Functions on G that are bi-invariant under K can (by restriction) be identified with functions f on $\mathbb{R}^n$ satisfying

$$f(k \cdot x) = f(x) \qquad (x \in \mathbb{R}^n,\ k \in K),$$

so with so-called *radial functions*, and the convolution product of two such functions corresponds with the ordinary convolution product on $\mathbb{R}^n$. Hence $C_c^\#(G)$ is a commutative convolution algebra.

(ii) Spherical functions

Denote by Δ the Laplacian on $\mathbb{R}^n$, i.e.

$$\Delta = \frac{\partial^2}{\partial x_1^2} + \cdots + \frac{\partial^2}{\partial x_n^2}.$$

This differential operator is invariant under the group G, which means that for any $g \in G$ and any function f of class C^2 on $\mathbb{R}^n$ one has

$$\Delta(L_g f) = L_g(\Delta f),$$

where

$$(L_g f)(x) = f(g^{-1} \cdot x).$$

If f_1 and f_2 are two functions of class C^2 on $\mathbb{R}^n$, f_1 with compact support, then

$$\Delta(f_1 * f_2) = \Delta f_1 * f_2 = f_1 * \Delta f_2.$$

Theorem 7.1.1. *Let φ be a bi-K-invariant function on G, considered as a radial function on $\mathbb{R}^n$. This function is a spherical function if and only if*

(1) *φ is C^∞,*

(2) *there exists a complex number λ such that $\Delta\varphi = \lambda\varphi$,*

(3) *$\varphi(0) = 1$.*

(a) Suppose first that φ is a spherical function. For all continuous radial functions f with compact support we have

$$f * \varphi = \chi(f)\,\varphi$$

where

$$\chi(f) = \int_{\mathbb{R}^n} \varphi(-x)\, f(x)\, dx.$$

If we take f of class C^∞ such that $\chi(f) \neq 0$, we see from this relation that φ is C^∞ and moreover

$$\Delta f * \varphi = \chi(\Delta f)\,\varphi = \chi(f)\,\Delta\varphi,$$

hence

$$\Delta\varphi = \lambda\varphi$$

with

$$\lambda = \frac{\chi(\Delta f)}{\chi(f)}.$$

(b) Let now φ be a C^∞ function on $\mathbb{R}^n$, radial and solution of the equation

$$\Delta\varphi = \lambda\varphi$$

for some $\lambda \in \mathbb{C}$. Considered as a function of $r = \|x\| = \sqrt{x_1^2 + \cdots + x_n^2}$, φ is a regular solution of the differential equation

$$\frac{d^2\varphi}{dr^2} + \frac{n-1}{r}\frac{d\varphi}{dr} = \lambda\varphi$$

and therefore (see [14]) it is proportional to the function $J_\lambda(r)$ defined by

$$J_\lambda(r) = \Gamma\Big(\frac{n}{2}\Big) \sum_{k=0}^{\infty} \frac{\lambda^k}{k!\,\Gamma(k+\frac{n}{2})} \Big(\frac{r}{2}\Big)^{2k} \tag{7.1.1}$$

$$= \Gamma\Big(\frac{n}{2}\Big) \left(\frac{\sqrt{\lambda} r}{2}\right)^{\frac{2-n}{2}} I_{\frac{n-2}{2}}(\sqrt{\lambda} r),$$

where I_ν is the modified Bessel function of index ν.

Observe that different λ give different solutions. Assume now, in addition, that we have $\varphi(0) = 1$. Let f be a radial function on $\mathbb{R}^n$, continuous and with compact support. The function $\psi = f * \varphi$ is then radial, C^∞ and again solution of

$$\Delta\psi = \lambda\psi,$$

hence $\psi = C\,\varphi$ for some constant C, depending on f, given by

$$C = \chi(f) = \int_{\mathbb{R}^n} \varphi(-x)\, f(x)\, dx,$$

hence

$$f * \varphi = \chi(f)\,\varphi,$$

so φ is a spherical function, by Proposition 6.1.6.

Let now s be a complex number and $\xi \in \mathbb{R}^n$ a vector of length equal to 1. The function f defined by

$$f(x) = \mathrm{e}^{s(\xi,x)}$$

is an eigenfunction of the Laplacian:

$$\Delta f = s^2 f.$$

Consider the integral

$$\varphi_s(x) = \int_{S^{n-1}} \mathrm{e}^{s(\xi,x)}\, d\sigma(\xi) \tag{7.1.2}$$

where S^{n-1} is the unit sphere in $\mathbb{R}^n$ and σ the normalized surface measure on S^{n-1}. The function φ_s is radial, C^∞ and satisfies

$$\Delta\varphi_s = s^2\varphi_s, \quad \varphi_s(0) = 1,$$

so that φ_s is a spherical function. Conversely, if φ is a spherical function then there exists $s \in \mathbb{C}$ such that $\varphi = \varphi_s$. Clearly $\varphi_s = \varphi_{-s}$.

The preceding integral can be made explicit using spherical coordinates:

$$\varphi_s(r) = \varphi(r,s) = \frac{\Gamma(\frac{n}{2})}{\sqrt{\pi}\Gamma(\frac{n-1}{2})} \int_0^\pi e^{sr\cos\theta} \sin^{n-2}\theta \, d\theta. \tag{7.1.3}$$

For $n = 3$ we get

$$\varphi(r,s) = \frac{\sinh(sr)}{sr},$$

while for general n, using the power series expansion of the exponential function, we obtain

$$\begin{aligned}\varphi(r,s) &= \Gamma\Big(\frac{n}{2}\Big) \sum_{k=0}^\infty \frac{1}{k!\,\Gamma(k+\frac{n}{2})} \Big(\frac{sr}{2}\Big)^{2k} \\ &= \Gamma\Big(\frac{n}{2}\Big) \Big(\frac{sr}{2}\Big)^{\frac{2-n}{2}} I_{\frac{n-2}{2}}(sr).\end{aligned}$$

Compare this expression with (7.1.1).

(iii) Bounded spherical functions

Proposition 7.1.2. *The spherical function φ_s is bounded if and only if* $\operatorname{Re} s = 0$.

If $\operatorname{Re} s = 0$ then it follows from (7.1.2) that $|\varphi_s(x)| \le 1$ for all $x \in G$. If $\operatorname{Re} s > 0$, we shall show that

$$\varphi(r,s) \sim \frac{\Gamma(\frac{n}{2})\, 2^{\frac{n-3}{2}}}{\sqrt{\pi}} \frac{e^{sr}}{(sr)^{\frac{n-1}{2}}} \qquad (r \to \infty).$$

From (7.1.3) we get, by an elementary substitution,

$$\varphi_s(r) = \varphi(r,s) = \frac{\Gamma(\frac{n}{2})}{\sqrt{\pi}\,\Gamma(\frac{n-1}{2})} \int_{-1}^1 e^{srt} (1-t^2)^{\frac{n-3}{2}} \, dt,$$

and, setting $t = 1 - \frac{u}{r}$, we obtain

$$\varphi(r,s) = \frac{\Gamma(\frac{n}{2})}{\sqrt{\pi}\,\Gamma(\frac{n-1}{2})} \frac{e^{sr}}{r^{\frac{n-1}{2}}} \int_0^{2r} e^{-su} u^{\frac{n-3}{2}} \Big(2 - \frac{u}{r}\Big)^{\frac{n-3}{2}} du.$$

For $\operatorname{Re} s > 0$ we get, using Lebesgue's dominated convergence theorem,

$$\lim_{r\to\infty} \int_0^{2r} e^{-su} u^{\frac{n-3}{2}} \Big(2 - \frac{u}{r}\Big)^{\frac{n-3}{2}} du = \frac{\Gamma(\frac{n-1}{2})\, 2^{\frac{n-3}{2}}}{s^{\frac{n-1}{2}}}.$$

(iv) Positive-definite spherical functions

Proposition 7.1.3. *The spherical function φ_s is positive-definite if and only if one has* $\operatorname{Re} s = 0$.

If φ_s is positive-definite then it is bounded and hence $\operatorname{Re} s = 0$ by Proposition 7.1.2.

Conversely, we have to show that $\varphi_{i\nu}$ ($\nu \in \mathbb{R}$) is positive-definite. Let us use the integral representation (7.1.2) of $\varphi_{i\nu}$. Then we get

$$\sum_{j,k=1}^{N} \varphi_{i\nu}(x_j - x_k)\, c_j\, \overline{c_k} = \int_{S^{n-1}} \Big| \sum_{j=1}^{N} c_j\, \mathrm{e}^{i\nu(\xi, x_j)} \Big|^2 d\sigma(\xi) \geq 0.$$

Notice that for Euclidean motion groups the set of bounded spherical functions coincides with the set of positive-definite spherical functions.

(v) Plancherel measure

Let f be a radial function in $L^1(\mathbb{R}^n)$. Then the Euclidean Fourier transform $\mathcal{F} f$ of f is again radial, so $\mathcal{F} f$ is a function of the radius s. We shall write $\mathcal{F} f(s)$, with abuse of notation. There is a nice relation with the spherical Fourier transform $\widehat{f}$ of f defined by

$$\hat{f}(s) = \int_{\mathbb{R}^n} f(x)\, \varphi_{2\pi i s}(-x)\, dx.$$

Indeed,

$$\begin{aligned} \widehat{f}(s) &= \int_{\mathbb{R}^n} f(x) \int_{S^{n-1}} \mathrm{e}^{-2\pi i s(x,\xi)} d\sigma(\xi)\, dx \\ &= \mathcal{F} f(s), \end{aligned}$$

where $s \in \mathbb{R}$, $s \geq 0$. This implies that for sufficiently regular radial functions f on $\mathbb{R}^n$ (e.g. $f = u * v$ with $u, v \in C_c(\mathbb{R}^n)$ and both radial)

$$f(0) = \int_{\mathbb{R}^n} \mathcal{F} f(x)\, dx = \frac{2\pi^{\frac{n}{2}}}{\Gamma(\frac{n}{2})} \int_0^\infty \hat{f}(s)\, s^{n-1}\, ds.$$

Hence the Plancherel measure for the Euclidean motion group, with respect to $\mathrm{SO}(n, \mathbb{R})$, is equal to

$$d\nu(s) = \frac{2\pi^{\frac{n}{2}}}{\Gamma(\frac{n}{2})} s^{n-1}\, ds,$$

where we have identified the dual space Z with $[0, \infty)$ via the correspondence $s \leftrightarrow \varphi_{2\pi i s}$. This identification is easily seen to be a homeomorphism. The proof is left to the reader.

(vi) Representations of class one

Let for $y \in \mathbb{R}^n$, $\chi_y(x) = \mathrm{e}^{-2\pi i(x,y)}$ $(x \in \mathbb{R}^n)$ be a character of $\mathbb{R}^n$. One easily sees that $\chi_y(k \cdot x) = \chi_{k^{-1}y}(x)$ for $x \in \mathbb{R}^n$, $k \in K$. So K acts on the characters of $\mathbb{R}^n$. Let us denote for $y \in \mathbb{R}^n$, K_y the stabilizer of χ_y, which coincides with the stabilizer of y; it is a closed subgroup of K. Let $d\dot{k}$ denote the normalized K-invariant measure on K/K_y.

We now define a unitary representation of G by inducing the representation $1 \otimes \chi_y$ from $K_y \ltimes \mathbb{R}^n$ to G.

The Hilbert space V_y consists of the measurable functions f on G satisfying

(a) $f(ga) = \chi_y(-a)f(g) \quad (g \in G; a \in \mathbb{R}^n)$,

(b) $f(gk) = f(g) \quad (g \in G; k \in K_y)$,

(c) $\int_{K/K_y} |f(k)|^2 d\dot{k} < \infty$.

The Hilbert norm is given by $\|f\| = (\int_{K/K_y} |f(k)|^2 d\dot{k})^{1/2}$. The representation π_y is defined by

$$(\pi_y(g)f)(g') = f(g^{-1}g') \qquad (g, g' \in G).$$

One easily verifies that π_y is a unitary representation of G on V_y, called the *representation induced by* $1 \otimes \chi_y$.

There exists (up to scalars) only one function fixed by K in V_y, namely

$$f_0(ka) = \chi_y(-a) \qquad (k \in K, a \in \mathbb{R}^n).$$

Observe that $\|f_0\| = 1$. So $\dim(V_y)_e = 1$.

The function f_0 is also cyclic in V_y. Indeed, assume that $(\pi_y(g)f_0, f) = 0$ for all $g \in G$, where $f \in V_y$. Then clearly, taking $g = a \in \mathbb{R}^n$,

$$\int_{K/K_y} \mathrm{e}^{-2\pi i(a,k\cdot y)}\overline{f(k)}\, d\dot{k} = 0 \quad \text{for all } a \in \mathbb{R}^n,$$

hence

$$\int_{K/K_y} \hat{h}(k \cdot y)\, \overline{f(k)}\, d\dot{k} = 0$$

for all $h \in L^1(\mathbb{R}^n)$, hence $f = 0$ (the functions $\hat{h}$ are dense in $\mathcal{C}_0(\mathbb{R}^n)$ with respect to the supremum norm).

Theorem 7.1.4. (a) *The unitary representations π_y $(y \in \mathbb{R}^n)$, induced by $1 \otimes \chi_y$ from $K_y \ltimes \mathbb{R}^n$ to G, are irreducible.*

(b) *Two representations π_y and $\pi_{y'}$ are equivalent if and only if $\|y\| = \|y'\|$.*

(c) *The representations π_y exhaust the set of irreducible unitary representations of class one (up to equivalence).*

Statement (a) follows from Lemma 6.2.3, (b) follows from Corollary 6.2.6, since for all $a \in \mathbb{R}^n$

$$
\begin{aligned}
(f_0, \pi_y(a) f_0) &= \int_{K/K_y} f_0(a^{-1}k) d\dot{k} = \int_{K/K_y} e^{2\pi i (a, k\cdot y)} d\dot{k} \\
&= \int_{S^{n-1}} e^{2\pi i (a,\xi)\|y\|} d\sigma(\xi) = \varphi_{2\pi i \|y\|}(a).
\end{aligned}
$$

Statement (c) is now an immediate consequence of this computation, Proposition 7.1.3 and Corollary 6.3.3.

7.2 The sphere

For $m \geq 2$, let $G = \mathrm{SO}(m)$, the real special orthogonal group of the quadratic form

$$
(x, y) = x_1 y_1 + \cdots + x_m y_m,
$$

where $x = (x_1, \ldots, x_m)$, $y = (y_1, \ldots, y_m) \in \mathbb{R}^m$. This group acts transitively on the sphere S^{m-1} in $\mathbb{R}^m$, and the stabilizer of the point $e_1 = (1, 0, \ldots, 0)$ is isomorphic to $\mathrm{SO}(m-1)$. We shall show that (G, K) is a Gelfand pair.

Let $x = (x_1, \ldots, x_m) = g \cdot e_1$ be an element of S^{m-1}. Clearly, we can find $k \in K$ such that $k \cdot x = (x_1, 0, \ldots, 0, y)$ with $y^2 = x_2^2 + \cdots + x_m^2$. Let A be the subgroup of G consisting of the matrices of the form

$$
\begin{pmatrix} \cos\theta & 0 & \sin\theta \\ 0 & I_{m-2} & 0 \\ -\sin\theta & 0 & \cos\theta \end{pmatrix}, \qquad 0 \leq \theta < 2\pi.
$$

Here I_{m-2} is the $(m-2) \times (m-2)$ identity matrix and the zeros denote the appropriate row and column vectors respectively. Then there is $a \in A$ such that $a \cdot (x_1, 0, \ldots, 0, y) = e_1$. Therefore $g \cdot e_1 = k^{-1} \cdot (a^{-1} \cdot e_1)$, or $g = k^{-1} a^{-1} l$ for some $l \in K$. In other words, $G = KAK$.

Now define the involution σ by

$$
\sigma(g) = JgJ \qquad (g \in G)
$$

where J is the matrix $J = \mathrm{diag}(-1, 1, \ldots, 1)$. Clearly, σ leaves G invariant, and also K, whereas $\sigma(a) = a^{-1}$ for $a \in A$. Therefore $\sigma(g) \in K g^{-1} K$ for all $g \in G$ and we conclude from Proposition 6.1.3 that (G, K) is a Gelfand pair.

To determine the spherical functions of the pair (G, K) we shall now develop, in a separate section, the theory of *spherical harmonics*. This will also give the complete set of class one representations and their dimensions, hence the Plancherel measure. Observe that each spherical function is positive-definite, since G is compact (Theorem 6.5.1).

7.3 Spherical harmonics

A suitable reference for this section is [44].

(i) Notations and main theorem

Let $S = S^{m-1}$ be, as before, the unit sphere in $\mathbb{R}^m$, $m \geq 2$. For $n \geq 0$, consider the linear space Π_n of polynomials $P(x) = P(x_1, \ldots, x_m)$, homogeneous of degree n, with complex coefficients. So

$$P(\lambda x) = \lambda^n P(x) \qquad (x \in \mathbb{R}^m, \lambda \in \mathbb{R}). \tag{7.3.1}$$

Writing for any multi-index $\alpha = (\alpha_1, \ldots, \alpha_m) \in \mathbb{Z}_+^m$

$$x^\alpha = x_1^{\alpha_1} \cdots x_m^{\alpha_m}$$

and $|\alpha| = \alpha_1 + \cdots + \alpha_m$, one thus has

$$P(x) = \sum_{|\alpha|=n} c_\alpha \, x^\alpha$$

for suitable $c_\alpha \in \mathbb{C}$.

Let $H_n = \{P \in \Pi_n : \Delta P = 0\}$, the subspace of *harmonic polynomials*, and set

$$\mathcal{H}_n = \{P|_S : P \in H_n\}, \tag{7.3.2}$$

the space of restrictions to S of polynomials in H_n, the space of *spherical harmonics of degree n*.

Observe that $H_0 = \Pi_0$, $H_1 = \Pi_1$. We shall regard the spaces $\mathcal{H}_n$ as linear subspaces of $L^2(S)$, the Hilbert space of L^2-functions on S with respect to the normalized G-invariant measure ds on S and scalar product

$$(f \mid g) = \int_S f(s)\,\overline{g(s)}\,ds \qquad (f, g \in L^2(S)).$$

Clearly, G acts on $L^2(S)$ and also on each $\mathcal{H}_n$, because Δ is a G-invariant differential operator, as follows from the equality

$$U(g)f(s) = f(g^{-1} \cdot s).$$

The mapping U is a (continuous) unitary representation of G on $L^2(S)$. Call U_n the restriction of U to $\mathcal{H}_n$, so $U_n(g) = U(g)|_{\mathcal{H}_n}$ $(g \in G)$.

For later reference we mention the following proposition.

Proposition 7.3.1. *The dimension of* Π_n *is equal to* $\binom{m+n-1}{n}$.

To see this, observe that Π_n is spanned by the monomials $x^\alpha = x_1^{\alpha_1} x_2^{\alpha_2} \cdots x_m^{\alpha_m}$ with $\alpha = (\alpha_1, \ldots, \alpha_m) \in \mathbb{Z}_+^m$, $|\alpha| = \alpha_1 + \cdots + \alpha_m = n$. Regard these monomials as coefficients in

$$\prod_{1 \le j \le m} (1 - x_j t)^{-1} = \prod_{1 \le j \le m} \sum_{k \ge 0} x_j^k t^k.$$

Then $\dim \Pi_n$ is the coefficient of t^n in $(1-t)^{-m}$. So

$$\dim \Pi_n = \frac{1}{n!} \frac{d^n}{dt^n} (1-t)^{-m}|_{t=0} = \frac{(n+m-1)!}{n!\,(m-1)!} = \binom{m+n-1}{n}.$$

We now come to the main theorem of this section.

Theorem 7.3.2. *Let* $m \ge 3$.

(1) *One has* $L^2(S) = \bigoplus_{n=0}^\infty \mathcal{H}_n$.

(2) *All spaces* $\mathcal{H}_n$ *are irreducible under* U_n.

(3) *The irreducible representations* U_n *on* $\mathcal{H}_n$ *are pairwise inequivalent.*

(4) *If* $m = 3$, *any irreducible representation of* SO(3) *is equivalent to one of the representations* U_n.

The proof of this theorem will be given in the following subsections.

(ii) We first show

Lemma 7.3.3. *For any pair* (n,k) *with* $n \ne k$, $\mathcal{H}_n$ *and* $\mathcal{H}_k$ *are orthogonal subspaces of* $L^2(S)$.

We shall use spherical coordinates in $\mathbb{R}^m$. Let $x \in \mathbb{R}^m$, $x \ne 0$, and write $x = (r,s)$ with $r = \|x\|$, $s = x/r \in S$. One has

$$\int_{\|x\| \le 1} f(x)\,dx = \omega_m \int_0^1 r^{m-1} dr \int_S f(r,s)\,ds$$

with $\omega_m = m \operatorname{vol}\{x : \|x\| \le 1\}$.

Now apply Green's formula

$$\int_{\|x\| \le 1} (u \Delta v - v \Delta u)\,dx = \omega_m \int_S \left(u \frac{\partial v}{\partial r} - v \frac{\partial u}{\partial r} \right) ds$$

and take $u = P \in H_n$, $v = Q \in H_k$. For $P \in H_n$ we of course have

$$P(x) = P(rs) = r^n P(s),$$

hence $\frac{\partial P}{\partial r} = nr^{n-1}\, P(s)$. Similarly for Q. So we get from Green's formula

$$\int_S (kPQ - nQP)\, ds = 0 \quad \text{or} \quad (k-n)\int_S PQ\, ds = 0.$$

Hence $\mathcal{H}_n \perp \mathcal{H}_k$ if $n \neq k$. Notice that with Q also $\overline{Q}$, the complex conjugate of Q, belongs to $\mathcal{H}_k$.

(iii) The next step is

Lemma 7.3.4. *$\sum_{n=0}^{\infty} \mathcal{H}_n$ is a dense subspace of $L^2(S)$.*

Let $\mathcal{P}$ denote the set of restrictions to S of arbitrary polynomials, so $\mathcal{P} \simeq \Pi = \sum_{n\geq 0} \Pi_n$. We shall show

Proposition 7.3.5. *One has $\mathcal{P} = \sum_{n=0}^{\infty} \mathcal{H}_n$.*

Lemma 7.3.4 then follows, since $\mathcal{P}$ is dense in $C(S)$ (in the topology of uniform convergence) by Stone–Weierstrass' theorem, and since $C(S)$ is dense in $L^2(S)$.

To show Proposition 7.3.5 we introduce some notions which are useful in their own right.

For $P \in \Pi$ we denote by $P(\partial)$ the differential operator

$$P(\partial) = P\Big(\frac{\partial}{\partial x_1}, \dots, \frac{\partial}{\partial x_m}\Big).$$

So, if P is in Π_k, i.e.

$$P = \sum_{|\alpha|=k} a_\alpha\, x^\alpha,$$

then $P(\partial) = \sum_{|\alpha|=k} a_\alpha\, (\frac{\partial}{\partial x})^\alpha$, where $(\frac{\partial}{\partial x})^\alpha = (\frac{\partial}{\partial x_1})^{\alpha_1} \dots (\frac{\partial}{\partial x_m})^{\alpha_m}$.

We introduce a Hermitean form on Π by

$$\langle P, Q\rangle = P(\partial)\overline{Q\,(0)}.$$

Clearly, $\langle P, Q\rangle = 0$ for $P \in \Pi_k$, $Q \in \Pi_l$ with $l \neq k$. Furthermore

$$\langle x^\alpha, x^\beta\rangle = \begin{cases} \alpha! & \text{if } \alpha = \beta, \\ 0 & \text{if } \alpha \neq \beta. \end{cases}$$

Here $\alpha! = \alpha_1! \cdots \alpha_m!$.

So, if $P, Q \in \Pi_k$, $P(x) = \sum_{|\alpha|=k} a_\alpha x^\alpha$, $Q(x) = \sum_{|\beta|=k} b_\beta x^\beta$, then

$$\langle P, Q\rangle = \sum_{|\alpha|=k} a_\alpha\, \overline{b_\alpha}\, \alpha!.$$

Hence $\langle P, Q\rangle$ defines a scalar product on Π.

Lemma 7.3.6. *For each $n \geq 2$ one has $\Pi_n = H_n \oplus \|x\|^2\, \Pi_{n-2}$.*

Let Δ be the usual Laplacian, $\Delta = \sum_{i=1}^{m} \frac{\partial^2}{\partial x_i^2}$. Then Δ maps Π_n into Π_{n-2} for all n. One has

$$\langle \|x\|^2 P, Q\rangle = P(\partial)\overline{\Delta Q(0)} = \langle P, \Delta Q\rangle$$

for $P, Q \in \Pi$. So Δ is the adjoint of multiplication by $\|x\|^2$. This holds in particular for $P \in \Pi_{n-2}$, $Q \in \Pi_n$.

The mapping $P \mapsto \|x|^2\, P$ from Π_{n-2} into Π_n is injective, thus $\Delta : \Pi_n \to \Pi_{n-2}$ is surjective. Now consider the mapping

$$P \mapsto |x\|^2 \Delta P$$

on Π_n. This operator is self-adjoint with kernel H_n and image $\|x\|^2 \Pi_{n-2}$, which are mutually orthogonal. So the lemma is proved.

Corollary 7.3.7. *For every $n \geq 0$ one has*

$$\Pi_n = \bigoplus_{o \leq k \leq [\frac{1}{2}n]} \|x\|^{2k}\, H_{n-2k}.$$

The proof is by induction on n.

The proof of Proposition 7.3.5 is now clear, since $\|x\|^2 = 1$ on S.

Corollary 7.3.8. *We have*

$$\dim \mathcal{H}_n = \frac{(n+m-3)!\,(m+2n-2)}{(m-2)!\,n!}.$$

Observe that $\dim \mathcal{H}_n = \dim \Pi_n - \dim \Pi_{n-2}$.

Remark 7.3.9. Let $P \in \Pi_k$. Then $P(\partial) : \Pi_k \to \Pi_{n-k}$ is surjective and $\Pi_n = \ker P(\partial) \oplus \overline{P(x)}\, \Pi_{n-k}$ (orthogonal direct sum). The proof is similar to the proof of Lemma 7.3.6.

(iv) Irreducibility of the representations U_n

We will now show that the spaces $\mathcal{H}_n$ are irreducible under U_n. To do this, we use the fact that any $\mathcal{H}_n$ contains a special simple function, a so-called *zonal function*, which spans the space of K-fixed vectors in $\mathcal{H}_n$ and is cyclic in $\mathcal{H}_n$. The result then follows from Lemma 6.2.3.

For $t \in S$ consider the linear form

$$f \mapsto f(t)$$

on $\mathcal{H}_n$. There is a unique function $Z_t \in \mathcal{H}_n$ with the property

$$(f \mid Z_t) = f(t) \qquad (f \in \mathcal{H}_n). \tag{7.3.3}$$

We introduce the notation

$$Z(s,t) = Z_t(s) \qquad (s,t \in S).$$

The function of two variables Z is called the *reproducing kernel* of $\mathcal{H}_n$, since any function $f \in \mathcal{H}_n$ is reproduced by means of (7.3.3).

It is easily checked that Z is G-invariant

$$Z(g \cdot s, g \cdot t) = Z(s,t)$$

for all $g \in G$ and $s,t \in S$. In particular, if we consider $Z(s,e_1) = Z_n(s)$, this function is in $\mathcal{H}_n$ and is K-fixed. Therefore Z_n only depends on s_1: there is a function $f_n : [-1,1] \to \mathbb{R}$ with

$$Z_n(x_1,\dots,x_m) = f_n(x_1)$$

for all $x = (x_1,\dots,x_m) \in S$. So Z_n is constant on the spheres $x_2^2 + \cdots + x_m^2 = c$ (c a constant), and is therefore called a *zonal function.*

Lemma 7.3.10. *The function f_n is a polynomial of degree at most n.*

The function Z_n is the restriction to S of a polynomial, homogeneous of degree n. For $-1 \le t \le 1$ we have $f_n(t) = Z_n(t,0,\dots,0,\sqrt{1-t^2})$, so

$$f_n(t) = \sum_{k=0}^{n} c_k \, (1-t^2)^{\frac{n-k}{2}} \, t^k$$

where $c_k \in \mathbb{C}$. Since we also have $f_n(t) = Z_n(t,0,\dots,0,-\sqrt{1-t^2})$ (notice that $n > 2$), we get $c_k = 0$ for $n-k$ odd. This completes the proof of the lemma.

We shall now use spherical coordinates on S in more detail:

$$\begin{aligned}
x_1 &= \cos\theta_1, \\
x_2 &= \sin\theta_1 \, \cos\theta_2, \\
x_3 &= \sin\theta_1 \, \sin\theta_2 \, \cos\theta_3, \\
&\vdots \\
x_{m-1} &= \sin\theta_1 \cdots \sin\theta_{m-2} \, \cos\theta_{m-1}, \\
x_m &= \sin\theta_1 \cdots \sin\theta_{m-2} \, \sin\theta_{m-1}
\end{aligned}$$

where $0 \le \theta_{m-1} < 2\pi$, $0 \le \theta_k < \pi$ for $k < m-1$.

Writing, for $x \in \mathbb{R}^m$, $x = rs$ where $r = \|x\|$, $s = x/r \in S$, one has

$$dx = r^{m-1} dr\, d\mu(s)$$

with $d\mu(s) = \sin^{m-2}\theta_1 \sin^{m-3}\theta_2 \cdots \sin\theta_{m-2}\, d\theta_1 d\theta_2 \cdots d\theta_{m-1}$.

The surface measure on S is G-invariant and the surface of S is equal to

$$\omega_m = \int_S d\mu(s) = \frac{2\pi^{\frac{m}{2}}}{\Gamma(\frac{m}{2})}.$$

For any $F \in C_c(\mathbb{R}^m)$ one has

$$\int_{\mathbb{R}^m} F(x)\, dx = \int_0^\infty \Big\{ \int_S F(rs)\, d\mu(s) \Big\}\, r^{m-1} dr,$$

and if φ is a continuous function on S only depending on s_1 and hence a zonal function, then $\varphi(s) = f(s_1)$ and

$$\int_S \varphi(s)\, d\mu(s) = c_m \int_0^\pi f(\cos\theta)\, \sin^{m-2}\theta\, d\theta,$$

where c_m is a positive constant not depending on the choice of φ. Taking $\varphi = 1$ one can easily compute the positive constant c_m. Observe that $d\mu(s) = \omega_m\, ds$.

Change of coordinates $t = \cos\theta$ gives

$$\int_0^\pi f(\cos\theta)\, \sin^{m-2}\theta\, d\theta = \int_{-1}^1 f(t)\,(1-t^2)^{\frac{m-3}{2}}\, dt.$$

So we have

Proposition 7.3.11. *Let φ be a continuous zonal function on S, $\varphi(s) = f(s_1)$. Then there is a positive constant c, not depending on the particular choice of φ, such that*

$$\int_S \varphi(s)\, ds = c \int_{-1}^1 f(t)\,(1-t^2)^{\frac{m-3}{2}}\, dt.$$

Proposition 7.3.12. *The polynomials f_n, $n \geq 0$, satisfy the condition*

$$\int_{-1}^1 f_n(t)\, \overline{f_k(t)}\,(1-t^2)^{\frac{m-3}{2}}\, dt = 0$$

if $n \neq k$.

From the orthogonality $\mathcal{H}_n \perp \mathcal{H}_k$ follows

$$\int_S Z_n(s)\, \overline{Z_k(s)}\, ds = 0 \qquad (n \neq k).$$

But the function $s \mapsto Z_n(s)\overline{Z_k(s)}$ is a zonal function corresponding to $f_n(t)\overline{f_k(t)}$. The proposition now follows from Proposition 7.3.11.

Corollary 7.3.13. *The degree of the polynomial f_n is precisely n.*

Use the Gram–Schmidt procedure for orthogonalization.

Observe that f_n is even if n is even, odd if n is odd.

Corollary 7.3.14. *The polynomials f_n, $n \geq 0$, form a complete orthogonal system in $L^2([-1,1], (1-t^2)^{\frac{m-3}{2}}\,dt)$.*

Proposition 7.3.15. *For any $n \geq 0$ one has $f_n(1) = \dim \mathcal{H}_n$.*

Observe that $f_n(1) = Z_n(e_1) = Z(e_1, e_1)$. But the reproducing kernel Z_n of $\mathcal{H}_n$ is G-invariant, so $Z(e_1, e_1) = Z(g \cdot e_1, g \cdot e_1)$ for all $g \in G$, so $Z(e_1, e_1) = Z(s, s)$ for all $s \in S$. Integrating over S we obtain

$$Z(e_1, e_1) = \int_S Z(s, s)\, ds.$$

Now select an orthonormal basis $\varphi_1, \ldots, \varphi_{d_n}$ in $\mathcal{H}_n$ where $d_n = \dim \mathcal{H}_n$. Then clearly

$$Z(s, t) = \sum_{i=1}^{d_n} \varphi_i(s)\overline{\varphi_i(t)}.$$

Hence

$$\int_S Z(s, s)\, ds = \sum_{i=1}^{d_n} \int_S |\varphi_i(s)|^2\, ds = d_n.$$

Remark 7.3.16. For $m = 3$ the polynomials $f_n/f_n(1)$ are precisely the *Legendre polynomials.*

For $m \geq 3$ the polynomials $f_n/f_n(1)$ are called *Gegenbauer polynomials* or *ultra-spherical polynomials.* See, e.g., [56].

Proposition 7.3.17. *Let $\varphi \in \mathcal{H}_n$ be a zonal function. Then φ is a scalar multiple of Z_n.*

Since $\varphi \in \mathcal{H}_n$ is zonal, we have $\varphi(s) = f(s_1)$. This function f satisfies, using Proposition 7.3.12,

$$\int_{-1}^{1} f(t)\overline{f_k(t)}\,(1-t^2)^{\frac{m-3}{2}}\, dt = 0$$

for all $k \neq n$, because $\mathcal{H}_n \perp \mathcal{H}_k$. Since the f_k form a complete orthogonal system, f must be a multiple of f_n, so $\varphi = cZ_n$ for some complex scalar c.

Proposition 7.3.18. *The zonal function Z_n is a cyclic vector in $\mathcal{H}_n$.*

Let $\varphi \in \mathcal{H}_n$ be such that

$$\int_S Z_n(g^{-1} \cdot s)\, \overline{\varphi(s)}\, ds = 0$$

for all $g \in G$. We have to show that $\varphi = 0$. Now, since $Z_n(g^{-1} \cdot s) = Z(s, g \cdot e_1) = Z(s,t)$ with $t = g \cdot e_1$, we get

$$\overline{\varphi(t)} = \int_S Z(s,t)\, \overline{\varphi(s)}\, ds = 0$$

for all $t \in S$, because Z is the reproducing kernel of $\mathcal{H}_n$.

Corollary 7.3.19. *The spaces $\mathcal{H}_n$ are irreducible under U_n.*

This clearly follows, as said before, from Propositions 7.3.17 and 7.3.18 and Lemma 6.2.3.

(v) Equivalence and more

Proposition 7.3.20. *The irreducible unitary representations U_n on $\mathcal{H}_n$ are pairwise inequivalent.*

This is most easily seen by comparing the dimensions of the spaces $\mathcal{H}_k$. Let $d_k = \dim \mathcal{H}_k$. Then a simple computation shows that $d_{k+1} > d_k$ for all $k \geq 0$.

For $m = 3$ we get $d_k = 2k + 1$. So the statement about SO(3) in Theorem 7.3.2 follows from the well-known representation theory of SO(3), or rather of SU(2), since $\mathrm{SO}(3) \simeq \mathrm{SU}(2)/\pm I$. We indeed get odd dimensions only. See, e.g., [24], [45]. The irreducible unitary representations of SU(2) are given by the natural action of SU(2) on the space V_n of homogeneous polynomials in two complex variables z_1 and z_2 of degree n ($n = 1, 2, \ldots$), so polynomials of the form $\sum_{k=0}^{n} c_k\, z_1^k\, z_2^{n-k}$. It is clear that only for even n the element $-I$ acts trivially. Therefore $\dim V_n = n + 1$ is odd.

This completes the proof of Theorem 7.3.2.

Theorem 7.3.21. *For any $n \geq 2$ one has*

$$\mathcal{H}_n = \mathrm{span}\{(\alpha_1 s_1 + \cdots + \alpha_m s_m)^n : \alpha_1^2 + \cdots + \alpha_m^2 = 0\}$$

where $(s_1, \ldots, s_m) \in S$ and $(\alpha_1, \ldots, \alpha_m) \in \mathbb{C}^m$.

For $m = 2$ the statement is clear, since $\dim \mathcal{H}_n = 2$ for $n \neq 0$. For $m > 2$ it is sufficient to observe that the right-hand side is a G-invariant subspace of $\mathcal{H}_n$, so equal to $\mathcal{H}_n$ because U_n is irreducible.

In the case of SO(2) we have $\dim \mathcal{H}_n = 2$ for $n \neq 0$, and these spaces are spanned by $(s_1 + is_2)^n$ and $(s_1 - is_2)^n$. Setting $\chi_n\left(\begin{smallmatrix} \cos\theta & \sin\theta \\ -\sin\theta & \cos\theta \end{smallmatrix}\right) = \mathrm{e}^{in\theta}$ we get $U_n = \chi_n \oplus \chi_{-n}$ for $n \neq 0$. The space $\mathcal{H}_0$ is one-dimensional and $U_0 = \chi_0$.

(vi) Radial parts

(1) Define the function $q : \mathbb{R}^m \backslash \{0\} \to S^{m-1}$ by

$$q(x_1, \ldots, x_m) = \Big(\frac{x_1}{r}, \ldots, \frac{x_m}{r}\Big)$$

where $r = \|x\|$.

The mapping q is a C^∞ mapping onto S^{m-1}, that is *submersive*. Let Δ be the Laplacian on $\mathbb{R}^m$ and let Ω be the Laplacian on S^{m-1}. Then for any C^∞ function F on S^{m-1} one has

$$\Delta(F \circ q) = (\Omega F) \circ q. \tag{7.3.4}$$

(2) Let $Q : S^{m-1} \to [-1, 1]$ be defined by

$$Q(s_1, \ldots, s_m) = s_1.$$

Then Q is a *submersive* C^∞ mapping from $\{s \in S^{m-1} : s_1 \neq \pm 1\}$ onto $(-1, 1)$. Let f be a C^∞ function on $(-1, 1)$ and let L be the radial part of Ω. Then

$$(Lf) \circ Q = \Omega(f \circ Q). \tag{7.3.5}$$

(3) *Computation of* Ω *on* $\mathcal{H}_n$

Clearly, for $m \geq 3$, Ω acts on $\mathcal{H}_n$ as a scalar since it commutes with the action of G (apply Schur's lemma): $\Omega F = \lambda F$, where F is the restriction to S^{m-1} of a *harmonic* polynomial P, homogeneous of degree n. So $\Delta P = 0$. We shall compute ΩF, or more precisely the eigenvalue λ, while our result below also holds for $m = 2$.

We have

$$F \circ q\,(x_1, \ldots, x_m) = P\Big(\frac{x_1}{r}, \ldots, \frac{x_m}{r}\Big) = r^{-n} P(x_1, \ldots, x_m) \qquad (r \neq 0).$$

A simple computation gives

$$\begin{aligned}\Delta(r^{-n} P(x)) &= n(n+1)\, r^{-n-2} P(x) - n(m-1)\, r^{-n-2}\, P(x) \\ &\quad - 2n\, r^{-n-2}\Big(x_1 \frac{\partial P}{\partial x_1} + \cdots + x_m \frac{\partial P}{\partial x_m}\Big) + r^{-n} \Delta P(x).\end{aligned}$$

Now

$$x_1 \frac{\partial P}{\partial x_1} + \cdots + x_m \frac{\partial P}{\partial x_m} = n P(x),$$

because P is homogeneous of degree n (Euler's equation). So finally we get

$$\Delta(r^{-n} P(x)) = -n(n+m-2) r^{-n-2} P(x).$$

Thus

$$\Omega F = -n(n+m-2)\, F$$

on $\mathcal{H}_n$.

(4) *Computation of L on $[-1,1]$*

Let f be a C^∞ function on $[-1,1]$. We shall compute an expression for Lf. We have $(Lf)\circ Q = \Omega(f\circ Q)$ and therefore

$$(Lf)\circ Q\circ q = \Delta(f\circ Q\circ q)$$

where q is defined in (1). Let us compute the right-hand side. Notice that $Q\circ q(x_1,\dots,x_m) = \frac{x_1}{r}$. We get

$$\frac{\partial}{\partial x_i} f\Big(\frac{x_1}{r}\Big) = \begin{cases} -\frac{x_i x_1}{r^3}\frac{df}{dt}\big(\frac{x_1}{r}\big) & \text{if } i\neq 1,\\ \frac{r^2-x_1^2}{r^3}\frac{df}{dt}\big(\frac{x_1}{r}\big) & \text{if } i = 1,\end{cases}$$

and

$$\frac{\partial^2}{\partial x_i^2} f\Big(\frac{x_1}{r}\Big) = \begin{cases} \frac{x_i^2x_1^2}{r^6}\frac{d^2f}{dt^2}\big(\frac{x_1}{r}\big) - \frac{x_1(r^2-3x_i^2)}{r^5}\frac{df}{dt}\big(\frac{x_1}{r}\big) & \text{if } i\neq 1,\\ \Big(\frac{r^2-x_1^2}{r^3}\Big)^2\frac{d^2f}{dt^2}\big(\frac{x_1}{r}\big) - 3\frac{(r^2-x_1^2)x_1}{r^5}\frac{df}{dt}\big(\frac{x_1}{r}\big) & \text{if } i = 1.\end{cases}$$

So

$$\Delta f\Big(\frac{x_1}{r}\Big) = \frac{r^2-x_1^2}{r^4}\frac{d^2f}{dt^2}\Big(\frac{x_1}{r}\Big) - (m-1)\frac{x_1}{r^4}\frac{df}{dt}\Big(\frac{x_1}{r}\Big).$$

Hence, taking $r=1$,

$$(Lf)(t) = (1-t^2)\frac{d^2f}{dt^2} - (m-1)t\frac{df}{dt}.$$

So the polynomial f_n defined in (iv) (cf. Lemma 7.3.10) satisfies the differential equation

$$(1-t^2)\frac{d^2u}{dt^2} - (m-1)t\frac{du}{dt} + n(m+n-2)u = 0.$$

It follows, see [56] for the notation, that this is the differential equation of the Gegenbauer polynomials, hence

$$f_n/f_n(1) = C_n^{\frac{m-2}{2}}.$$

7.4 Spherical functions on spheres

We continue what we did in Section 7.2. Let $m\geq 3$. Recall Z_n, the zonal function in $\mathcal{H}_n$, and set $\varphi_n = Z_n/Z_n(e_1)$. We know that $Z_n(e_1) = d_n = \dim\mathcal{H}_n$. The function φ_n can be considered as a bi-K-invariant function on G.

Theorem 7.4.1. *The spherical functions of the Gelfand pair (G,K) are precisely the functions φ_n, $n\in\mathbb{N}$.*

Let us first show that φ_n is a spherical function. Let f be a bi-K-invariant, continuous function on G and set $\psi = f * \varphi_n$. This function is bi-K-invariant again and, considered as a function on S, an element of $\mathcal{H}_n$, and moreover a zonal function, so ψ is a multiple of φ_n, i.e.

$$\psi = f * \varphi_n = \chi_n(f)\varphi_n$$

with

$$\chi_n(f) = \int_G f(g)\,\varphi(g^{-1})\,dg.$$

Hence, by Proposition 6.1.6, φ_n is a spherical function. Since the functions φ_n constitute a complete orthonormal system in the space of zonal functions in $L^2(S)$, the functions φ_n are the only spherical functions.

Since the group G is compact, the functions φ_n are positive-definite. Moreover, the representations U_n of G on $\mathcal{H}_n$ exhaust the set of class one representations, up to equivalence. We leave this fact as an exercise.

Consider the function f on $\mathbb{R}^m$ given by

$$f(x) = (x_1 + ix_2)^n,$$

which is a harmonic polynomial, homogeneous of degree n (by Theorem 7.3.21). Restrict this function to S. Then $f \in \mathcal{H}_n$. Consider now

$$\psi(s) = \int_K f(k \cdot s)\,dk \qquad (s \in S).$$

This is a zonal function in $\mathcal{H}_n$ with $\psi(e_1) = 1$, hence $\psi = \varphi_n$. Applying spherical coordinates, we have the following explicit integral formula:

$$\varphi_n(s) = C_n^{\frac{m-2}{2}}(\cos\varphi) = \frac{\Gamma(\frac{m-1}{2})}{\sqrt{\pi}\,\Gamma(\frac{m-2}{2})} \int_0^\pi (\cos\varphi + i\,\sin\varphi\,\cos\theta)^n\,\sin^{m-3}\theta\,d\theta,$$

where we substitute $s_1 = \cos\varphi$ in $s = (s_1, \dots, s_m)$.

7.5 Real hyperbolic spaces

In this section we follow mainly [15].

(i) Introduction

For $n \in \mathbb{N}$, $n \geq 2$, let $\mathrm{SO}(1,n)$ be the special orthogonal group of the quadratic form

$$[x, y] = x_0 y_0 - x_1 y_1 - \cdots - x_n y_n,$$

where $x = (x_0, x_1, \dots, x_n)$, $y = (y_0, y_1, \dots, y_n)$ are elements of $\mathbb{R}^{n+1}$. Let J be the matrix $\operatorname{diag}(1, -1, \dots, -1)$. Then clearly $\mathrm{SO}(1,n)$ consists of the real $(n+1) \times (n+1)$ matrices g with $\det g = 1$ and

$$JgJ = {}^t g^{-1}.$$

So we evidently have ${}^t g \in \mathrm{SO}(1,n)$ for all $g \in \mathrm{SO}(1,n)$. It is an exercise to show that the set of all $g \in \mathrm{SO}(1,n)$ satisfying $g_{00} = [ge_0, e_0] > 0$ is a subgroup G of $\mathrm{SO}(1,n)$, where $e_0 = (1, 0, \dots, 0) \in \mathbb{R}^{n+1}$.

Indeed, it is therefore sufficient to show that $[x, y] > 0$ for all $x, y \in \mathbb{R}^{n+1}$ with $x_0 > 0$, $y_0 > 0$ and $[x,x] = [y,y] = 1$. The group G is clearly an open subgroup of $\mathrm{SO}(1,n)$ of index 2.

Let $X = \{x \in \mathbb{R}^{n+1} : x_0 > 0,\ [x,x] = 1\}$. It is called a one-sheeted hyperboloid. By a similar reasoning as above one shows that G acts on X. Let $K = \mathrm{SO}(n)$ be the stabilizer of e_0 in G and define the subgroup A of G by

$$A = \left\{ a_t = \begin{pmatrix} \cosh t & 0 & \sinh t \\ 0 & I_{n-1} & 0 \\ \sinh t & 0 & \cosh t \end{pmatrix} : \quad t \in \mathbb{R} \right\}.$$

Let $x = (x_0, x_1, \dots, x_n) \in X$. There is a $k \in K$ such that $k \cdot x = (x_0, 0, \dots, 0, \alpha)$ where $\alpha^2 = x_1^2 + \dots + x_n^2$. Notice that $x_0 > 0$ and $x_0^2 - \alpha^2 = 1$. Then we can find $t \in \mathbb{R}$ with

$$a_t \cdot e_0 = (x_0, 0, \dots, 0, \alpha).$$

So $x = k^{-1} a_t \cdot e_0$ and consequently

$$G = KAK$$

and G acts transitively on X. Moreover, G turns out to be connected. Since G is also open in $\mathrm{SO}(1,n)$, it coincides with the connected component $\mathrm{SO}_0(1,n)$ of the identity of $\mathrm{SO}(1,n)$, thus $G = \mathrm{SO}_0(1,n)$.

Let w be the matrix $\operatorname{diag}(1, -1, -1, \dots, -1)$ in K. Then $w a_t w^{-1} = a_{-t}$ for all $t \in \mathbb{R}$. So we actually have

Proposition 7.5.1 (Cartan decomposition). *The group $G = \mathrm{SO}_0(1,n)$ can be written as*

$$G = K \bar{A}^+ K$$

where $\bar{A}^+ = \{a_t \in A : t \geq 0\}$.

Consider now the Cartan involution θ on G defined by

$$\theta(g) = {}^t g^{-1} = JgJ \qquad (g \in G).$$

Clearly, θ leaves K fixed and $\theta(a) = a^{-1}$ for all $a \in A$. Therefore, because $G = KAK$, we have

$$\theta(g) \in Kg^{-1}K$$

for all $g \in G$. So we have by Proposition 6.1.3:

Proposition 7.5.2. *The pair* $(\mathrm{SO}_0(1,n), \mathrm{SO}(n))$ *is a Gelfand pair.*

By Proposition 6.1.2 we may conclude that G is unimodular: its left Haar measure dg is also right-invariant. Moreover $X \simeq G/K$ admits a G-invariant measure dx. Clearly, one can normalize dg such that, fixing dx and taking the normalized Haar measure on K,

$$\int_G f(g)\,dg = \int_X \left\{ \int_K f(gk)\,dk \right\} d\dot{g}$$

for all $f \in C_c(G)$. Here $d\dot{g}$ is the notation for the G-invariant measure on $X \simeq G/K$, so $d\dot{g} \simeq dx$. See Section 4.5.

To write down an explicit form for dx, we use "hyperbolic" coordinates (similar to spherical coordinates). One gets

$$\int_G f(g)\,dg = 2\frac{\pi^{\frac{n}{2}}}{\Gamma(\frac{n}{2})} \int_K \int_0^\infty \int_K \sinh^{n-1} t\, f(ka_tk')\,dk\,dt\,dk' \tag{7.5.1}$$

for $f \in C_c(G)$. Observe that the factor in front of the integral is the measure of the sphere S^n. So for $f \in C_c^\#(G)$ one has

$$\int_G f(g)\,dg = \int_0^\infty A(t)\, f(a_t)\,dt, \tag{7.5.2}$$

where $A(t) = 2\frac{\pi^{\frac{n}{2}}}{\Gamma(\frac{n}{2})} \sinh^{n-1} t$.

(ii) Iwasawa decomposition

Let Ξ be the cone

$$\Xi = \{\xi = (\xi_0, \xi_1, \ldots, \xi_n) \in \mathbb{R}^{n+1} : [\xi, \xi] = 0,\ \xi_0 > 0\}.$$

One clearly has $\Xi = KA \cdot \xi^0$ with $\xi^0 = (1, 0, \ldots, 0, 1)$. Moreover G acts on Ξ. It is therefore sufficient to show that

$$[g \cdot \xi^0, e_0] > 0$$

for all $g \in G$. We have

$$[g \cdot \xi^0, e_0] = [\xi^0, g^{-1} \cdot e_0] = x_0 + x_n$$

if $g^{-1} \cdot e_0 = x \in X$, $x = (x_0, x_1, \ldots, x_n)$. Since $x_0^2 = 1 + x_1^2 + \cdots + x_n^2$, we obtain $x_0^2 > x_n^2$, and thus $x_0 + x_n > 0$ because $x_0 > 0$. So G acts transitively on Ξ.

To determine the stabilizer of ξ^0 in G one has to perform an easy computation. Define the following subgroups of G:

$$N = \left\{ n = n_z = \begin{pmatrix} 1 + \frac{1}{2}\|z\|^2 & z^t & -\frac{1}{2}\|z\|^2 \\ z & I_{n-1} & -z \\ \frac{1}{2}\|z\|^2 & z^t & 1 - \frac{1}{2}\|z\|^2 \end{pmatrix} : z \in \mathbb{R}^{n-1} \right\},$$

$$M \simeq \mathrm{SO}(n-1) = \left\{ \begin{pmatrix} 1 & 0 & 0 \\ 0 & l & 0 \\ 0 & 0 & 1 \end{pmatrix} : l \in \mathrm{SO}(n-1) \right\}.$$

Observe that M is compact, N abelian (isomorphic to $\mathbb{R}^{n-1}$) and

$$m\, n_z\, m^{-1} = n_{m(z)} \qquad (m \in M,\ n_z \in N).$$

So MN is a subgroup of G and it is unimodular. Moreover $\operatorname{Stab} \xi^0 = MN$ and thus Ξ admits a G-invariant measure. We have $G = KMAN = KAN$ since M and A commute. We also have

$$a_t\, n_z\, a_{-t} = n_{e^t z} \qquad (a_t \in A,\ n_z \in N).$$

Theorem 7.5.3 (Iwasawa decomposition). *The mapping $(k, a, n) \mapsto kan$ from $K \times A \times N$ into G is a diffeomorphism.*

Let $g = k a_t n_z$. Then we have

(a) $[g \cdot \xi^0, e_0] = [a_t \cdot \xi^0, e_0] = e^t$,

(b) $[g \cdot e_i, e_0] = [a_t n_z \cdot e_i, e_0] = [a_t n_z a_{-t} \cdot e_i, e_0] = [n_{e^t z} \cdot e_i, e_0] = e^t z_{i-2}$ for $3 \le i \le n+1$.

So we may conclude that t, z and k depend in a C^∞ way on g. In addition we see that the decomposition of g is unique. This proves the theorem.

According to Section 4.5 (special cases of (4.5.10)) we can normalize the invariant measure dg on G such that

Proposition 7.5.4. *For all $f \in C_c(G)$ we have*

$$\int_G f(g)\, dg = \int_K \int_{-\infty}^{\infty} \int_{\mathbb{R}^{n-1}} f(k a_t n_z)\, e^{(n-1)t}\, dk\, dt\, dz.$$

Let now $f \in C_c(\Xi)$. Then we get by Proposition 7.5.4 that

$$\mu : f \mapsto \int_K \int_0^\infty f(\lambda k \cdot \xi^0)\, \lambda^{n-1} \frac{d\lambda}{\lambda}\, dk$$

is a G-invariant measure on Ξ. Here dk is as usual the normalized Haar measure on K. Notice that the integral can actually be taken over K/M instead of K. Setting $B = K \cdot \xi^0 \simeq K/M$ we might also write for the invariant measure μ:

$$\mu(f) = \int_B \int_0^\infty f(\lambda b)\lambda^{n-1} \frac{d\lambda}{\lambda}\, d\sigma(b) \qquad (f \in C_c(\Xi)),$$

with $d\sigma(b) \simeq d\dot{k}$ ($\dot{k} \in K/M$). We also have $B = \{\xi \in \Xi : \xi_0 = 1\}$, so $B \simeq S^{n-1}$, the unit sphere in $\mathbb{R}^n$, and $d\sigma$ can be seen as the (normalized) surface measure on S^{n-1}.

(iii) Spherical functions

We define the Laplace–Beltrami (or Laplace) operator Δ on X in the same way as the operator Ω in Section 7.3 (vi) (1).

Let q be the mapping $q(x) = \frac{x}{r}$ defined on $\{x \in \mathbb{R}^{n+1} : [x,x] > 0,\ x_0 > 0\}$, where $r = [x,x]^{1/2}$. Then for any C^∞ function F on X one has

$$-\Box\,(F \circ q) = (\Delta F) \circ q$$

where $\Box = \frac{\partial^2}{\partial x_0^2} - \frac{\partial^2}{\partial x_1^2} - \cdots - \frac{\partial^2}{\partial x_n^2}$.

The differential operator Δ is G-invariant:

$$\Delta(L_g F) = L_g(\Delta F)$$

where $(L_g F)(x) = F(g^{-1} \cdot x)$ ($g \in G$; $x \in X$).

Let f_1, f_2 be two C^∞ functions on G, bi-K-invariant and f_1 with compact support. These functions may also be viewed as functions on X and one has

$$\Delta(f_1 * f_2) = (\Delta f_1) * f_2 = f_1 * (\Delta f_2).$$

This follows from the G-invariance of Δ and the commutativity of $C_c^\#(G)$.

Similar to Section 7.3 (vi) (2) we define the radial part L of Δ. So let $Q : X \to [1,\infty)$ be defined by

$$Q(x_0, x_1, \ldots, x_n) = x_0.$$

Then for any C^∞ function f on $[1,\infty)$ we have

$$(Lf) \circ Q = \Delta(f \circ Q).$$

In a similar way as in Section 7.3 (vi) (4) we can compute L and we obtain

$$L = (u^2 - 1)\frac{d^2}{du^2} + n\,u\,\frac{d}{du}.$$

Setting $u = \cosh t$ $(t \geq 0)$, we get

$$L = \frac{d^2}{dt^2} + \frac{A'(t)}{A(t)}\frac{d}{dt}$$

where, we recall, $A(t) = 2\,\frac{\pi^{\frac{n}{2}}}{\Gamma(\frac{n}{2})}\sinh^{n-1} t$.

So if φ is a bi-K-invariant C^∞ function on G satisfying $\Delta\varphi = \lambda\varphi$, then

$$\frac{d^2\Phi}{dt^2} + \frac{A'(t)}{A(t)}\frac{d\Phi}{dt} = \lambda\Phi$$

if $\Phi(t) = \varphi(a_t), t \geq 0$.

Theorem 7.5.5. *Let φ be a bi-K-invariant function on G. We may also consider φ as a K-invariant function on X. In order that φ is a spherical function it is necessary and sufficient that φ satisfies*

(1) *φ is C^∞,*

(2) *there is a complex number λ such that $\Delta\varphi = \lambda\varphi$,*

(3) *$\varphi(e_0) = 1$.*

The proof is similar to that of Theorem 7.1.1 for Euclidean motion groups and makes use of the fact that the space of solutions of the differential equation

$$\frac{d^2\Phi}{dt^2} + \frac{A'(t)}{A(t)}\frac{d\Phi}{dt} = \lambda\Phi$$

that are regular at $t = 0$ is one-dimensional.

To find an integral form for the spherical functions, we combine the methods from Section 7.1 (ii) and Section 7.4.

Let $f(x) = [x, \xi^0]^{s-\rho} = (x_0 + x_n)^{s-\rho}$ be defined on X, where $s \in \mathbb{C}$ and $\rho = \frac{n-1}{2}$. Since $[x, \xi^0] > 0$ for all $x \in X$, f is clearly well-defined and a C^∞ function for all $s \in \mathbb{C}$. Moreover, an easy computation gives $\Delta f = (s^2 - \rho^2) f$; compare with Section 7.3 (vi) (3). Set

$$\begin{aligned} \varphi_s(x) &= \int_K f(k \cdot x)\,dk \\ &= \int_K [x, k \cdot \xi^0]^{s-\rho}\,dk \qquad (x \in X). \end{aligned}$$

Then φ_s is K-invariant, C^∞, $\varphi_s(e_0) = 1$ and it satisfies again the differential equation $\Delta\varphi_s = (s^2 - \rho^2)\,\varphi_s$, by K-invariance of Δ. So φ_s is a spherical function and each spherical function is of this form. Notice that $\varphi_s = \varphi_{-s}$.

One easily verifies that

$$\varphi_s(a_t) = \frac{\Gamma(\frac{n}{2})}{\sqrt{\pi}\,\Gamma(\frac{n-1}{2})} \int_0^\pi (\cosh t - \sinh t\, \cos\theta)^{s-\rho} \sin^{n-2}\theta\, d\theta$$

for $t \geq 0$.

Compare this expression with the form of a spherical function on the sphere in Section 7.4.

Notice that φ_s, as a function on G, can also be written as

$$\varphi_s(g) = \int_K \mathrm{e}^{(s-\rho)t(g^{-1}k)}\, dk \qquad (g \in G) \tag{7.5.3}$$

if we write $g = k\, a_{t(g)}\, n$ (Iwasawa decomposition) for any $g \in G$.

(iv) Bounded spherical functions

Proposition 7.5.6. *The spherical function φ_s is bounded if and only if*

$$-\rho \leq \operatorname{Re} s \leq \rho.$$

(a) We first show that $|\varphi_s(g)| \leq 1$ if $-\rho \leq \operatorname{Re} s \leq \rho$. From the integral representation (7.5.3) for φ_s follows

$$|\varphi_s(g)| \leq \varphi_\sigma(g) \qquad (g \in G)$$

if $\sigma = \operatorname{Re} s$. So it is sufficient to consider the case where s is real. From the same integral representation it follows that the real-valued function $s \mapsto \varphi_s(g)$ is convex for any $g \in G$. Since $\varphi_\rho(g) = \varphi_{-\rho}(g) = 1$, the result follows.

(b) If $\operatorname{Re} s > 0$ we are going to show that

$$\varphi(a_t) \sim c(s)\, \mathrm{e}^{(s-\rho)t} \qquad (t \to \infty)$$

with

$$c(s) = \frac{2^{\rho-s}\,\Gamma(\frac{n}{2})}{\sqrt{\pi}\,\Gamma(\frac{n-1}{2})} \int_0^\pi (1 - \cos\theta)^{s-\rho} \sin^{n-2}\theta\, d\theta$$
$$= 2^{n-2}\, \frac{\Gamma(\frac{n}{2})\,\Gamma(s)}{\sqrt{\pi}\Gamma(s+\rho)}.$$

Indeed, we have for $t \geq 0$

$$(\cosh t)^{\rho-s}\, \varphi(a_t) = \frac{\Gamma(\frac{n}{2})}{\sqrt{\pi}\,\Gamma(\frac{n-1}{2})} \int_0^\pi (1 - \tanh t\, \cos\theta)^{s-\rho} \sin^{n-2}\theta\, d\theta,$$

and

- if $0 < \theta < \pi/2$ and $\operatorname{Re} s \geq \rho$, then $|(1 - \tanh t \ \cos\theta)^{s-\rho}| \leq 1$,
- if $0 < \operatorname{Re} s \leq \rho$, then $|(1 - \tanh t \ \cos\theta)^{s-\rho}| \leq (1 - \cos\theta)^{\operatorname{Re} s - \rho}$.

Similar considerations hold for $\pi/2 < \theta < \pi$. So applying Lebesgue's dominated convergence theorem we get the result.

Remark 7.5.7. From (b) we obtain that φ_s is bounded if $0 < \operatorname{Re} s \leq \rho$ and unbounded if $\operatorname{Re} s > \rho$. Since $\varphi_s = \varphi_{-s}$ we could also get the complete result of Proposition 7.5.6 if we knew that φ_s is bounded for imaginary s. This important fact follows from Theorem 7.5.9, where we show that $\varphi_{i\nu}$ ($\nu \in \mathbb{R}$) is positive-definite.

(v) Positive-definite spherical functions

(a) We begin with some preliminaries on positive-definite kernels. Let X be a set. A *kernel* Φ on X is a function on $X \times X$. It is said be positive-definite if for every N-tuple of elements $x_1, \dots, x_N$ in X and complex numbers $\alpha_1, \dots, \alpha_N$ one has

$$\sum_{i,j=1}^{N} \Phi(x_i, x_j)\, \alpha_i \overline{\alpha_j} \geq 0.$$

If X is a group and φ a function on X, then φ is positive-definite if and only if the kernel Φ defined by

$$\Phi(x, y) = \varphi(x^{-1} y)$$

is positive-definite.

In case of a locally compact group this definition agrees with Definition 5.1.5 if φ is continuous (see Remarks 5.1.7).

Let G be a locally compact group and K a compact subgroup, and set $X = G/K$. A kernel Φ on X is said to be *invariant* if

$$\Phi(gx, gy) = \Phi(x, y)$$

for all $g \in G$ and $x, y \in X$.

There is a bijective correspondence between the invariant kernels Φ on X and the bi-K-invariant functions φ on G. This correspondence is defined by the relation

$$\Phi(gK, hK) = \varphi(g^{-1}h) \qquad (g, h \in G).$$

The kernel Φ is continuous (C^∞ respectively) if and only if φ is. So the spherical functions φ_s on $G = \mathrm{SO}_0(1, n)$ correspond to kernels Φ_s on the hyperboloid of one sheet X.

(b) To determine the positive-definite spherical functions we apply the invariant measure μ on $\Xi \simeq G/MN$, defined after Proposition 7.5.4:

$$\mu(f) = \int_B \int_0^\infty f(\lambda b)\, \lambda^{n-1} \frac{d\lambda}{\lambda}\, d\sigma(b),$$

where σ is the normalized surface measure on the sphere $B = \{\xi \in \Xi : \xi_0 = 1\}$.

Let $\mathcal{H}_\alpha(\Xi)$ denote the space of continuous functions f on Ξ, homogeneous of degree α, i.e.

$$f(\lambda\xi) = \lambda^\alpha f(\xi) \qquad (\xi \in \Xi,\ \lambda > 0,\ \alpha \in \mathbb{C}).$$

The group G acts on $\mathcal{H}_\alpha(\Xi)$. Let ℓ be the linear form defined on $\mathcal{H}_{1-n}(\Xi)$ by

$$\ell(f) = \int_B f(b)\, d\sigma(b).$$

Proposition 7.5.8. *The linear form ℓ is G-invariant: if $f \in \mathcal{H}_{1-n}(\Xi)$, then*

$$\ell(L_y f) = \ell(f)$$

for all $g \in G$.

Let $\varphi \in C_c(\Xi)$ and set

$$f(\xi) = \int_0^\infty \varphi(\lambda\xi)\, \lambda^{n-1} \frac{d\lambda}{\lambda}.$$

Then f is continuous and homogeneous of degree $1-n$ and the mapping $\varphi \mapsto f$ commutes with G. Moreover

$$\ell(f) = \mu(\varphi).$$

Since any $f \in \mathcal{H}_{1-n}(\Xi)$ can be obtained in this way, the proposition follows.

Theorem 7.5.9. *The spherical function φ_s is positive-definite in the following two cases:*

(1) $s = i\nu$, $\nu \in \mathbb{R}$,

(2) *s is real, $-\rho \leq s \leq \rho$.*

The proof of this theorem will be given in this and the next subsection.

(a) Suppose that φ_s is positive-definite. Since φ_s is bi-K-invariant we have

$$\varphi_s(g^{-1}) = \varphi_s(g),$$

and since φ_s is positive-definite we also have

$$\varphi_s(g^{-1}) = \overline{\varphi_s(g)}.$$

So φ_s is real-valued, and thus $s^2 = \Delta\varphi_s(0) + \rho^2$ is real, so s is imaginary or real. Because φ_s, being positive-definite, is also bounded, we get either $s \in \mathbb{R}$, $-\rho \leq s \leq \rho$ or $s \in i\mathbb{R}$.

(b) We will now show that $\varphi_{i\nu}$ ($\nu \in \mathbb{R}$) is positive-definite. Let P_s be the function defined on $X \times \Xi$ by

$$P(x, \xi) = [x, \xi]^{s-\rho}.$$

This function is called the *Poisson kernel* of X and is characterized by the properties

(1) $P_s(gx, g\xi) = P_s(x, \xi)$ for all $g \in G$,

(2) $P_s(x, \lambda\xi) = \lambda^{s-\rho} P(x, \xi)$ for all $\lambda > 0$,

(3) $P_s(e_0, b) = 1$ for all $b \in B$.

Notice that $[x, \xi] > 0$ for all $x \in X$, $\xi \in \Xi$. Consider the kernel H_s on X defined by

$$H_s(x, y) = \int_B P_s(x, b)\, P_{-s}(y, b)\, d\sigma(b).$$

By the invariance property (1) of the Poisson kernel, the homogeneity property (2) and Proposition 7.5.8 we see that H_s is G-invariant. Moreover $H_s(x, e_0) = \varphi_s(x)$. So we get $H_s = \Phi_s$:

$$\Phi_s(x, y) = \int_B P_s(x, b)\, P_{-s}(y, b)\, d\sigma(b).$$

From this integral representation we easily deduce that the kernel $\Phi_{i\nu}$, ν real, is positive-definite, because

$$\sum_{i,j=1}^{N} \Phi_{i\nu}(x_i, x_j)\, \alpha_i \overline{\alpha_j} = \int_B \Big| \sum_{i=1}^{N} \alpha_i\, P_{i\nu}(x_i, b) \Big|^2 d\sigma(b) \geq 0.$$

(vi) Positive-definite spherical functions (continued)

(c) Note that $[\xi, \xi'] \geq 0$ for $\xi, \xi' \in \Xi$. Set $W_s(\xi, \xi') = [\xi, \xi']^{s-\rho}$. For $\operatorname{Re} s > 0$ we have for all $b \in B$

$$\gamma(s) = \int_B W_s(b, b')\, d\sigma(b') = \frac{\Gamma(\frac{n}{2})}{\sqrt{\pi}\,\Gamma(\frac{n-1}{2})} \int_0^{\pi} (1 - \cos\theta)^{s-\rho} \sin^{n-2}\theta\, d\theta$$

$$= 2^{s+\rho-1} \frac{\Gamma(\frac{n}{2})}{\sqrt{\pi}} \frac{\Gamma(s)}{\Gamma(s+\rho)}.$$

The following relation permits to switch between $P_{-s}(x, \xi)$ and $P_s(x, \xi)$ for $\operatorname{Re} s \neq 0$:

$$P_s(x, \xi) = \frac{1}{\gamma(s)} \int_B P_{-s}(x, b')\, W_s(\xi, b')\, d\sigma(b') \qquad (x \in X,\ \xi \in \Xi).$$

The proof of this intertwining relation uses the properties (1), (2), (3) of Poisson kernels and Proposition 7.5.8. We obtain for $\operatorname{Re} s > 0$

$$\Phi_s(x, y) = \frac{1}{\gamma(s)} \int_B \int_B P_{-s}(x, b)\, P_{-s}(y, b')\, W_s(b, b')\, d\sigma(b) d\sigma(b').$$

From this integral representation we are going to derive that Φ_s is positive-definite for $0 < s < \rho$.

Lemma 7.5.10. *Let* $0 < s < \rho$. *For all* $f \in C(B)$ *one has*

$$\int_B \int_B f(b)\overline{f(b')}\, W_s(b,b')\, d\sigma(b)\, d\sigma(b') \geq 0.$$

The positive-definiteness of Φ_s follows immediately from this lemma, for

$$\sum_{i,j=1}^{N} \Phi_s(x_i, x_j)\,\alpha_i\overline{\alpha_j} = \frac{1}{\gamma(s)} \int_B \int_B f(b)\overline{f(b')}\, W_s(b,b')\, d\sigma(b)\, d\sigma(b')$$

with

$$f(b) = \sum_{i=1}^{N} \alpha_i\, P_{-s}(x_i, b).$$

Let us prove the lemma. It is based on the following property of positive-definite kernels: if H is a positive-definite kernel and $\|H\|_\infty < 1$, then for $\beta > 0$, $(1-H)^{-\beta}$ is also a positive-definite kernel, since

$$(1-H)^{-\beta} = 1 + \sum_{k=1}^{\infty} \frac{\beta(\beta+1)\cdots(\beta+k-1)}{k!} H^k$$

and the product of two positive-definite kernels is again positive-definite (exercise!).

Set for $0 < \alpha < 1$,

$$W_s^\alpha(b,b') = [1 - \alpha(b_1b_1' + \cdots + b_nb_n')]^{s-\rho}.$$

For $s < \rho$, W_s^α is a positive-definite kernel, hence for all $f \in C(B)$,

$$\int_B \int_B f(b)\overline{f(b')}\, W_s^\alpha(b,b')\, d\sigma(b)\, d\sigma(b') \geq 0.$$

If $0 < s < \rho$ we can take the limit for α tending to 1, and we obtain

$$\int_B \int_B f(b)\overline{f(b')}\, W_s(b,b')\, d\sigma(b)\, d\sigma(b') \geq 0.$$

(vii) Representations of the spherical principal series

Until now we considered only representations on a Hilbert space, in particular unitary representations. We shall now broaden our scope and consider also representations on a *Banach space*. The definition is the same as in Section 5.1. A *Banach space representation* is a pair $(\pi, \mathcal{H})$ of a Banach space $\mathcal{H}$ and a homomorphism $\pi : G \to \mathrm{GL}(\mathcal{H})$ such that the mapping $(x, v) \mapsto \pi(x)v$ from $G \times \mathcal{H}$ to $\mathcal{H}$ is continuous.

The definitions of irreducibility and equivalence of representations are the same as for Hilbert space representations. The definitions of $\pi(f)$ for $f \in C_c(G)$ and of $\pi(h)$ for $h \in C(K)$ are a little more involved, but it turns out that the same conclusions hold as in the case of a Hilbert space representation. So, in particular, $\pi(f) \in \mathrm{End}(\mathcal{H})$ and $\pi(h) \in \mathrm{End}(\mathcal{H})$, where $\mathrm{End}(\mathcal{H})$ is the space of continuous endomorphisms of $\mathcal{H}$.

A next step would be to consider representations on Fréchet spaces or even on complete locally convex spaces. We postpone this to Chapter 8.

Spherical principal series representations are those representations that are obtained by induction from one-dimensional representations of the group MAN.

Let s be a complex number and let $\mathcal{H}_s$ be the space of continuous functions f on G satisfying

$$f(xma_tn) = \mathrm{e}^{(s-\rho)t} f(x)$$

with $x \in G, m \in M, n \in N, a_t \in A, \rho = \frac{n-1}{2}$. Let π_s be the representation of G on $\mathcal{H}_s$ defined by

$$\pi_s(g)\, f(x) = f(g^{-1}x).$$

We clearly may identify $\mathcal{H}_s$ with the space $\mathcal{H}_{s-\rho}(\Xi)$ from (v) since $\Xi \simeq G/MN$, and also with the space $C(B)$, by restricting the functions in $\mathcal{H}_{s-\rho}(\Xi)$ to B. Provided with the supremum norm (uniform topology), $C(B)$ is a Banach space and π_s a Banach space representation.

Let ℓ be the G-invariant linear form on $\mathcal{H}_{-\rho}$ defined in (v). Denote by $\langle\,,\rangle$ the bilinear form defined on $\mathcal{H}_s \times \mathcal{H}_{-s}$ by

$$\langle f, g\rangle = \int_B f(b)\, g(b)\, d\sigma(b).$$

This form is continuous, non-degenerate and G-invariant:

$$\langle \pi_s(x) f, \pi_{-s}(x)\, g\rangle = \langle f, g\rangle \qquad (x \in G),$$

as follows from the invariance of ℓ.

Let u_s be the function on G defined by

$$u_s(ka_tn) = \mathrm{e}^{(s-\rho)t} \qquad (k \in K,\ t \in \mathbb{R}, n \in N).$$

The function u_s belongs to $\mathcal{H}_s$ and is K-invariant. Moreover, as is easily seen from (7.5.3),

$$\varphi_s(g) = \langle u_s, \pi_{-s}(g)\, u_{-s}\rangle \qquad (g \in G).$$

If s is imaginary, say $s = i\nu$, then $\mathcal{H}_{i\nu}$ can be provided with a pre-Hilbert space structure with squared norm

$$\|f\|^2 = \int_B |f(b)|^2\, d\sigma(b),$$

and $\pi_{i\nu}(x)$ is then unitary for all $x \in G$. Completing the space $\mathcal{H}_{i\nu}$ and extending $\pi_{i\nu}(x)$ to this completion gives a unitary representation on a Hilbert space, which we again denote by $\pi_{i\nu}$. Observe that the unitary representation $\pi_{i\nu}$ is irreducible as soon as the Banach space representation $\pi_{i\nu}$ on $\mathcal{H}_{i\nu}$ is: if V is a non-trivial closed invariant subspace of the Hilbert space, then $V \cap \mathcal{H}_s$ is a non-trivial closed invariant subspace of $\mathcal{H}_s$.

The spherical function $\varphi_{i\nu}$ is clearly positive-definite.

Now we want to make a closer study of the representations π_s on $\mathcal{H}_s$ with $s \in i\mathbb{R}$, and $s \in \mathbb{R}$ satisfying $-\rho < s < \rho$.

(1) Irreducibility

Because of the duality between π_s and π_{-s} it is obvious that π_s and π_{-s} are irreducible if u_s and u_{-s} are cyclic vectors in $\mathcal{H}_s$ and $\mathcal{H}_{-s}$ respectively.

Lemma 7.5.11. *The vector u_s is cyclic if $s \neq \rho + k$, $k = 0, 1, 2, \ldots$.*

Notice that $u_s(g) = [e_0, g \cdot \xi^0]^{s-\rho}$ $(g \in G)$. Let μ be a measure on B (a continuous linear form on $C(B)$) such that

$$\int_B [g \cdot e_0, b]^{s-\rho} d\mu(b) = 0$$

for all $g \in G$. Then clearly all K-translates of μ satisfy the same condition, and we may thus replace μ by $f(b)\, d\sigma(b)$ for some $f \in \mathcal{H}_{-s}$ by taking a suitable convolution product $f(b) = g * \mu(b) = \int_K g(k)\,(L_k\mu)(b) dk$ with $g \in C(K)$. If we can show that the condition implies $f = 0$, then clearly $\mu = 0$ and u_s is cyclic in $\mathcal{H}_s$.

So let $f \in \mathcal{H}_s$ be such that $\int_B [g \cdot e_0, b]^{s-\rho} f(b) d\sigma(b) = 0$ for all $g \in G$. Then again the same holds for $\pi_{-s}(g)f$ for all $g \in G$. Taking $g = a_t$ and differentiating the expression in $t = 0$ several times, we get for $s \neq \rho + k$ $(k = 0, 1, 2, \ldots)$

$$\int_B b_n^l\, f(b)\, d\sigma(b) = 0 \tag{7.5.4}$$

for $l = 0, 1, 2, \ldots$, so f is orthogonal to all M-fixed vectors in $C(B)$. Since the K-translates of these vectors span $L^2(B)$ (see Section 7.3), and (7.5.4) also holds for the K-translates of f, we get $f = 0$. Hence u_s is a cyclic vector in $\mathcal{H}_s$.

Corollary 7.5.12. *The representations π_s on $\mathcal{H}_s$ are irreducible for $s \neq \pm(\rho + k)$, $k = 0, 1, 2, \ldots$. In particular, π_s is irreducible for $s \in i\mathbb{R}$, and $s \in \mathbb{R}$ satisfying $-\rho < s < \rho$. For $s = \rho$ the trivial one-dimensional representation is a sub-representation of π_ρ.*

(2) Intertwining operators

Define for $\operatorname{Re} s > 0$ the mapping A_s on $\mathcal{H}_{-s}$ as follows:

$$(A_s f)(\xi) = \int_B W_s(\xi, b')\, f(b')\, d\sigma(b') \qquad (\xi \in \Xi). \tag{7.5.5}$$

Here W_s is as in (iv). Since $\int_B W_s(b,b')\,d\sigma(b') = \gamma(s)$ exists for $\operatorname{Re} s > 0$, one easily sees that A_s is well-defined:

$$|(A_s f)(b)| \leq \gamma(s)\,\|f\|_\infty \qquad (f \in \mathcal{H}_{-s}, b \in B), \tag{7.5.6}$$

where $\|\cdot\|_\infty$ denotes the supremum norm on $C(B)$.

Because of the G-invariance of the form ℓ (see (v) (b)), it follows that $A_s f$ is continuous, so in $\mathcal{H}_{-s}$. By (7.5.6) A_s is a continuous mapping from $\mathcal{H}_{-s}$ to $\mathcal{H}_s$. Finally,

$$\pi_s(x)\,A_s = A_s\,\pi_{-s}(x)$$

for all $x \in G$. The mapping A_s is called an *intertwining operator.*

Let now s be real, $0 < s < \rho$. Then π_{-s} is irreducible and

$$\langle f \mid A_s g\rangle = \int_B \int_B W_s(b,b')\,f(b)\,\overline{g(b')}\,d\sigma(b)\,d\sigma(b') \qquad (f, g \in \mathcal{H}_{-s}) \tag{7.5.7}$$

is a G-invariant positive-definite form on $\mathcal{H}_{-s}$ according to Lemma 7.5.10. It is even a scalar product. We also have

$$\langle f \mid A_s f\rangle \leq \gamma(s)\,\|f\|_\infty^2 \qquad (f \in \mathcal{H}_{-s}). \tag{7.5.8}$$

From (7.5.7) we obtain a pre-unitary structure on $\mathcal{H}_{-s}$. Completing $\mathcal{H}_{-s}$ and extending $\pi_{-s}(x)$ continuously to this completion for all $x \in G$, π_{-s} becomes a unitary representation. Moreover, π_{-s} remains irreducible as a unitary representation. Indeed, if V is a non-trivial closed invariant subspace, then $V \cap C(B)$ is non-trivial and closed in the uniform topology, by (7.5.8).

The irreducible unitary representations π_{-s} with $0 < s < \rho$, thus obtained, are called representations of the *complementary series*.

Clearly, $\varphi_{-s}(x) = \frac{1}{\gamma(s)}\langle u_{-s} \mid A_s\pi_{-s}(x)u_{-s}\rangle$ $(x \in G)$, see (vi). Notice that $A_s u_{-s} = \gamma(s)\,u_s$, which gives an alternative proof of this expression for φ_{-s}.

(3) Class one representations

We can now conclude that the class one representations are given by

- the unitary spherical principal series $\pi_{i\nu}$ $(\nu \in \mathbb{R},\ \pi_{i\nu} \sim \pi_{-i\nu})$,
- the complementary series π_{-s} $(0 < s < \rho)$,
- the trivial representation.

The associated positive-definite spherical functions are $\varphi_{i\nu}$, φ_s and φ_ρ, respectively.

(vii) Plancherel formula

In this subsection we shall determine the Plancherel measure for the pair (G, K). We start with a summary of some (advanced) functional analysis, see [11, Chapter XII] or [46, Chapter VI].

(1) Spectral theory of self-adjoint operators on a Hilbert space

Let $\mathcal{H}$ be a Hilbert space with scalar product denoted by $(\ |\ )$ and A a linear operator defined on some subspace of $\mathcal{H}$, called the domain of A and denoted by $D(A)$,

$$A : D(A) \to \mathcal{H}.$$

We shall assume throughout that $D(A)$ is dense in $\mathcal{H}$. The adjoint A^* of A is defined as follows: its domain $D(A^*)$ is the set of elements $f \in \mathcal{H}$ such that the linear form

$$g \mapsto (Ag \mid f) \qquad (g \in D(A))$$

is continuous. For such f an element $f^* \in \mathcal{H}$ exists such that

$$(Ag \mid f) = (g \mid f^*) \qquad (g \in D(A))$$

and one sets

$$A^* f = f^*.$$

Clearly, $D(A^*)$ is a linear subspace of $\mathcal{H}$, not necessarily a dense subspace.

An operator $(D(A), A)$ is said to be *self-adjoint* if $D(A) = D(A^*)$ and $A = A^*$.

Let A be a linear operator with domain $D(A)$. The *resolvent set* of A is the set of complex numbers z such that $zI - A$ is a bijection of $D(A)$ onto $\mathcal{H}$ and such that the inverse $R_z = (zI - A)^{-1}$ is a continuous linear operator on $\mathcal{H}$. The operator R_z is called the *resolvent* of A. If z and z' are two complex numbers of the resolvent set of A, then

$$R_z - R_{z'} = (z - z')R_z R_{z'}$$

(resolvent equation). The resolvent set is open and $z \mapsto R_z$ is an analytic function on this open set. The complement of the resolvent set is called the *spectrum* of A.

If A is self-adjoint, then its spectrum is real. Moreover, if $\operatorname{Im} z \neq 0$ then

$$\|R_z\| \leq \frac{1}{|\operatorname{Im} z|}.$$

A *spectral function* on $\mathbb{R}$ is a mapping from $\mathbb{R}$ into the set of orthogonal projections of the Hilbert space $\mathcal{H}$

$$E : \lambda \mapsto E_\lambda$$

such that

(a) $\lambda \mapsto E_\lambda$ is increasing: $E_{\lambda_1} E_{\lambda_2} = E_{\lambda_1}$ if $\lambda_1 \leq \lambda_2$,

(b) $\lim_{\lambda\to\infty} E_\lambda f = f$, $\lim_{\lambda\to-\infty} E_\lambda f = 0$ for all $f \in \mathcal{H}$.

Observe that $\lambda \mapsto (E_\lambda f \mid g)$ is a function of bounded variation on $\mathbb{R}$, so it yields a Riemann–Stieltjes integral or measure, called the *spectral measure* associated to the spectral function. For all bounded continuous functions ϕ on $\mathbb{R}$ one defines

$$E(\phi) = \int_{-\infty}^{\infty} \phi(\lambda)\, dE_\lambda,$$

which means

$$(E(\phi) f \mid g) = \int_{-\infty}^{\infty} \phi(\lambda)\, d(E_\lambda f \mid g)$$

for all $f, g \in \mathcal{H}$.

This is a bounded operator on $\mathcal{H}$, and

$$E(\overline{\phi}) = E(\phi)^*.$$

If ϕ is an arbitrary continuous function on $\mathbb{R}$, one can again define $E(\phi)$ with domain $\{f \in \mathcal{H} : \int |\phi(\lambda)|^2 d(E_\lambda f \mid f) < \infty\}$. This domain is dense in $\mathcal{H}$. If ϕ is real-valued, then $E(\phi)$ is self-adjoint.

Let now A be a self-adjoint operator with domain $D(A)$. There exists a spectral function E on $\mathbb{R}$ such that

$$A = \int_{-\infty}^{\infty} \lambda\, dE_\lambda.$$

This function is unique if we assume that $\lambda \mapsto E_\lambda f$ is left continuous for all $f \in \mathcal{H}$, which we can. If R_z is the resolvent of A, then

$$R_z = \int_{-\infty}^{\infty} \frac{dE_\lambda}{z-\lambda}.$$

From this we obtain, if ε and μ are real numbers, then

$$\operatorname{Im}(R_{\mu+i\varepsilon} f \mid f) = -\int_{-\infty}^{\infty} \frac{\varepsilon}{(\mu-\lambda)^2+\varepsilon^2} d(E_\lambda f \mid f).$$

Now $\psi_\varepsilon(\mu) = \frac{1}{\pi}\frac{\varepsilon}{(\mu-\lambda)^2+\varepsilon^2}$ is well known to be an approximation of the delta-function at λ, i.e.

$$\lim_{\varepsilon\to 0} \int_{-\infty}^{\infty} \phi(\mu)\, \psi_\varepsilon(\mu)\, d\mu = \phi(\lambda) \qquad (\phi \in C_c(\mathbb{R})).$$

So we get

$$\lim_{\varepsilon \downarrow 0} \int_{-\infty}^{\infty} \operatorname{Im}(R_{\lambda+i\varepsilon} f \mid f)\, \phi(\lambda)\, d\lambda = -\pi \int_{-\infty}^{\infty} \phi(\lambda)\, d(E_\lambda f \mid f) \tag{7.5.9}$$

for all $f \in \mathcal{H}$ and $\phi \in C_c(\mathbb{R})$. The latter formula is important, for it helps in many cases to determine the spectral measure associated with a self-adjoint operator.

(2) Application: the Laplace–Beltrami operator

Let Δ be the Laplace–Beltrami operator on the manifold X (see (iii)) with domain $D(\Delta) = C_c^\infty(X)$, the space of C^∞ functions on X with compact support. This domain is a dense subspace of $L^2(X)$. We obtain a densely defined linear operator, which is known to be *symmetric*, i.e.

$$(\Delta f \mid g) = (f \mid \Delta g)$$

for all $f, g \in D(\Delta)$. Otherwise formulated: $\Delta \subset \Delta^*$. This rather plausible fact follows from another view on the Laplace–Beltrami operator coming from the theory of Lie groups and Lie algebras. We shall give a summary here.

We shall make use of the theory of Lie groups and Lie algebras to some extent. We refer to [55] and [24]; see also [21].

Let $G = \mathrm{SO}_0(1, n)$ be as before and let $\mathfrak{g}$ be its Lie algebra consisting of the $(n+1) \times (n+1)$ matrices Y satisfying

$$J\, {}^tY J = -Y.$$

The Cartan involution θ of G induces an involution on $\mathfrak{g}$: $\theta(Y) = -\,{}^tY = JYJ$, and its ± 1-eigenspaces are given by $\mathfrak{k}$ and $\mathfrak{p}$ respectively,

$$\mathfrak{g} = \mathfrak{k} \oplus \mathfrak{p},$$

where $\mathfrak{k}$ is the Lie algebra of K consisting of the anti-symmetric matrices and $\mathfrak{p}$ consisting of the symmetric matrices

$$Y = \begin{pmatrix} 0 & y_1 & \dots & y_n \\ y_1 & 0 & \dots & 0 \\ \vdots & 0 & \dots & 0 \\ y_n & 0 & \dots & 0 \end{pmatrix}$$

with $y_1, \dots, y_n$ in $\mathbb{R}$. The Lie algebra $\mathfrak{g}$ is semisimple and its Killing form is given by

$$B(Y, Z) = (n-1)\, \operatorname{tr} YZ.$$

This form is negative-definite on $\mathfrak{k}$ and positive-definite on $\mathfrak{p}$. Moreover, it is invariant under θ and $\langle Y, Z \rangle = -B(Y, \theta Z)$ is a scalar product on $\mathfrak{g}$.

Any $Y \in \mathfrak{g}$ corresponds to both a left- and a right-invariant differential operator on G:

$$(Yf)(x) = \frac{d}{dt}\Big|_{t=0} f(xe^{tY}) \quad \text{(left-invariant case)},$$

$$(Yf)(x) = \frac{d}{dt}\Big|_{t=0} f(e^{tY}x) \quad \text{(right-invariant case)}$$

($x \in G, f \in C^\infty(G)$). This correspondence can easily be extended to $U(\mathfrak{g})$, the universal enveloping algebra of $\mathfrak{g}$; for example in the left-invariant case

$$(Y_1 \dots Y_n) f(x) = \frac{\partial^n}{\partial t_1 \dots \partial t_n} f(x\, e^{t_1 Y_1} \dots e^{t_n Y_n})\Big|_{t_1=\dots=t_n=0}.$$

The Casimir operator ω is defined as follows. Select a basis $Y_1, \dots, Y_m$ of $\mathfrak{g}$ and a dual basis $Y_1', \dots, Y_m'$ of $\mathfrak{g}$: $B(Y_i, Y_j') = 1$ if $i = j$, $B(Y_i, Y_j') = 0$ if $i \neq j$. Then

$$\omega = Y_1 Y_1' + \dots + Y_m Y_m',$$

where $m = \dim \mathfrak{g}$. This differential operator does not depend on the special choice of the bases and it is left- and right-invariant under G.

Let $Z_1, \dots, Z_m$ be another basis of $\mathfrak{g}$ and $Z_1', \dots, Z_m'$ the unique dual basis. Then we may write

$$Y_i = \sum_j s_{ij}\, Z_j \qquad \text{with} \quad s_{ij} = B(Y_i, Z_j'),$$

$$Y_l' = \sum_k t_{lk}\, Z_k' \qquad \text{with} \quad t_{lk} = B(Y_l', Z_k).$$

Moreover,

$$Z_i = \sum_j u_{ij}\, Y_j \qquad \text{with} \quad u_{ij} = B(Z_i, Y_j') = t_{ji},$$

$$Z_l' = \sum_k v_{lk}\, Y_k' \qquad \text{with} \quad v_{lk} = B(Z_l', Y_k) = s_{kl}.$$

Now $B(Z_i, Z_l') = \sum_j u_{ij}\, v_{lj}$. This expression is equal to 1 if $i = l$, and equal to 0 if $i \neq l$. So the same holds for $\sum_j t_{ji}\, s_{jl}$. We finally have

$$\sum_{i=1}^m Y_i Y_i' = \sum_{i,j,k} s_{ij}\, t_{ik} Z_j Z_k' = \sum_{j=1}^m Z_j Z_j'.$$

Because the Killing form is $\mathrm{Ad}(G)$-invariant, the Casimir operator is too, so ω is left- and right-invariant under G.

Selecting orthonormal bases $Y_1, \dots, Y_l$ of $\mathfrak{k}$ ($l = \dim \mathfrak{k}$) and $Y_{l+1}, \dots, Y_{l+n}$ of $\mathfrak{p}$ with respect to the above scalar product, we have

$$\omega = -Y_1^2 - \dots - Y_l^2 + Y_{l+1}^2 + \dots + Y_{l+n}^2.$$

If f is a C^∞ function on G, right-invariant under K (or actually a C^∞ function on X), then

$$\omega f = \omega_{\mathfrak{p}} f$$

where $\omega_{\mathfrak{p}} = Y_{l+1}^2 + \dots + Y_{l+n}^2$. One also denotes $\omega_{\mathfrak{k}} = -Y_1^2 - \dots - Y_l^2$. Both $\omega_{\mathfrak{k}}$ and $\omega_{\mathfrak{p}}$ are bi-invariant under K. Observe that $\omega_{\mathfrak{p}}$ can be viewed as a left-G-invariant differential operator on X of the second order. Such an operator is unique up to complex constants, so $\omega_{\mathfrak{p}} = c\Delta$ with c a *real* constant here.

Any left-G-invariant differential operator D on X is completely determined by its behaviour at e_0:

$$f \mapsto Df(e_0) \qquad (f \in C^\infty(X)).$$

Choosing local coordinates $y_1, \dots, y_n$ near e_0, where

$$(y_1, \dots, y_n) \leftrightarrow \mathrm{e}^{Y(y_1,\dots,y_n)} \cdot e_0 \in X,$$

we have $Df(e_0) = p(\frac{\partial}{\partial y_1}, \dots, \frac{\partial}{\partial y_n}) f(e_0)$, where p is a polynomial. Since D acts on $C^\infty(X) \simeq C^\infty(G/K)$, the polynomial p must be K-invariant, hence a one-variable polynomial in $y_1^2 + \dots + y_n^2$. So $\omega_{\mathfrak{p}}$ and Δ, both being second order invariant differential operators on X, are proportional.

Let $f, g \in C_c^\infty(X)$. Then, considering $Y \in \mathfrak{g}$ as a right-invariant differential operator, we easily get

$$(Yf \mid g) = -(f \mid Yg)$$

and hence

$$(\omega_{\mathfrak{p}} f \mid g) = (f \mid \omega_{\mathfrak{p}} g).$$

So Δ is a symmetric operator.

Any symmetric operator A has a closure, denoted by $\overline{A}$, its graph is the closure of the graph of A. One can show ([46]) that $\overline{A} = A^{**}$. Clearly, $\overline{A} \subset A^*$, since A^* is a closed operator and $\overline{A}$ is symmetric again.

Proposition 7.5.13. *The operator $(D(\Delta), \Delta)$ is essentially self-adjoint, that means $(D(\overline{\Delta}), \overline{\Delta})$ is self-adjoint. One even has $\overline{\Delta} = \Delta^*$.*

The domain $D(\overline{\Delta})$ consists of functions $u \in L^2(X)$ for which there exists a sequence of functions $f_m \in C_c^\infty(X)$ satisfying

$$\lim_{m\to\infty} f_m = u \quad \text{and} \quad \lim_{m\to\infty} \Delta f_m = v$$

for some $v \in L^2(X)$. Then $\overline{\Delta} u = v$. The convergence has to be considered in the $L^2(X)$-topology.

In the proof of the proposition we shall use the following properties: if $f \in C_c^\infty(X)$ and $h \in C_c^\infty(G)$, then

$$\Delta(h * f) = h * \Delta f = \omega' h * f$$

where $\omega' = \omega/c$, and if in addition $u \in L^2(X)$, then

$$(h * u \mid f) = (u \mid \widetilde{h} * f)$$

where $\widetilde{h}(x) = \overline{h(x^{-1})}$ $(x \in G)$.

Let us now prove the proposition.

(a) If $u \in L^2(X)$ and $h \in C_c^\infty(G)$, then $h * u$ belongs to the domain $D(\overline{\Delta})$ and moreover

$$\overline{\Delta}(h * u) = \Delta(h * u) = \omega' h * u.$$

Indeed, let $\{f_m\}$ be a sequence of functions in $C_c^\infty(X)$ such that

$$\lim_{m\to\infty} f_m = u \quad (\text{in } L^2(X)).$$

Then

$$\lim_{m\to\infty} h * f_m = h * u$$

and

$$\begin{aligned}\lim_{m\to\infty} \Delta\,(h * f_m) &= \lim_{m\to\infty} \omega'\, h * f_m \\ &= \omega'\, h * u = \Delta\,(h * u).\end{aligned}$$

(b) Let now $u \in L^2(X)$ belong to $D(\Delta^*)$: for all $f \in C_c^\infty(X)$ one has

$$(u \mid \Delta\, f) = (\Delta^* u \mid f).$$

Let $\{h_m\}$ be an approximate unit (see Section 4.6 (vii)) consisting of a sequence of functions $h_m \in C_c^\infty(G)$, $h_m \geq 0$, $\int_G h_m(x)\,dx = 1$. Then

$$\lim_{m\to\infty} h_m * u = u \quad (\text{in } L^2(X)).$$

Moreover, $h_m * u \in D(\overline{\Delta}) \subset D(\Delta^*)$ for all $m = 1, 2, 3, \ldots$ and we have for all $f \in C_c^\infty(X)$

$$\begin{aligned}(\overline{\Delta}\,(h_m * u) \mid f) &= (h_m * u \mid \Delta f) \quad (\text{since } \overline{\Delta} \text{ is symmetric}) \\ &= (u \mid \widetilde{h}_m * \Delta f) = (u \mid \Delta(\widetilde{h}_m * f)) \\ &= (\Delta^* u \mid \widetilde{h}_m * f) = (h_m * \Delta^* u \mid f).\end{aligned}$$

Hence $\overline{\Delta}\,(h_m * u) = h_m * \Delta^* u$ for all m, and therefore

$$\lim_{m\to\infty} \overline{\Delta}\,(h_m * u) = \Delta^* u.$$

We may conclude that $u \in D(\overline{\Delta})$, so $D(\overline{\Delta}) = D(\Delta^*)$ and $\overline{\Delta} = \Delta^*$.

Remark 7.5.14. The domain of $\overline{\Delta} = \Delta^*$ consists of all $u \in L^2(X)$ such that Δu "in distribution sense" is a regular distribution belonging to $L^2(X)$. This is easily seen by those readers who are familiar with the theory of distributions (see, e.g., [41] and [43]). Others may skip this remark. The theory of distributions will play a major role in Chapter 8.

We are now going to determine the resolvent of $\overline{\Delta}$ and then its spectral measure. We need some preparation.

(3) Asymptotic behaviour of the spherical functions

Set for $a_t \in A$, $\varphi_s(a_t) = \varphi(s,t)$. The function $\varphi(s,.)$ is a solution of the differential equation

$$\frac{d^2y}{dt^2} + (n-1)\frac{\cosh t}{\sinh t}\frac{dy}{dt} = (s^2 - \rho^2)\,y. \tag{7.5.10}$$

Setting $z = -\sinh^2 t$ we obtain

$$4z(z-1)\frac{d^2y}{dz^2} + 2[(n+1)z - n]\frac{dy}{dz} = (s^2 - \rho^2)\,y. \tag{7.5.11}$$

The hypergeometric function ${}_2F_1(a,b;c;z)$ satisfies the differential equation

$$z(z-1)\frac{d^2y}{dz^2} + [(a+b+1)z - c]\frac{dy}{dz} + aby = 0 \tag{7.5.12}$$

(see [13]). Thus we can express the function $\varphi(s,t)$ as

$$\varphi(s,t) = {}_2F_1\Big(\frac{s+\rho}{2}, \frac{-s+\rho}{2}; \frac{n}{2}; -\sinh^2 t\Big).$$

We shall determine the asymptotic behaviour of the solutions of (7.5.10) for $t \to \infty$ and for $t \downarrow 0$.

(a) For $t \to \infty$ we make use of the following relation for hypergeometric functions (see [13]):

$$\begin{aligned}{}_2F_1(a,b;c;z) = {} & \frac{\Gamma(c)\,\Gamma(b-a)}{\Gamma(b)\,\Gamma(c-a)}(-z)^{-a}\,{}_2F_1\Big(a, 1-c+a; 1-b+a; \frac{1}{z}\Big) \\ & + \frac{\Gamma(c)\,\Gamma(a-b)}{\Gamma(a)\,\Gamma(c-b)}(-z)^{-b}\,{}_2F_1\Big(b, 1-c+b; 1-a+b; \frac{1}{z}\Big) \\ & \qquad (|\arg(-z)| < \pi).\end{aligned}$$

Both functions on the right-hand side are solutions of (7.5.11). Let us now take $a = \frac{s+\rho}{2}$, $b = \frac{-s+\rho}{2}$, $c = -\frac{n}{2}$, $z = -\sinh^2 t$ and set

$$\Phi(s,t) = |\sinh t|^{s-\rho}\,{}_2F_1\Big(\frac{-s+\rho}{2}, \frac{-s-\rho+1}{2}; 1-s; \frac{-1}{\sinh^2 t}\Big).$$

Then we have for s not an integer

$$\varphi(s,t) = c(s)\,\Phi(s,t) + c(-s)\,\Phi(-s,t)$$

with

$$c(s) = 2^{n-2}\frac{\Gamma(\frac{n}{2})}{\sqrt{\pi}}\,\frac{\Gamma(s)}{\Gamma(s+\rho)}.$$

This c-function coincides with the one considered in (iv). To see this, apply the duplication formula of the Γ-function.

(b) For t near 0, we apply the following method. First of all, $\varphi(s,.)$ is a regular solution at $t = 0$ of equation (7.5.10). To find a second solution, linearly independent of $\varphi(s,.)$, we apply the method of quadrature.

So set $y(t) = \varphi(s,t)\,z(t)$. Then we get

$$\varphi(s.t)\,z''(t) + \left((n-2)\frac{\cosh t}{\sinh t}\,\varphi(s,t) + 2\,\frac{d\varphi}{dt}(s,t)\right)\,z'(t) = 0,$$

and hence z satisfies

$$z'(t) = \frac{c}{\varphi(s,t)}\,\exp\Big(-\int_{t_0}^{t}(n-2)\frac{\cosh u}{\sinh u}\,du\Big).$$

We have

$$\frac{\cosh u}{\sinh u} = \frac{1}{u}\,(1 + a_2u^2 + a_4u^4 + \cdots)$$

near $u = 0$.

Hence (7.5.10) has a solution Ψ such that near $t = 0$

$$\Psi(t) \sim t^{2-n}, \qquad \Psi'(t) \sim (2-n)t^{1-n} \qquad (n \neq 2),$$

$$\Psi(t) \sim \log t, \qquad \Psi'(t) \sim \frac{1}{t} \qquad (n = 2).$$

Clearly, Ψ is linearly independent of $\varphi(s,.)$.

(4) Formulation and proof of the Plancherel formula

For f a continuous function on G, bi-K-invariant and with compact support, we define its spherical Fourier transform by

$$\widehat{f}(s) = \int_G f(x)\,\varphi_s(x)\,dx$$

where s is a complex parameter. Notice that this definition differs slightly from that in (6.4.4). Clearly, $\widehat{f}$ is an entire function of s. In view of (7.5.2) we may also write

$$\widehat{f}(s) = \int_0^\infty f(a_t)\,\varphi(s,t)\,A(t)\,dt.$$

Theorem 7.5.15. *Let f be a continuous function on G, bi-K-invariant and with compact support. Then*

$$\int_G |f(x)|^2 dx = \frac{1}{2\pi\kappa}\int_0^\infty |\widehat{f}(i\nu)|^2 \frac{d\nu}{|c(i\nu)|^2},$$

where c is as in (3),

$$c(s) = 2^{n-2}\frac{\Gamma(\frac{n}{2})}{\sqrt{\pi}}\frac{\Gamma(s)}{\Gamma(s+\rho)},$$

and κ is the constant

$$\kappa = \lim_{t\to\infty} \mathrm{e}^{(n-1)t} A(t) = 2^{2-n}\frac{\pi^{\frac{n}{2}}}{\Gamma(\frac{n}{2})}.$$

(a) A formula for partial integration. Recall that $L = \frac{d^2}{dt^2} + \frac{A'(t)}{A(t)}\frac{d}{dt}$ is the radial part of the Laplace–Beltrami operator (see (iii)). For two functions F and G of class C^1 on $(0,\infty)$ we set

$$[F,G](t) = A(t)\,[F'(t)G(t) - F(t)G'(t)].$$

By partial integration we get, if F and G are C^2 on $(0,\infty)$ and $0<\alpha<\beta$,

$$\int_\alpha^\beta (LF\cdot G - F\cdot LG)(t)\,A(t)dt = [F,G](\beta) - [F,G](\alpha).$$

So if F and G are two eigenfunctions of L with the same eigenvalue, then $[F,G]$ is constant. Consider in particular the functions $\varphi(s,t)$ and $\Phi(-s,t)$, introduced in (3). From the asymptotic behaviour of these functions for $t\to\infty$ we obtain

$$[\varphi(s,.),\Phi(-s,.)] = 2\,s\,\kappa\,c(s).$$

The equation $Lu = (s^2-\rho^2)\,u$ also has a solution Ψ with asymptotics near $t=0$ as described in (3). The functions $\varphi(s,t)$ and $\Psi(t)$ form a fundamental system of solutions of $Lu = (s^2-\rho^2)\,u$, hence there are constants c_1 and c_2 (depending on s) such that

$$\Phi(-s,t) = c_1\,\varphi(s,t) + c_2\,\Psi(t).$$

Let F be an even C^2 function on $\mathbb{R}$. If $F(0)=0$, then it follows from the asymptotic behaviour of $\Phi(-s,t)$ near $t=0$ that

$$\lim_{t\to 0}\,[F,\Phi(-s,.)](t) = 0,$$

and if $F(0)\neq 0$,

$$[F,\Phi(-s,.)] = [F - F(0)\varphi(s,.),\Phi(-s,.)] + [F(0)\,[\varphi(s,.),\Phi(-s,.)],$$

hence

$$\lim_{t\to 0}\,[F,\Phi(-s,.)](t) = 2\kappa\,s\,c(s)\,F(0).$$

If, in addition, F has compact support, we get

$$\int_0^\infty (zF - LF)(t)\,\Phi(-s,t)\,A(t)\,dt = 2\kappa\,s\,c(s)\,F(0), \tag{7.5.13}$$

with $z = s^2 - \rho^2$.

(b) The resolvent of the closure of the Laplace–Beltrami operator. Let K_s be the function on G defined on the complement of K by

$$K_s(ka_tk') = \frac{1}{2\kappa\,s\,c(s)}\,\Phi(-s,t) \qquad (t \neq 0),$$

where k and k' are elements of the group K. If $\operatorname{Re} s > \rho$, the function K_s is integrable. This follows easily from the asymptotic behaviour of $\Phi(-s,t)$ near $t = 0$ and for $t \to \infty$. If $\operatorname{Re} s > 0$, then the function K_s is in $L^2(G)^\#$ at infinity, i.e.

$$\int_{C_\delta} |K_s(x)|^2\,dx < \infty \tag{7.5.14}$$

for some $\delta > 0$, where $C_\delta = \{Ka_tK : t > \delta\}$.

Let f be a function in $C_c^\infty(X)$ and let u be the function on G defined by

$$u = f * K_s.$$

Obviously u is right K-invariant and can thus be viewed as a function on X; it is again C^∞ and in $L^2(X)$ for $\operatorname{Re} s > 0$, by (7.5.14). It is even in $D(\overline{\Delta})$ for $\operatorname{Re} s > 0$ (see (2), Remark 7.5.14).

For $\operatorname{Re} s > \rho$ the mapping $f \mapsto f * K_s$ can be extended to a *continuous* linear operator on $L^2(X)$. The function u satisfies

$$zu - \Delta u = f \quad (z = s^2 - \rho^2,\ \operatorname{Re} s > 0).$$

This follows from (7.5.13). We may conclude that the resolvent R_z of $\overline{\Delta}$ is given by

$$R_z f = f * K_s \quad (z = s^2 - \rho^2)$$

if $\operatorname{Re} s > \rho$. Since s is an analytic function of z for $\operatorname{Re} s > 0$, provided $z \notin (-\infty, -\rho^2]$, and since for the above operator R_z the function $(R_z f \mid f)$ is still analytic for these values of z, we conclude that the support of the spectral measure (see (2)) is contained in $(-\infty, -\rho^2]$, and our R_z is indeed the resolvent everywhere.

(c) The spectral measure of the Laplace–Beltrami operator. Let $f \in C_c^\infty(G)^\#$. If $\lambda = -\rho^2 - \nu^2$, $\nu \geq 0$, then

$$\lim_{\varepsilon \downarrow 0} (R_{\lambda + i\varepsilon} f \mid f) = \int_G K_{i\nu}(x)\,(\widetilde{f} * f)(x)\,dx,$$

where the limit is uniform for λ in compact subsets of $(-\infty, -\rho^2]$. Therefore we obtain for any $h \in C_c(\mathbb{R})$

$$\lim_{\varepsilon \downarrow 0} \int_{-\infty}^{\infty} \mathrm{Im}(R_{\lambda+i\varepsilon} f \mid f)\, h(\lambda)\, d\lambda$$
$$= \int_0^{\infty} \int_G \mathrm{Im}(K_{i\nu}(x))\, (\widetilde{f} * f)(x)\, h(-\rho^2 - \nu^2)\, 2\nu\, dx d\nu.$$

From the relation

$$\varphi(s,t) = c(s)\, \Phi(s,t) + c(-s)\, \Phi(-s,t)$$

we obtain

$$\mathrm{Im}\, K_{i\nu}(x) = -\frac{1}{4\kappa} \frac{\varphi(i\nu, t)}{|c(i\nu)|^2},$$

hence, if dE_λ denotes the spectral measure of $\overline{\Delta}$,

$$\int_{-\infty}^{\infty} h(\lambda)\, d(E_\lambda f \mid f) = \frac{1}{2\pi\kappa} \int_0^{\infty} |\widehat{f}(i\nu)|^2\, h(-\rho^2 - \nu^2)\, \frac{d\nu}{|c(i\nu)|^2}$$

(see (2)). The total measure of $d(E_\lambda f \mid f)$ is equal to $\| f \|^2$, hence

$$\int_G |f(x)|^2 dx = \frac{1}{2\pi\kappa} \int_0^{\infty} |\widehat{f}(i\nu)|^2 \frac{d\nu}{|c(i\nu)|^2}.$$

Chapter 8
Theory of Generalized Gelfand Pairs

Literature: [49], [53].

In this chapter we shall extend the theory of Gelfand pairs (G, K) to the case where the subgroup K is not necessarily compact. The examples that we have in mind are $G = \mathrm{O}(1,n) \ltimes \mathbb{R}^{n+1}$, $H = \mathrm{O}(1,n)$ and $G = \mathrm{O}(1,n)$, $H = \mathrm{O}(1,n-1)$. We shall introduce the theory of generalized Gelfand pairs and apply it to these examples, later on in Chapter 9.

We assume here explicitly some familiarity with the theory of Lie groups and Lie algebras (see [55], [21], [24]), as well as some knowledge of distribution theory (see [41], [43]) and of the theory of locally convex topological vector spaces (see [4], [50]).

About the theory of generalized Gelfand pairs we shall sometimes be rather sketchy, often giving references rather than proofs. The examples (in Chapter 9) will however be treated in detail.

8.1 C^∞ vectors of a representation

Let G be a Lie group with finitely many connected components, $\mathfrak{g}$ its Lie algebra and $U(\mathfrak{g}_c)$ its universal enveloping algebra over $\mathbb{C}$, realized as the algebra of left-invariant differential operators on G. Any basis of $\mathfrak{g}$ as a real vector space generates the algebra $U(\mathfrak{g}_c)$. Fix a left-invariant Haar measure dx on G. Let π be a representation of G on the Fréchet space $\mathcal{H}$. Recall that a Fréchet space is a complete locally convex space with a countable set of semi-norms. The definition of representation is the same as in Section 7.5 (vii). The reason for extending from Hilbert spaces to Fréchet spaces will be become clear soon. A vector $v \in \mathcal{H}$ is called a C^∞ vector if the mapping

$$x \mapsto \pi(x)v \qquad (x \in G)$$

from G to $\mathcal{H}$ is C^∞. Denote by $\mathcal{H}_\infty$ the subspace of all C^∞ vectors in $\mathcal{H}$. Given $f \in C_c^\infty(G)$ denote, as usual, by $\pi(f)$ the continuous linear operator on $\mathcal{H}$ defined by

$$\pi(f)v = \int_G f(x)\,\pi(x)v\,dx \qquad (v \in \mathcal{H}).$$

The definition of the operator $\pi(f)$ is more involved than in the case of a Hilbert space, but in principle it goes along the same route.

For all $f \in C_c^\infty(G)$ and all $v \in \mathcal{H}$ we have $\pi(f)v \in \mathcal{H}_\infty$. The space of all linear combinations of such C^∞ vectors is called the *Gårding subspace* of π; it is a dense subspace of $\mathcal{H}$. Indeed, let $\{f_n\}_{n\geq 1}$ be an *approximate unit*, that is a sequence of elements of $C_c^\infty(G)$ satisfying

(a) $f_n \geq 0$, $\int_G f_n(x)\,dx = 1$,

(b) there exists a decreasing sequence $(O_n)_{n\geq 1}$ of compact neighbourhoods of the unit element e such that Supp $f_n \subset O_n$ for all n and $\bigcap_{n\geq 1} O_n = \{e\}$.

Then for each $v \in \mathcal{H}$, $\pi(f_n)v \to v$ as $n \to \infty$. In particular, $\mathcal{H}_\infty$ is dense in $\mathcal{H}$.

Given $v \in \mathcal{H}_\infty$, set $\widetilde{v}(x) = \pi(x)v$ $(x \in G)$. So $\widetilde{v} \in C^\infty(G, \mathcal{H})$. For $D \in U(\mathfrak{g}_c)$ and $v \in \mathcal{H}_\infty$ we define

$$\pi(D)v = D\widetilde{v}(e).$$

In particular, if $X \in \mathfrak{g}$,

$$\pi(X)v = \frac{d}{dt}\Big|_{t=0} \pi(\exp tX)v,$$

where "exp" denotes the exponential mapping from $\mathfrak{g}$ to G. Obviously, we have $\pi(D)\mathcal{H}_\infty \subset \mathcal{H}_\infty$, since $\pi(x)\pi(D)v = D\widetilde{v}(x)$ $(v \in \mathcal{H}_\infty,\ x \in G)$. In addition,

$$\pi(D_1)\pi(D_2)v = \pi(D_1D_2)v \qquad (D_1, D_2 \in U(\mathfrak{g}_c),\ v \in \mathcal{H}_\infty).$$

We have obtained a representation of the algebra $U(\mathfrak{g}_c)$ on $\mathcal{H}_\infty$.

The space $\mathcal{H}_\infty$ is also G-invariant: if $v \in \mathcal{H}_\infty$, then $\pi(x)v \in \mathcal{H}_\infty$ for all $x \in G$; moreover, we get a representation of G on $\mathcal{H}_\infty$ in this way, which we call π_∞. If we provide $\mathcal{H}_\infty$ with the topology of $C^\infty(G, \mathcal{H})$, via the injection $v \mapsto \widetilde{v}$, $\pi_\infty(x)$ becomes continuous for all $x \in G$. The space $\mathcal{H}_\infty$, provided with this topology, can be considered as a *closed* subspace of $C^\infty(G, \mathcal{H})$, so as a Fréchet space. We leave the proof to the reader. Finally, π_∞ is a (continuous) representation of G on $\mathcal{H}_\infty$.

The topology on $\mathcal{H}_\infty$ can alternatively be defined by means of a set of semi-norms. Let, for example, $\mathcal{H}$ be a Hilbert space and let $X_1, \ldots, X_n$ be a basis of $\mathfrak{g}$ (as a real vector space). Then the set of semi-norms $\|\cdot\|_m$ is given by the formula

$$\|v\|_m^2 = \sum_{|\alpha|\leq m} \|\pi(X_1)^{\alpha_1}\cdots\pi(X_n)^{\alpha_n}v\|^2$$

with $|\alpha| = \alpha_1 + \cdots + \alpha_n$, α_i non-negative integers, and $v \in \mathcal{H}_\infty$. The topology does not depend on the choice of the basis of $\mathfrak{g}$.

Examples

1. Let G be a Lie group with finitely many connected components. Then $C^\infty(G)$ is in a natural way a Fréchet space. The group acts from the left and for this action every function in $C^\infty(G)$ is a C^∞ vector.

2. Let $G = \mathrm{SO}_0(1,n)$ and let π_s ($s \in \mathbb{C}$) be a spherical principal series representation on $C(B)$, see Section 7.5 (vii). Then clearly $C^\infty(B)$ consists of C^∞ vectors. On the other hand, if $f \in C(B)$ is a C^∞ vector for π_s, then $k \mapsto f(k^{-1}b)$ is a C^∞ function on K for all $b \in B$, so $f \in C^\infty(B)$.

3. Let again $G = \mathrm{SO}_0(1,n)$ and consider $\pi_{i\nu}$ ($\nu \in \mathbb{R}$), the unitary spherical principal series on $L^2(B)$. Then again $C^\infty(B)$ consists of C^∞ vectors since the injection of $C^\infty(B)$ into the Hilbert space $L^2(B)$ is continuous. The converse is also true: each C^∞ vector for $\pi_{i\nu}$ belongs to $C^\infty(B)$. For the proof, see Example 4. It follows also and easily from the following useful and remarkable lemma, due to Dixmier and Malliavin [9]:

Lemma 8.1.1 (Decomposition lemma). *Let G be a real Lie group.*

(a) *Any function $f \in C_c^\infty(G)$ can be written as a finite sum of functions of the form $g * h$ ($g, h \in C_c^\infty(G)$), where 'star' means convolution product.*

(b) *If $\mathcal{H}$ is a Fréchet space and π a representation of G on $\mathcal{H}$, then $\mathcal{H}_\infty$ coincides with the Gårding subspace of π.*

Since the injection $C^\infty(B) \to L^2(B)_\infty$ is continuous, the space $C^\infty(B)$ also coincides topologically with $L^2(B)_\infty$, by the closed graph theorem.

4. Let again $G = \mathrm{SO}_0(1,n)$ and consider now π_{-s} ($0 < s < \rho$), the complementary series of G. These representations are realized on the Hilbert space $\overline{\mathcal{H}_{-s}}$, being the completion of $\mathcal{H}_{-s} \simeq C(B)$ taken with respect to the scalar product

$$(f \mid g)_s = \langle f \mid A_s g \rangle \qquad (f, g \in \mathcal{H}_{-s}),$$

see Section 7.5 (vi). Write $\|f\|_s$ for the norm of $f \in \overline{\mathcal{H}_{-s}}$. In order to determine the space of C^∞ vectors in this case, we need some preparation. Clearly, $C^\infty(B)$ consists of C^∞ vectors, since again the injection of $C^\infty(B)$, with its usual topology, into $\overline{\mathcal{H}_{-s}}$ is continuous, by formula (7.5.8). The converse is also true: $C^\infty(B)$ is precisely the space of C^∞ vectors for π_{-s} (also topologically). To show this we study the *K-decomposition* for π_{-s}.

Restricting π_{-s} to K, we get a unitary representation of K which is the orthogonal direct sum of representations U_l on $\mathcal{H}_l$ (see Section 7.3 for notation), each occurring once, so

$$\overline{\mathcal{H}_{-s}} = \bigoplus_{l=0}^{\infty} \mathcal{H}_l.$$

The scalar product on $\mathcal{H}_l$, inherited from $\overline{\mathcal{H}_{-s}}$, is proportional to the standard one. To find the positive factor, it is sufficient to compute, for each l, the integral

$$A_s\varphi_l(\xi^o) = \int_B W_s(\xi^0, b)\,\varphi_l(b)\,db,$$

where φ_l is the spherical function in $\mathcal{H}_l$, which, as we have seen before, is a Gegenbauer polynomial. The above integral has been evaluated in [56]:

$$\gamma(s,l) = \gamma(s)\,\frac{\Gamma(-s+\rho+l)}{\Gamma(s+\rho+l)}\,\frac{\Gamma(s+\rho)}{\Gamma(-s+\rho)}.$$

By the asymptotics of the Γ-function: $\frac{\Gamma(z+a)}{\Gamma(z+b)} \sim z^{a-b}\,(1+\mathcal{O}(\frac{1}{z}))$ $(|\arg z| < \pi)$, we get

$$\gamma(s,l)^{-1} \le \text{const.}\, l^{2s} \qquad (l \to \infty),$$

the constant depending only on s.

Let us decompose the C^∞ vector $f \in \overline{\mathcal{H}_{-s}}$ as

$$f = \sum_{l=0}^{\infty} f_l,$$

where $f_l \in \mathcal{H}_l$. Then we have $f_l = \sum_{i=1}^{d_l}(f\,|\,e_i)_s\, e_i$ with $d_l = \dim \mathcal{H}_l$ and $e_1,\dots,e_{d_l}$ an orthonormal basis in $\mathcal{H}_l$ with respect to the scalar product of $\overline{\mathcal{H}_{-s}}$. Let $\omega_{\mathfrak{k}}$ be the Casimir operator on K, considered as a left-invariant differential operator on G; $\omega_{\mathfrak{k}} = -X_1^2 - \cdots - X_k^2$ in the notation of Section 7.5 (vii), where $X_1,\dots,X_k$ is an orthonormal basis of the Lie algebra $\mathfrak{k}$ of K with respect to $-B(X,Y)$, B being the Killing form of the Lie algebra $\mathfrak{g}$ of G. Then $\pi_{-s}(\omega_{\mathfrak{k}})$ is a scalar on $\mathcal{H}_l$, equal to a positive constant (not depending on l) times the action of $-\Omega$ on $\mathcal{H}_l$, the latter being $a_l = l(l+n-2)$ times the identity (see Section 7.3 (v)). We shall ignore the constant here. Now notice that for each $m = 0,1,2,\dots$ and $b \in B$ one has

$$\begin{aligned}\sum_{i=1}^{d_l} |(f\,|\,e_i)_s\, e_i(b)| &= \sum_{i=1}^{d_l} |(\pi_{-s}(\omega_{\mathfrak{k}}^m) f\,|\,e_i)_s|\, a_l^{-m}\, |e_i(b)| \\ &\le \|\pi_{-s}(\omega_{\mathfrak{k}}^m) f\|_s\, a_l^{-m} \Big(\sum_{i=1}^{d_l} |e_i(b)|^2\Big)^{1/2} \\ &\le \|\pi_{-s}(\omega_{\mathfrak{k}}^m) f\|_s\, a_l^{-m}\, \gamma(s,l)^{-1/2}\, d_l^{1/2}.\end{aligned}$$

Hence $\sum_{l=0}^{\infty} f_l(b)$ converges uniformly to a continuous function, and this function equals f. Even stronger, for each non-negative integer p, $\sum_{l=0}^{\infty}(\omega_{\mathfrak{k}}^p f_l)(b)$ converges uniformly too.

To show that $f \in C^\infty(B)$, we have to prove that $\sum_{l=0}^{\infty}(Df_l)(b)$ converges uniformly on B for each left-invariant differential operator D on K. So we have to relate D to $\omega_{\mathfrak{k}}$.

Let $D = X_j$, a basis element occurring in the definition of $\omega_{\mathfrak{k}}$. Then we have for $f \in \mathcal{H}_l$ and $b \in B$

$$\begin{aligned} |X_j f(b)|^2 &\leq d_l \, \|X_j f\|_2^2 \quad (\text{norm of } L^2(B)) \\ &\leq d_l \, (-X_j^2 f \mid f) \\ &\leq d_l \, (\omega_{\mathfrak{k}} f \mid f) = d_l \, a_l \, \|f\|_2^2, \end{aligned}$$

and hence

$$|X_j X_{j'} f(b)|^2 \leq d_l \, a_l^2 \, \|f\|_2^2,$$

and therefore

$$|Df(b)|^2 \leq c_q \, d_l \, a_l^q \, \|f\|_2^2,$$

if $q =$ order of D, c_q being a constant. So we finally get for $1 \leq i \leq d_l$,

$$|De_i(b)|^2 \leq c_q \, \gamma(s,l)^{-1} \, d_l \, a_l^q.$$

We may conclude that $\sum_{l=0}^{\infty}(Df_l)(b)$ converges uniformly to a continuous function on B, and hence $f \in C^\infty(B)$.

Observe that a similar reasoning may be used for the unitary spherical principal series.

8.2 Invariant Hilbert subspaces

Let again G be a Lie group with finitely many connected components and let H be a closed subgroup of G. Assume throughout this chapter both G and H to be unimodular, so that G/H admits an invariant measure (cf. Section 4.5). Fix Haar measures dg on G, dh on H and a G-invariant measure dx on G/H such that, formally, $dg = dx\, dh$.

Let π be a (continuous) unitary representation of G on a Hilbert space $\mathcal{H}$. Endow $\mathcal{H}_\infty$ with its topology defined in Section 8.1. Denote by $\mathcal{H}_{-\infty}$ the anti-dual of $\mathcal{H}_\infty$ endowed with the strong topology (uniform convergence on bounded subsets of the locally convex space $\mathcal{H}_\infty$). Any $a \in \mathcal{H}_{-\infty}$ is a continuous, anti-linear form on $\mathcal{H}_\infty$ and $\mathcal{H}_{-\infty}$ is again a Fréchet space.

The inclusion $\mathcal{H}_\infty \subset \mathcal{H}$ (continuous injection) and the isomorphism of the anti-dual of a Hilbert space with itself, yield a continuous inclusion $\mathcal{H} \subset \mathcal{H}_{-\infty}$, so that

$$\mathcal{H}_\infty \subset \mathcal{H} \subset \mathcal{H}_{-\infty}.$$

The group G acts on $\mathcal{H}_{-\infty}$ and the corresponding representation is called $\pi_{-\infty}$.

As in [41] and [43], denote by $D(G)$, $D(G/H)$ the spaces of C^∞ functions with compact support on G and G/H respectively. These spaces were previously denoted by $C_c^\infty(G)$ and $C_c^\infty(G/H)$, but we shall adopt Schwartz's notation from now on. Endow both spaces with their usual topology. Let $D'(G)$, $D'(G/H)$ be the topological anti-dual of $D(G)$ and $D(G/H)$ respectively, provided with the strong topology. Both $D(G/H)$ and $D'(G/H)$ are reflexive, they are each other's dual (see [43]). For $v \in \mathcal{H}_\infty$, $a \in \mathcal{H}_{-\infty}$ we set $\langle a, v\rangle = \overline{a(v)}$ and we also write $\langle v, a\rangle$ instead of $\overline{\langle a, v\rangle}$. Similarly we set $\langle T, \varphi\rangle = \overline{\langle \varphi, T\rangle} = T(\varphi)$ for either $\varphi \in D(G)$, $T \in D'(G)$ or $\varphi \in D(G/H)$, $T \in D'(G/H)$. Denote by $\varphi_0 \mapsto \varphi$ the canonical projection mapping $D(G) \to D(G/H)$ given by

$$\varphi(x) = \int_H \varphi_0(gh)\, dh \qquad (x = gH \in G/H). \tag{8.2.1}$$

For $\varphi_0 \in C_c(G)$ this mapping was considered already in Section 4.3; for $\varphi_0 \in D(G)$ the treatment is similar: the function φ is seen to be C^∞ by applying local coordinates, while the surjectivity follows as in Section 4.3.

For any $a \in \mathcal{H}_{-\infty}$ and $\varphi_0 \in D(G)$ we define

$$\pi_{-\infty}(\varphi_0)\, a = \int_G \varphi_0(g)\, \pi_{-\infty}(g)\, a\, dg.$$

Then $\pi_{-\infty}(\varphi_0)\, a \in \mathcal{H}_\infty$. Indeed, it is easily seen that $\pi_{-\infty}(\varphi_0)\, a \in \mathcal{H}$. We leave it to the reader to show this. Applying then the decomposition lemma (Lemma 8.1.1) the result follows.

For the representations π, π_∞ and $\pi_{-\infty}$ we have the following properties:

(i) *π is irreducible if and only if π_∞ is;*

(ii) *π is irreducible if and only if $\pi_{-\infty}$ is.*

Let us show (ii). If π is irreducible, then $\pi_{-\infty}$ is: if V is a closed G-invariant subspace of $\mathcal{H}_{-\infty}$, then $V \cap \mathcal{H}$ is closed in $\mathcal{H}$, so $V \cap \mathcal{H} = \{0\}$ or $V \cap \mathcal{H} = \mathcal{H}$. If $V \cap \mathcal{H} = \{0\}$ then $V = \{0\}$ since $V \cap \mathcal{H}$ is dense in V. Indeed, with $v \in V$, $v \neq 0$ there is a $\varphi_0 \in D(G)$ with $\pi_{-\infty}(\varphi_0)v \neq 0$ and this vector is in $\mathcal{H}$ and in V. If $V \cap \mathcal{H} = \mathcal{H}$ then $V = \mathcal{H}_{-\infty}$.

Let now $\pi_{-\infty}$ be irreducible and let V be a non-trivial invariant closed subspace of $\mathcal{H}$. Then V is dense in $\mathcal{H}_{-\infty}$. Let $\varphi_0 \in D(G)$. If $v \in \mathcal{H}_{-\infty}$ is the limit of $v_n \in V$ in $\mathcal{H}_{-\infty}$, then v and v_n satisfy $\pi(\varphi_0)v_n \to \pi_{-\infty}(\varphi_0)v$ in $\mathcal{H}$, by the closed graph theorem applied to the mapping $\pi_{-\infty}(\varphi_0) : \mathcal{H}_{-\infty} \to \mathcal{H}$. Therefore $\pi_{-\infty}(\varphi_0)v \in V$, hence $\mathcal{H}_\infty \subset V$, so $V = \mathcal{H}$. Hence π is irreducible.

The proof of (i) is along the same lines.

A vector $a \in \mathcal{H}_{-\infty}$ is called *cyclic* if the space of all vectors of the form $\pi_{-\infty}(\varphi_0)\, a$ ($\varphi_0 \in D(G)$), is dense in $\mathcal{H}$.

An equivalent definition requires that $\mathrm{span}(\pi_{-\infty}(G)a)$ is dense in $\mathcal{H}$. In the present context we prefer the definition given above.

Define now

$$\mathcal{H}_{-\infty}^{H} = \{a \in \mathcal{H}_{-\infty} \,:\, \pi_{-\infty}(h)\, a = a \text{ for all } h \in H\}.$$

We shall say that π can be realized on a *Hilbert subspace* of $D'(G/H)$ if there is a continuous linear injection $j : \mathcal{H} \to D'(G/H)$ such that

$$j\,\pi(g) = L_g\, j$$

for all $g \in G$ (L_g denotes left-translation by g). The space $j(\mathcal{H})$ is called an (invariant) *Hilbert subspace* of $D'(G/H)$. We have the following theorem.

Theorem 8.2.1. *The unitary representation $(\pi, \mathcal{H})$ can be realized on a Hilbert subspace of $D'(G/H)$ if and only if $\mathcal{H}_{-\infty}^{H}$ contains cyclic vectors. There is a one-to-one correspondence between the cyclic vectors of $\mathcal{H}_{-\infty}^{H}$ and the continuous linear injections $j : \mathcal{H} \to D'(G/H)$ satisfying $j\,\pi(g) = L_g\, j$ ($g \in G$). To a cyclic vector a in $\mathcal{H}_{-\infty}^{H}$ corresponds the injection j for which $j^* : D(G/H) \to \mathcal{H}$ is given by $j^*(\varphi) = \pi_{-\infty}(\varphi_0)\, a$.*

The proof is based on the decomposition lemma. Let $\{f_n\}$ be an approximate unit as in Section 8.1. Let j^* be the dual mapping of j, so $j^* : D(G/H) \to \mathcal{H}$, and define $j_0^* : D(G) \to \mathcal{H}$ by $j_0^*(\varphi_0) = j^*(\varphi)$, where φ and φ_0 correspond as in (8.2.1). Then clearly j_0^* is continuous and $j_0^* L_g = \pi(g)\, j_0^*$ for all $g \in G$. Let $v \in \mathcal{H}_\infty$ be of the form $v = \pi(\psi_0)\, w$ for some $\psi_0 \in D(G)$, $w \in \mathcal{H}$. Then we have

$$\begin{aligned}
\langle j_0^*(f_n), v\rangle &= \langle j_0^*(f_n), \pi(\psi_0) w\rangle \\
&= \langle \pi(\widetilde{\psi}_0) j_0^*(f_n), w\rangle \\
&= \langle j_0^*(\widetilde{\psi}_0 * f_n), w\rangle .
\end{aligned}$$

This latter expression tends to $\langle j_0^*(\widetilde{\psi}_0), w\rangle$ when $n \to \infty$. So $\lim_{n\to\infty}\langle j_0^*(f_n), v\rangle$ exists and defines, by the special properties of the topology of $\mathcal{H}_{-\infty}$, an element a of $\mathcal{H}_{-\infty}$: $\langle a, v\rangle = \lim_{n\to\infty}\langle j_0^*(f_n), v\rangle$, $v \in \mathcal{H}_\infty$ (weak convergence of a sequence of elements in $\mathcal{H}_{-\infty}$ defines an element of $\mathcal{H}_{-\infty}$, by the closed graph theorem). Furthermore

$$\begin{aligned}
\langle j_0^*(\varphi_0), v\rangle &= \lim_{n\to\infty} \langle j_0^*(\varphi_0 * f_n), v\rangle \\
&= \lim_{n\to\infty} \langle j_0^*(f_n), \pi(\widetilde{\varphi}_0) v\rangle \\
&= \langle a, \pi(\widetilde{\varphi}_0) v\rangle \qquad (\varphi_0 \in D(G),\ v \in \mathcal{H}_\infty).
\end{aligned}$$

Hence $j_0^*(\varphi_0) = \pi_{-\infty}(\varphi_0)\, a$, and thus $j^*(\varphi) = \pi_{-\infty}(\varphi_0)\, a$. Clearly, the vector a is cyclic since j is injective. Given j, the vector $a \in \mathcal{H}_{-\infty}$ is unique, and clearly H-fixed. Indeed, just replace φ_0 by $R_h\varphi_0$, where for $h \in H$, $R_h\varphi$ is defined by $R_h\varphi(g) = \varphi_0(gh)$ ($g \in G$). The other statements of the theorem are obvious.

Let π be a unitary representation realized on a Hilbert subspace of $D'(G/H)$ and $j : \mathcal{H} \to D'(G/H)$ the corresponding injection. Denote by ξ_π the cyclic vector in $\mathcal{H}_{-\infty}^H$ defined by Theorem 8.2.1. Then we set

$$\langle T, \varphi_0 \rangle = \langle \xi_\pi, \ \pi_{-\infty}(\varphi_0)\, \xi_\pi \rangle \qquad (\varphi_0 \in D(G)).$$

The anti-linear functional T is a left and right H-invariant distribution on G. We call T the *reproducing distribution* of π (or $\mathcal{H}$); it is positive-definite, bi-H-invariant and satisfies

$$\|j^*\varphi\|^2 = \langle T, \widetilde{\varphi}_0 * \varphi_0 \rangle \tag{8.2.2}$$

for all $\varphi_0 \in D(G)$. Given a positive-definite, bi-H-invariant distribution T on G, formula (8.2.2) shows the way to define a G-invariant Hilbert subspace of $D'(G/H)$ with T as reproducing distribution. Indeed, let V be the space $D(G/H)$ provided with the scalar product

$$(\varphi \mid \psi) = \langle T, \widetilde{\varphi}_0 * \psi_0 \rangle.$$

Let V_0 be the subspace of V consisting of the elements of length zero and define $\mathcal{H}$ to be the completion of V/V_0 and j^* the natural projection $D(G/H) \to \mathcal{H}$. Then clearly

$$\|j^*\varphi\|^2 = \langle T, \widetilde{\varphi}_0 * \varphi_0 \rangle$$

for all $\varphi_0 \in D(G)$.

An easy calculation shows that jv is a C^∞ function for all $v \in \mathcal{H}_\infty$. Actually

$$jv(x) = \langle \pi(g^{-1})v, \xi_\pi \rangle \qquad (x = gH \in G/H). \tag{8.2.3}$$

Note that j can be extended to $\mathcal{H}_{-\infty}$ (as anti-dual of $j^* : D(G/H) \to \mathcal{H}_\infty$). Then $j(\xi_\pi)$ is precisely equal to the reproducing distribution T, considered as an H-invariant element of $D'(G/H)$. One has

$$jj^*(\varphi) = \varphi_0 * T \tag{8.2.4}$$

for all $\varphi_0 \in D(G)$.

Let π_1 and π_2 be two unitary representations on $\mathcal{H}_1$ and $\mathcal{H}_2$ respectively, realized on $D'(G/H)$ by means of the G-equivariant injections j_1 and j_2. Denote the associated reproducing distributions by T_1 and T_2. Assume that $T_1 = T_2$. Then we have $\|j_1^*\varphi\|^2 = \|j_2^*\varphi\|^2$ for all $\varphi \in D(G/H)$, thus U given by $U(j_1^*\varphi) = j_2^*\varphi$ is well-defined and can be extended to a unitary operator from $\mathcal{H}_1$ onto $\mathcal{H}_2$, commuting with the actions of G. Moreover, $j_1 = j_2 \circ U$. We say that the triples $(\pi_1, \mathcal{H}_1, j_1)$ and $(\pi_2, \mathcal{H}_2, j_2)$ are *equivalent*. We shall alternatively call the Hilbert subspaces $j_1(\mathcal{H}_1)$ and $j_2(\mathcal{H}_2)$ equivalent.

Summarizing we have

Proposition 8.2.2. *The correspondence $\mathcal{H} \leftrightarrow T$ that associates to each invariant Hilbert subspace its reproducing distribution is a bijection between the set of equivalence classes of G-invariant Hilbert subspaces of $D'(G/H)$ and the set of bi-H-invariant positive-definite distributions on G.*

A bi-H-invariant positive-definite distribution T is said to be *extremal* if for any bi-H-invariant positive-definite distribution T_1 with $T_1 \leq T$ one has $T_1 = \lambda T$ for some $\lambda \geq 0$. Here $T_1 \leq T$ means $\langle T_1, \widetilde{\varphi}_0 * \varphi_0 \rangle \leq \langle T, \widetilde{\varphi}_0 * \varphi_0 \rangle$ for all $\varphi_0 \in D(G)$.

One has the following proposition.

Proposition 8.2.3. *Let π be a unitary representation, realized on a Hilbert subspace $j(\mathcal{H})$ of $D'(G/H)$ and let T be its reproducing distribution. Then the following two statements are equivalent:*

(i) *π is irreducible,*

(ii) *T is extremal.*

Let π be irreducible and let T_1 be a positive-definite bi-H-invariant distribution on G with $T_1 \leq T$. Then T_1 is the reproducing distribution of an invariant Hilbert subspace of $D'(G/H)$, thus of a unitary representation on a Hilbert space $\mathcal{H}_1$ realized on $D'(G/H)$. Call the natural G-equivariant injection j_1. The mapping A given by $A(j^*\varphi) = j_1^*\varphi$ is well-defined and continuous since $T_1 \leq T$ implies $\|j_1^*\varphi\|_{\mathcal{H}_1} \leq \|j^*\varphi\|_{\mathcal{H}}$ for all $\varphi \in D(G/H)$. Extend A to $\mathcal{H}$. Then $Aj^* = j_1^*$, hence $jA^*Aj^* = j_1 j_1^*$. Now A^*A is in $\mathrm{End}(\mathcal{H})$ and commutes with $\pi(G)$, so it is a positive scalar λ, by Schur's lemma. Hence $j_1 j_1^* = \lambda j j^*$ and thus $T_1 = \lambda T$.

Conversely, let T be extremal. Suppose that $\mathcal{H}_1 \subset \mathcal{H}$ is a closed invariant subspace and let P denote the orthogonal projection of $\mathcal{H}$ onto $\mathcal{H}_1$. Then $j_1^*\varphi = P\, j^*\varphi$ where j_1 is the injection j restricted to $\mathcal{H}_1$ and $\varphi \in D(G/H)$. Hence

$$\|j_1^*\varphi\|^2 = \|P\, j^*\varphi\|^2 \leq \|j^*\varphi\|^2,$$

or $\langle T_1, \widetilde{\varphi}_0 * \varphi_0 \rangle \leq \langle T, \widetilde{\varphi}_0 * \varphi_0 \rangle$ for all $\varphi_0 \in D(G)$. Here T_1 is the reproducing distribution of $\mathcal{H}_1$. Because T is extremal, we get $T_1 = \lambda\, T$ for some $\lambda \geq 0$, hence $\|P\, j^*\varphi\|^2 = \lambda\, \|j^*\|^2$, so $P = 0$ or $P = I$. So $\mathcal{H}_1 = \{0\}$ or $\mathcal{H}_1 = \mathcal{H}$. Hence π is irreducible.

Denote by Γ_G the cone of bi-H-invariant positive-definite distributions and by $\mathrm{ext}(\Gamma_G)$ the subset of the extremal distributions. It is easily seen that this set can be identified with the set of extremal rays of Γ_G.

We can choose a so-called admissible parametrization $s \mapsto T_s$ $(s \in S)$ of $\mathrm{ext}(\Gamma_G)$, where S is a topological Hausdorff space, see [49]. Fix such a space S.

Proposition 8.2.4 ([49, Proposition 9]). *For any $T \in \Gamma_G$ there exists a (not necessarily unique) Radon measure m on S such that*

$$\langle T, \varphi_0 \rangle = \int_S \langle T_s, \varphi_0 \rangle \, dm(s)$$

for all $\varphi_0 \in D(G)$.

The measure m occurring in Proposition 8.2.4 is a special kind of a so-called Borel measure, being a measure on the Borel subsets of S (see [43]). This measure concept is more general than the one we introduced for locally compact spaces in Section 4.1. We do not elaborate on it; it is a topic in regular courses on measure theory, see, e.g., [40]. The result of this proposition, except for the fixed parametrization independent of T, has been obtained by L. Schwartz and K. Maurin. See [49] for references. The proof of the proposition is also sketched there. A preliminary version was obtained by Thomas who applied Choquet's theorem, a generalization of Krein–Milman's theorem. The result of Proposition 8.2.4 however is more precise.

We shall be mainly interested here in the decomposition of the distribution $T \in \Gamma_G$ given by

$$\langle T, \varphi_0 \rangle = \int_H \varphi_0(h) \, dh \qquad (\varphi_0 \in D(G)),$$

which corresponds to the δ-distribution at the origin of G/H. This decomposition might be called a Plancherel formula for G/H.

We conclude this section by showing that any $T \in \mathrm{ext}(\Gamma_G)$ is an *eigendistribution of all bi-G-invariant differential operators* on G. We begin with a summary of some additional well-known facts from functional analysis.

Let A be a densely defined linear operator from a Hilbert space $\mathcal{H}_1$ into a Hilbert space $\mathcal{H}_2$ and let by A^* be its adjoint. Then one has:

(1) A^* is a closed operator. A has a closure if and only if the domain $D(A^*)$ of A^* is dense.

(2) If A is a closed operator, then $D(A^*A)$ is dense and A^*A is a positive self-adjoint operator.

(3) If A is a closed operator, then it has a unique polar decomposition: $A = U(A^*A)^{1/2}$, where U is a bounded operator $\mathcal{H}_1 \to \mathcal{H}_2$, a so-called partial isometry (cf. [34]).

Let now $(\pi, \mathcal{H})$ be a unitary representation of G on a Hilbert space $\mathcal{H}$ with scalar product denoted by $(\,|\,)$. For $X \in \mathfrak{g}$ and $v, w \in \mathcal{H}_\infty$ one has

$$(\pi(\exp tX)v \mid w) = (v \mid \exp(-tX)w),$$

hence the domain of $\pi_\infty(X)^*$ contains $\mathcal{H}_\infty$ and $\pi_\infty(X)^* w = \pi_\infty(-X)w$ for $w \in \mathcal{H}_\infty$. From this we easily see that for any $D \in U(\mathfrak{g}_c)$ one has

$$\pi_\infty(D)^* w = \pi_\infty(\widetilde{D})w \qquad (w \in \mathcal{H}_\infty)$$

for some well-defined $\widetilde{D} \in U(\mathfrak{g}_c)$. Hence $\pi_\infty(D)$ has a closed extension and $\pi_\infty(D)^* \pi_\infty(D)$ is essentially self-adjoint.

Lemma 8.2.5. *Let $(\pi, \mathcal{H})$ be an irreducible unitary representation of G and $z \in U(\mathfrak{g}_c)$ a bi-G-invariant differential operator on G. Then $\pi_\infty(z)$ is a scalar on $\mathcal{H}_\infty$.*

The operator $\pi_\infty(\widetilde{z})\pi_\infty(z) = A$ is defined on $\mathcal{H}_\infty$, it is essentially self-adjoint and commutes with $\pi_\infty(g)$ for all $g \in G$. The same holds for the closure of A. The projections of the spectral decomposition of the closure of A commute with all $\pi(g)$ $(g \in G)$, are thus equal to zero or I. Hence A is a positive scalar. From the polar decomposition of the closure of $\pi_\infty(z)$ we see now that $\pi_\infty(z)$ is a bounded operator, commuting with all $\pi(g)$ $(g \in G)$, so $\pi_\infty(z)$ is a scalar operator.

Let now $T \in \operatorname{ext}(\Gamma_G)$ and let $\mathcal{H} \subset D'(G/H)$ be the associated minimal invariant Hilbert subspace. Then $T \in \mathcal{H}_{-\infty}^H$ and thus $\pi_{-\infty}(\widetilde{z})\,T = \lambda\,T$ for z as in Lemma 8.2.5 and λ a complex scalar. So $\widetilde{z}\,T = \lambda\,T$, since the representation π is just left-translation. Hence T is an eigendistribution for all bi-G-invariant differential operators on G, since with z also $\widetilde{z}$ is bi-G-invariant. So we have shown:

Proposition 8.2.6. *Any $T \in \operatorname{ext}(\Gamma_G)$ is an eigendistribution of all bi-G-invariant differential operators on G.*

8.3 Generalized Gelfand pairs

We are now going to generalize the classical notion of a Gelfand pair. We adopt the notations of Section 8.2.

Definition 8.3.1. The pair (G, H) is called a *generalized Gelfand pair* if for each irreducible unitary representation π on a Hilbert space $\mathcal{H}$ one has $\dim \mathcal{H}_{-\infty}^H \leq 1$.

If the subgroup H is compact, then $\mathcal{H}^H$ is dense in $\mathcal{H}_{-\infty}^H$, so the condition $\dim \mathcal{H}^H \leq 1$ is equivalent to $\dim \mathcal{H}_{-\infty}^H \leq 1$. A classical Gelfand pair is thus also a generalized Gelfand pair. Clearly, $C_c(G)^\#$ is reduced to zero if H is non-compact, so that the condition that $C_c(G)^\#$ should be abelian, is automatically satisfied. But how to find then a criterion in order that the condition in the definition is satisfied? The following result is proved in [49].

Proposition 8.3.2. *The following statements are equivalent:*

(i) *(G, H) is a generalized Gelfand pair.*

(ii) *Any unitary representation $(\pi, \mathcal{H})$ that can be realized on a Hilbert subspace of $D'(G/H)$ is multiplicity free, i.e. the commutant of $\pi(G) \subset \mathrm{End}(\mathcal{H})$ is abelian.*

(iii) (Bochner–Schwartz–Godement theorem) *For every $T \in \Gamma_G$ there exists a unique Radon measure m on S such that*

$$\langle T, \varphi_0 \rangle = \int_S \langle T_s, \varphi_0 \rangle \, dm(s)$$

for all $\varphi_0 \in D(G)$.

We shall only show the implication (ii) $\Rightarrow$ (i).

Let $(\pi, \mathcal{H})$ be an irreducible unitary representation and assume that $\dim \mathcal{H}_{-\infty}^{H} > 1$. Then there exist at least two linearly independent G-equivariant injections from $\mathcal{H}$ to $D'(G/H)$. Denote these injections by j_1 and j_2 and set $\mathcal{H}_1 = j_1(\mathcal{H}), \mathcal{H}_2 = j_2(\mathcal{H})$. Provide $\mathcal{H}_1$ and $\mathcal{H}_2$ with the Hilbert space structure inherited from $\mathcal{H}$ by j_1 and j_2 respectively. We clearly have $\mathcal{H}_1 \neq \mathcal{H}_2$ (as subsets of $D'(G/H)$). Indeed, suppose that $\mathcal{H}_1 = \mathcal{H}_2$ and consider $j_1^{-1} j_2$. This is a closed linear mapping, hence, by the closed graph theorem, a continuous mapping commuting with $\pi(G)$. By Schur's lemma we then get $j_2 = \lambda j_1$ for some $\lambda \in \mathbb{C}$ and we obtain a contradiction. Now consider the intersection $\mathcal{H}_1 \cap \mathcal{H}_2$. We shall show that it contains only zero. Provide this space with the Hilbert space structure with squared norm $\|v\|^2 = \|v\|_{\mathcal{H}_1}^2 + \|v\|_{\mathcal{H}_2}^2$. Then G acts unitarily on $\mathcal{H}_1 \cap \mathcal{H}_2$ again. Denote by i the identification $\mathcal{H}_1 \cap \mathcal{H}_2 \subset \mathcal{H}_1$. Then i is a continuous mapping and $ii^* : \mathcal{H}_1 \to \mathcal{H}_1$ commutes with the action of G, so $ii^* = \lambda I$ for some positive scalar λ, hence $i^* = \lambda I$, so $\mathcal{H}_1 \cap \mathcal{H}_2 \supset \mathcal{H}_1$ if $\lambda \neq 0$. The same then holds for $\mathcal{H}_2$, so $\mathcal{H}_1 = \mathcal{H}_2$. This contradicts our assumption, so $\lambda = 0$, and thus, as said before, $\mathcal{H}_1 \cap \mathcal{H}_2 = \{0\}$. Consider finally the Hilbert subspace $\mathcal{H}_1 + \mathcal{H}_2$, provided with the scalar product with squared norm $\|v\|^2 = \|v_1\|_{\mathcal{H}_1}^2 + \|v_2\|_{\mathcal{H}_2}^2$ if $v = v_1 + v_2$ ($v_1 \in \mathcal{H}_1$, $v_2 \in \mathcal{H}_2$). Since $\mathcal{H}_1$ and $\mathcal{H}_2$ are irreducible and equivalent under the unitary G-action, the commutant of $\mathcal{H}_1 + \mathcal{H}_2$ is clearly isomorphic to the algebra $\mathrm{M}_2(\mathbb{C})$ of 2×2 matrices over $\mathbb{C}$, which is not abelian. So $\dim \mathcal{H}_{-\infty}^{H} \leq 1$.

We now give a *criterion*, which will turn out to be very useful, to ensure condition (ii) of Proposition 8.3.2. It is due to Thomas.

Proposition 8.3.3 ([49]). *Let $J : D'(G/H) \to D'(G/H)$ be an anti-automorphism. If $J(\mathcal{H}) = \mathcal{H}$ for all G-invariant or all minimal G-invariant Hilbert subspaces of $D'(G/H)$, then (G, H) is a generalized Gelfand pair.*

The hypothesis $J\mathcal{H} = \mathcal{H}$ means: if j is the injection $\mathcal{H} \to D'(G/H)$, then there exists a anti-unitary mapping $U : \mathcal{H} \to \mathcal{H}$ such that $Jj = jU$ on $\mathcal{H}$. In that case we have $Jjj^*J^* = jj^*$. The converse is also true: if $Jjj^*J^* = jj^*$

then $\|j^*J^*\varphi\|^2 = \|j^*\varphi\|^2$ for all $\varphi \in D(G/H)$, hence $j^*\varphi \mapsto j^*J^*\varphi$ is a well-defined mapping, which can be extended to an anti-unitary mapping $U^* : \mathcal{H} \to \mathcal{H}$, so $U^*j^* = j^*J^*$ and therefore $jU = Jj$ and $J(\mathcal{H}) = \mathcal{H}$.

See [49, Theorem E] for the proof. We reproduce it here. If the minimal invariant spaces are invariant under J, then so are the others, for instance by Proposition 8.2.4. Let $\mathcal{H}$ be a G-invariant Hilbert subspace and let $\mathcal{A}$ be the commutant of $\pi(G)$. Denote by j the inclusion of $\mathcal{H}$ into $D'(G/H)$. Let $A \in \mathcal{A}$ be a positive operator, $\mathcal{H}_1 = (\ker A)^{\perp} \subset \mathcal{H}$ and $A^{1/2}$ the continuous injection $\mathcal{H}_1 \to \mathcal{H}$. The space $\mathcal{H}_1$ is a $\pi(G)$-invariant Hilbert subspace, $j \circ A^{1/2}$ the inclusion mapping. By hypothesis $Jj \circ A \circ j^*J^* = jAj^*$, so $jU \circ A \circ U^{-1}j^* = jAj^*$ for an anti-unitary mapping $U : \mathcal{H} \to \mathcal{H}$, so $UAU^{-1} = A$ on $\mathcal{H}$. Thus for any $A \in \mathcal{A}$ we have $A^* = UAU^{-1}$, since U is anti-unitary. Thus for $A, B \in \mathcal{A}$ we have $(AB)^* = A^*B^*$, hence $\mathcal{A}$ is abelian.

An interesting way to construct such an anti-automorphism J is as follows. Denote by $D'(G,H)$ the space of right-H-invariant distributions on G, with the relative topology of $D'(G)$. It is well known that $D'(G/H)$ can be identified with $D'(G,H)$. To see this, apply therefore the same method as in Section 4.5, for example. Let τ be an involutive automorphism of G that leaves H stable, i.e. $\tau(H) = H$. Then define $JT = \overline{T}^{\tau}$ on $D'(G,H)$. To check that the hypothesis of Proposition 8.3.3 is satisfied, it suffices to show that $JT = T$ for all (extremal) positive-definite bi-H-invariant distributions T on G. So we have:

Corollary 8.3.4 ([49], [53]). *Let τ be an involutive automorphism of G that leaves H stable, i.e. $\tau(H) = H$. Define $JT = \overline{T}^{\tau}$ for all $T \in D'(G,H)$. If $JT = T$ for all (extremal) positive-definite bi-H-invariant distributions on G, then (G,H) is a generalized Gelfand pair.*

Some examples of generalized Gelfand pairs

(1) The pair $(G \times G, \operatorname{diag}(G \times G))$ with G again unimodular, is a generalized Gelfand pair. Indeed, apply Corollary 8.3.4 with $\tau(x,y) = (y,x)$. Identifying $G \times G/\operatorname{diag}(G \times G)$ with G by means of $(x,y) \mapsto xy^{-1}$, we have to check that $\widetilde{T} = T$ for any $\operatorname{Ad}(G)$-invariant positive-definite distribution on G. However, this property is already satisfied for any positive-definite distribution on G. Indeed, for such a distribution T one has

$$\langle T, \widetilde{\varphi}_0 * \varphi_0 \rangle = \langle \widetilde{T}, \widetilde{\varphi}_0 * \varphi_0 \rangle$$

because these expressions are real-valued, and then, by writing

$$\begin{aligned} 4\,(\widetilde{\varphi}_0 * \psi_0) &= [\widetilde{(\varphi_0 + \psi_0)} * (\varphi_0 + \psi_0) - \widetilde{(\varphi_0 - \psi_0)} * (\varphi_0 - \psi_0)] \\ &= -i\,[\widetilde{(\varphi_0 + i\psi_0)} * (\varphi_0 + i\psi_0) - \widetilde{(\varphi_0 - i\psi_0)} * (\varphi_0 - i\psi_0)], \end{aligned}$$

we get

$$\langle T, \widetilde{\varphi}_0 * \psi_0 \rangle = \langle \widetilde{T}, \widetilde{\varphi}_0 * \psi_0 \rangle$$

for all $\varphi_0, \psi_0 \in D(G)$, and thus

$$\langle \varphi_0 * T, \psi_0 \rangle = \langle \varphi_0 * \widetilde{T}, \psi_0 \rangle$$

and hence $\varphi_0 * T = \varphi_0 * \widetilde{T}$ for all $\varphi_0 \in D(G)$. These expressions are C^∞ functions, and evaluating them at the unit element we get $T = \widetilde{T}$.

(2) The pair (G, H) where G is the semi-direct product $G = H \ltimes N$, N a closed abelian normal subgroup, e.g. $G = \mathrm{O}(1, n) \ltimes \mathbb{R}^{n+1}$. Take $\tau(hn) = hn^{-1}$.

(3) Let G and H be as usual, but now with H compact. Choose τ as in Proposition 6.1.3 (so $\tau = \theta$). We know already that (G, H) is a (generalized) Gelfand pair. It can also be checked that the criterion (Corollary 8.3.4) applies.

Let (G, H) be a generalized Gelfand pair and $\mathcal{H}$ an invariant Hilbert subspace of $D'(G/H)$, with reproducing distribution T given by

$$\langle T, \varphi \rangle = \langle \xi_\pi, \pi_{-\infty}(\varphi_0)\xi_\pi \rangle.$$

For any $\mathrm{Ad}(H)$-invariant $D \in U(\mathfrak{g}_c)$ one clearly gets $\pi_{-\infty}(D)\xi_\pi = \lambda\, \xi_\pi$ for some complex scalar λ. Then one easily get that T is an *eigendistribution* for all left G- and right H-invariant differential operators D on G.

Here is a result of a practical nature. Compare it with Lemma 6.2.3.

Proposition 8.3.5. *Let π be a unitary representation of G on $\mathcal{H}$ that can be realized on $D'(G/H)$. If $\dim \mathcal{H}_{-\infty}^H = 1$ then π is irreducible.*

Let $A \in \mathrm{End}(\mathcal{H})$ commute with $\pi(G)$. Clearly, A and A^* leave $\mathcal{H}_\infty$ invariant and are also continuous linear operators with respect to the topology of $\mathcal{H}_\infty$. So A acts on $\mathcal{H}_{-\infty}$ as a continuous operator, commuting with $\pi_{-\infty}(G)$. Hence A leaves $\mathcal{H}_{-\infty}^H$ invariant, and thus acts as a scalar λ there. We now have

$$A\pi_{-\infty}(\varphi_0)\, a = \pi_{-\infty}(\varphi_0) A\, a = \lambda\, \pi_{-\infty}(\varphi_0)\, a$$

for all $\varphi_0 \in D(G)$, $a \in \mathcal{H}_{-\infty}^H$. hence $A = \lambda I$, because all $a \in \mathcal{H}_{-\infty}^H$, $a \neq 0$ are cyclic. Schur's lemma now implies that π is irreducible.

Some results on the Bochner–Schwartz–Godement theorem

(1) **Theorem** (Bochner–Schwartz, see [43, p. 276]). *A distribution T on $\mathbb{R}^n$ is positive-definite if and only if it is the Fourier transform of a positive tempered measure μ:*

$$\langle T, \varphi \rangle = \int_{\mathbb{R}^n} (\mathcal{F}\varphi)(-y)\, d\mu(y) \quad (\varphi \in D(\mathbb{R}^n)),$$

where $\mathcal{F}$ denotes the Euclidean Fourier transform on $\mathbb{R}^n$.

(2) A similar statement for *Fourier series* is easily shown, since any distribution on $\mathbb{T}$ (or $\mathbb{T}^n$) is compactly supported. If T is positive-definite and $a_n = \langle T, e^{2\pi i n x}\rangle$, then $\{a_n\}$ is a tempered sequence and $a_n \geq 0$ for all n. The converse is clear.

(3) The Bochner–Schwartz–Godement theorem for the *Euclidean motion group* $G = SO(n) \ltimes \mathbb{R}^n$ reads as follows.

Let T be an $SO(n)$-invariant, radial, positive-definite distribution on $\mathbb{R}^n$, then we have

$$\langle T, \varphi\rangle = \int_0^\infty \widehat{\varphi}(s)\, s^{n-1}\, d\mu(s)$$

with μ an even, positive, tempered measure on $\mathbb{R}$, and conversely. This follows easily from Section 7.1.

(4) For the case $(SO_0(1,n), SO(n))$ we have to introduce complicated spaces like $\mathcal{S}(G)$, the Schwartz space of $G = SO_0(1,n)$, which is outside the scope of this book.

8.4 Invariant Hilbert subspaces of $L^2(G/H)$

We keep the notation of the previous section, but for the time being (G, H) need not be a generalized Gelfand pair. Let π be a unitary representation of G on a Hilbert space $\mathcal{H}$ that can be realized on an invariant Hilbert subspace of $D'(G/H)$, and let $j : \mathcal{H} \to D'(G/H)$ be the corresponding injection. Define ξ_π, T, j^* as before. Let $\varphi \in D(G/H)$ and $\varphi_0 \in D(G)$ be related as in (8.2.1).

Proposition 8.4.1. *The following conditions are equivalent:*

(i) $j(\mathcal{H}) \subset L^2(G/H)$.

(ii) *There exists a constant* $c > 0$ *such that, for all* $\varphi_0 \in D(G)$,

$$|\langle T, \widetilde{\varphi}_0 * \varphi_0\rangle| \leq c\, \|\varphi\|_2^2.$$

(i) $\Rightarrow$ (ii). The mapping $j : \mathcal{H} \to L^2(G/H)$ is closed and everywhere defined on $\mathcal{H}$, hence continuous by the closed graph theorem. This implies that $j^* : D(G/H) \to \mathcal{H}$ is continuous in the L^2-topology of $D(G/H)$, so (ii) follows.

(ii) $\Rightarrow$ (i). Clearly, (ii) implies that j^* is continuous with respect to the L^2-topology on $D(G/H)$. Write $\langle j^*\varphi, v\rangle = \langle \varphi, jv\rangle$ with $\varphi \in D(G/H)$, $v \in \mathcal{H}$, and observe that $jv \in L^2(G/H)$, and hence $j(\mathcal{H}) \subset L^2(G/H)$.

Observe that condition (i) implies that j, considered as a mapping from $\mathcal{H}$ to $L^2(G/H)$, is continuous. We shall say that π belongs to the *relative discrete series* of G (with respect to H) if π is irreducible and satisfies condition (i), i.e. if π can be realized on a Hilbert subspace of $L^2(G/H)$. We shall occasionally use the terminology: π *is square-integrable* mod H.

Proposition 8.4.2. *Let π be an irreducible unitary representation of G on $\mathcal{H}$ that can be realized on an invariant Hilbert subspace of $D'(G/H)$. Let $j : \mathcal{H} \to D'(G/H)$ be the corresponding G-equivariant injection. The following statements are equivalent:*

(i) *π is square-integrable mod H.*

(ii) *$j(\mathcal{H})$ is a closed linear subspace of $L^2(G/H)$.*

(iii) *$j(v) \in L^2(G/H)$ for at least one non-zero element $v \in \mathcal{H}$.*

It is sufficient to prove the implication (iii) $\Rightarrow$ (ii). Let $V = \{w \in \mathcal{H} : j(w) \in L^2(G/H)\}$. Clearly, V is a G-stable and non-zero linear subspace of $\mathcal{H}$, hence dense in $\mathcal{H}$. Now observe that $j : V \to L^2(G/H)$ is a closed linear operator: if $w_k \to w$ ($w_k \in V, w \in \mathcal{H}$) and $jw_k \to f$ in $L^2(G/H)$, then obviously $jw \in D'(G/H)$ is equal to f as a distribution. Polar decomposition of j and applying Schur's lemma yields: j can be extended to a continuous linear operator $\mathcal{H} \to L^2(G/H)$ (a scalar times a partial isometry) with closed image.

Remark 8.4.3. It also follows that there is a constant $c > 0$ such that $\|jv\|_2 = c\,\|v\|$ for all $v \in \mathcal{H}$.

One has the following orthogonality relations.

Proposition 8.4.4. *Let π, π' be irreducible unitary representations on $\mathcal{H}$ and $\mathcal{H}'$ respectively, both belonging to the relative discrete series. Define T, T' and $\xi_\pi, \xi_{\pi'}$ as usual. Then one has:*

(i) *$\int_{G/H} \langle \pi(x^{-1})v, \xi_\pi\rangle \overline{\langle \pi'(x^{-1})v', \xi_{\pi'}\rangle}\, dx = 0$ for all $v \in \mathcal{H}_\infty$, $v' \in \mathcal{H}'_\infty$ if π is not equivalent to π'.*

(ii) *There exists a constant $d_\pi > 0$, only depending on T, such that*

$$\int_{G/H} \langle \pi(x^{-1})v, \xi_\pi\rangle \overline{\langle \pi(x^{-1})v', \xi_\pi\rangle}\, dx = d_\pi^{-1}\, \langle v, v'\rangle$$

for all $v, v' \in \mathcal{H}_\infty$.

Consider the invariant sesqui-linear form on $\mathcal{H}_\infty \times \mathcal{H}'_\infty$ given by

$$(v, v') = \int_{G/H} \langle \pi(x^{-1})v, \xi_\pi\rangle \overline{\langle \pi'(x^{-1})v', \xi_{\pi'}\rangle}\, dx.$$

This form is continuous with respect to the topology on $\mathcal{H} \times \mathcal{H}'$, so there is a bounded linear operator $A : \mathcal{H} \to \mathcal{H}'$, commuting with the G-action, such that $(v, v') = \langle Av, v'\rangle$ ($v \in \mathcal{H}_\infty$, $v' \in \mathcal{H}'_\infty$), so $A = 0$ in case (i). In case (ii), A is a positive (non-zero) scalar, say d_π^{-1}, by Schur's lemma. The "only" dependency of d_π on T follows from the formula: $\|jj^*\varphi_0\|_2^2 = d_\pi^{-1}\, \|j^*\varphi_0\|^2$, so $\|\varphi_0 * T\|_2^2 = d_\pi^{-1}\langle T, \widetilde{\varphi}_0 * \varphi_0\rangle$ for all $\varphi_0 \in D(G)$.

Remark 8.4.5. Observe that $\|jv\|_2 = d_\pi^{-1/2}\,\|v\|$ for all $v \in \mathcal{H}_\infty$. So the constant c in Remark 8.4.3 is equal to $d_\pi^{-1/2}$.

The constant d_π is called the *formal degree* of π. It depends on the choice of j (or T). Once a canonical choice of j (or T) is possible, d_π has a more realistic meaning.

Special cases

1. *The group case:* $G \times G/\operatorname{diag}(G \times G)$

We begin with some elementary functional analysis on *Hilbert–Schmidt* and *trace-class* operators on a Hilbert space (see [36, p. 99]).

Let $\mathcal{H}$ be a separable Hilbert space and let $\operatorname{End}(\mathcal{H})$, as usual, denote the set of continuous linear operators on $\mathcal{H}$.

Lemma 8.4.6. *Let* $\{e_j\}$ *and* $\{f_k\}$ *be complete orthonormal sets in* $\mathcal{H}$. *Given* $A \in \operatorname{End}(\mathcal{H})$, *set*

$$T_e(A) = \sum_{j=1}^{\infty} \|Ae_j\|^2, \quad T_f(A) = \sum_{k=1}^{\infty} \|Af_k\|^2$$

($0 \le T_e(A), T_f(A) \le \infty$). *Then we have* $T_e(A) = T_e(A^*) = T_f(A)$.

Observe that

$$T_e(A) = \sum_{j=1}^{\infty} \|Ae_j\|^2 = \sum_{j=1}^{\infty}\left(\sum_{k=1}^{\infty} |(Ae_j \mid f_k)|^2\right)$$
$$= \sum_{k=1}^{\infty}\left(\sum_{j=1}^{\infty} |(A^* f_k \mid e_j)|^2\right) = T_f(A^*).$$

Therefore $T_e(A) = T_e(A^*)$ and $T_e(A^*) = T_f(A)$.

Definition 8.4.7. The operator $A \in \operatorname{End}(\mathcal{H})$ is called a *Hilbert–Schmidt operator* if $\sum_{j=1}^{\infty} \|Ae_j\|^2 < \infty$ for some complete orthonormal set $\{e_j\}$ in $\mathcal{H}$. We set

$$\|A\|_{\mathrm{HS}}^2 = \sum_{j=1}^{\infty} \|Ae_j\|^2,$$

and call $\|A\|_{\mathrm{HS}}$ the *Hilbert–Schmidt norm* of A.

By Lemma 8.4.6, the definition of the Hilbert–Schmidt norm does not depend on the particular choice of the set $\{e_j\}$. We list here a few properties of Hilbert–Schmidt operators.

Proposition 8.4.8. (a) $\|A\| \leq \|A\|_{\mathrm{HS}}$ *for all Hilbert–Schmidt operators* A.

(b) *Hilbert–Schmidt operators are compact operators.*

(c) *The set of all Hilbert–Schmidt operators is a two-sided* $*$*-ideal in the algebra* $\mathrm{End}(\mathcal{H})$.

(d) $\|AT\|_{\mathrm{HS}} \leq \|A\|_{\mathrm{HS}}\|T\|$ *and* $\|TA\|_{\mathrm{HS}} \leq \|T\|\|A\|_{\mathrm{HS}}$ *for all Hilbert–Schmidt operators* A *and all* $T \in \mathrm{End}(\mathcal{H})$.

Definition 8.4.9. $A \in \mathrm{End}(\mathcal{H})$ is called a *trace-class operator* if A is a finite sum of operators of the form BC, where B and C are Hilbert–Schmidt operators.

Lemma 8.4.10. *Let* A *be a trace-class operator. Then* A *can be written as a finite sum*

$$A = \sum_j c_j A_j^* A_j$$

where the A_j *are Hilbert–Schmidt operators and the* c_j *are complex scalars.*

This follows from the relation

$$\begin{aligned} 4A^*B = {} & [(A+B)^*(A+B) - (A-B)^*(A-B)] \\ & - i[(A+iB)^*(A+iB) - (A-iB)^*(A-iB)] \end{aligned}$$

for $A, B \in \mathrm{End}(\mathcal{H})$.

Let A be of trace-class. By Lemma 8.4.10, the series

$$\sum_{i=1}^{\infty} (Ae_i \mid e_i)$$

is *absolutely convergent* and its sum does not depend on the particular choice of the complete orthonormal set $\{e_i\}$. We call this sum the *trace* of A and denote it by $\operatorname{tr} A$.

We list here some properties of trace-class operators and their trace.

Proposition 8.4.11. (a) *The set* J *of operators of trace-class is a two-sided* $*$*-ideal in* $\mathrm{End}(\mathcal{H})$. *The positive operators in* J *generate* J *as a vector space over* $\mathbb{C}$.

(b) *The functional* tr *is linear on* J. *Furthermore, given* $A \in J$ *and* $T \in \mathrm{End}(\mathcal{H})$, *one has* $\operatorname{tr} TA = \operatorname{tr} AT$.

Property (b) is easily shown as follows. Obviously $\operatorname{tr} UAU^{-1} = \operatorname{tr} A$ for unitary operators U. Hence $\operatorname{tr} AU = \operatorname{tr} UA$ for unitary U. Now let $T \in \mathrm{End}(\mathcal{H})$. For the proof we may assume that T is self-adjoint and $-I \leq T \leq I$, where I is the identity operator. Then we can write $T = \frac{1}{2}(U + U^*)$, where $U = T + i(I - T^2)^{1/2}$ is unitary. Therefore $\operatorname{tr} AT = \operatorname{tr} TA$.

Examples of Hilbert–Schmidt operators and trace-class operators are given by *kernel operators*. We mention here two theorems, without proof.

Let X be a locally compact space, satisfying the second axiom of countability and μ a positive measure on X. Set $\mathcal{H} = L^2(X, \mu)$.

Theorem 8.4.12 (E. Schmidt). *Suppose A is a kernel operator on $\mathcal{H}$ with kernel $A(x, y) \in L^2(X \times X, \mu \otimes \mu)$, i.e.*

$$Af(x) = \int_X A(x, y)\, f(y)\, d\mu(y) \qquad (f \in \mathcal{H}).$$

Then A is a Hilbert–Schmidt operator and

$$\|A\|_{\mathrm{HS}}^2 = \int_X \int_X |A(x, y)|^2\, d\mu(x)\, d\mu(y).$$

The proof of this theorem is straightforward and is left to the reader.

Theorem 8.4.13 (J. Mercer, [36, p. 117] or [60, p. 536, Theorem 4]). *Assume that* $\operatorname{Supp}\mu = X$. *Let A be a positive operator in* $\operatorname{End}(\mathcal{H})$ *with a continuous kernel $A(x, y)$. Then A is of trace-class if and only if $A(x, x) \in L^1(X, \mu)$. This being the case, one has*

$$\operatorname{tr} A = \int_X A(x, x)\, d\mu(x).$$

Let now G be as usual and let π be a unitary representation of G on a separable Hilbert subspace $\mathcal{H}$. We shall say that π has a *character* if $\pi(\varphi_0)$ is of trace class for all $\varphi_0 \in D(G)$. The linear form

$$\Theta_\pi : \varphi_0 \mapsto \operatorname{tr} \pi(\varphi_0)$$

on $D(G)$ is called the character of π. Fixing a complete orthonormal set $\{e_i\}$ in $\mathcal{H}$, we obtain

$$\Theta_\pi(\varphi_0) = \lim_{n\to\infty} \int_G \varphi_0(x)\, \chi_n(x)\, dx$$

where $\chi_n(x) = \sum_{i=1}^n (\pi(x)\, e_i \,|\, e_i)$. Consequently $\Theta_\pi$is the pointwise limit of the sequence of distributions χ_n, and thus, since $D(G)$ is a barrelled locally convex space, by the closed graph theorem, a distribution (cf. [4, Chapter 3] or [43]). Notice that the character of π already exists as soon as $\pi(\varphi_0)$ is a Hilbert–Schmidt operator for all $\varphi_0 \in D(G)$. This follows immediately from the decomposition lemma (Lemma 8.1.1 (a)).

We now return to our special case. Let G_1 be as usual and set $G = G_1 \times G_1$ and $H = \operatorname{diag}(G)$. Let π be an irreducible unitary representation of G. Then π

can be realized on a Hilbert subspace of $D'(G/H) = D'(G_1)$ if π is of the form $\pi_1 \widehat{\otimes}_2 \overline{\pi_1}$ where π_1 is an irreducible unitary representation of G_1 on $\mathcal{H}_1$ whose character Θ_{π_1} exists [26]. The converse is also true if G_1 is a type I group (the groups of our examples are all type I groups) [26]. Actually, the reproducing distribution T associated with π can be taken equal to Θ_{π_1}. This is a canonical choice. The injection $j : \mathcal{H}_1 \widehat{\otimes}_2 \overline{\mathcal{H}_1} \to D'(G_1)$ has the form

$$j(v \otimes w)(x) = \langle \pi(x^{-1})v \,|\, w \rangle \qquad (x \in G_1; v, w \in \mathcal{H}_1).$$

In this case Propositions 8.4.2 and 8.4.4 yield the well-known properties of square-integrable representations of G_1. In fact, if π is square-integrable mod diag(G), then $\pi = \pi_1 \widehat{\otimes}_2 \overline{\pi_1}$ (G of type I) with π_1 a square-integrable representation of G_1 (mod H, with $H = \{e\}$), see [2]. The converse is also true. Recall that any square-integrable representation of G_1 has a character ([2, Section 5.2.3], applying the decomposition lemma).

2. *The case H normal*

Consider now the case of a pair (G, H) with H a normal subgroup. Given an irreducible unitary representation of G on a Hilbert space $\mathcal{H}$, the space $\mathcal{H}_{-\infty}^H$ is a closed subspace of $\mathcal{H}_{-\infty}$ which is G-invariant, and hence, since $\pi_{-\infty}$ is irreducible, either zero or the whole space. In the latter case the restriction of π to H is the identity representation on $\mathcal{H}$, hence π can be regarded as an irreducible unitary representation of the group G/H. Let us call this representation $\widetilde{\pi}$. We may conclude: there is a one-to-one correspondence between irreducible unitary representations $\widetilde{\pi}$ of the group G/H and irreducible unitary representations π of G which can be realized on $D'(G/H)$ by left translations. If $\widetilde{\pi}$ is square-integrable, then π is square-integrable mod H and conversely.

We now continue the general theory. Let us assume that (G, H) is a generalized Gelfand pair. Denote by $E_2(G/H)$ the set of equivalence classes of (irreducible) square-integrable representations mod H. Fix a representation π in each class, together with its realization j_π on a Hilbert subspace of $L^2(G/H)$, and call this set of representations S. Denote by T_π the reproducing distribution and by d_π the formal degree of π. Let $\mathcal{H}_\pi$ be the representation space of π. Define $\mathcal{H}_d = \bigoplus j_\pi(\mathcal{H}_\pi)$ and let E be the orthogonal projection of $L^2(G/H)$ onto $\mathcal{H}_d$. Then one has the following (partial) Plancherel formula for the relative discrete series.

Proposition 8.4.14. *For all $\varphi_0 \in D(G)$ one has*

$$\|E\varphi\|_2^2 = \sum_{\pi \in S} d_\pi \, \langle T_\pi, \widetilde{\varphi}_0 * \varphi_0 \rangle.$$

Notice that $E\varphi \in C^\infty(G/H)$ for all $\varphi \in D(G/H)$ (apply, for instance, the decomposition lemma). So the formula in Proposition 8.4.15 is equivalent to

$$(E\varphi)(eH) = \sum_{\pi \in S} d_\pi \, \langle T_\pi, \varphi_0 \rangle \qquad (\varphi_0 \in D(G)).$$

The above formulae do not depend on the particular choice of the set S. Indeed, $d_\pi \, T_\pi$ is independent of the particular choice of π in its equivalence class and of the choice of j_π. In fact, $d_\pi \, \langle T_\pi, \widetilde{\varphi}_0 * \varphi_0 \rangle = \| E_\pi \varphi \|_2^2$, where E_π is the orthogonal projection of $L^2(G/H)$ onto $j_\pi(\mathcal{H}_\pi)$. Choose therefore an orthonormal basis $\{e_i\}$ in $\mathcal{H}_\pi$. Then $\{d_\pi^{1/2} \, j(e_i)\}$ is an orthonormal basis for $j_\pi(\mathcal{H}_\pi)$ and

$$\begin{aligned} \| E_\pi \varphi \|_2^2 &= \sum_i d_\pi \, |\langle j e_i \,|\, \varphi \rangle|^2 = \sum_i d_\pi \, |\langle e_i \,|\, j^* \varphi \rangle|^2 \\ &= d_\pi \, \| j^* \varphi \|^2 = d_\pi \, \langle T_\pi, \widetilde{\varphi}_0 * \varphi_0 \rangle. \end{aligned}$$

We shall now apply our theory, developed so far, to the two cases mentioned at the beginning: $G = \mathrm{O}(1,n) \ltimes \mathbb{R}^{n+1}$, $H = \mathrm{O}(1,n)$ and $G = \mathrm{O}(1,n)$, $H = \mathrm{O}(1, n-1)$. Their Euclidean analogs were treated in Chapter 7. They gave rise to classical Gelfand pairs and the decomposition of $L^2(G/K)$ had no discrete series. There exist however classical Gelfand pairs (G, K) admitting square-integrable unitary representations mod K, or, equivalently, square-integrable positive-definite spherical functions, even if G/K is not compact; see [30]. We shall see in the next chapter that in the non-Euclidean example $G = \mathrm{O}(1,n)$, $H = \mathrm{O}(1, n-1)$ square-integrable representations mod H do occur in the decomposition of $L^2(G/H)$.

Chapter 9

Examples of Generalized Gelfand Pairs

Literature: [15], [28], [42].

9.1 Non-Euclidean motion groups

(i) Introduction

Let $n \geq 1$ be a natural number. The non-Euclidean motion group on $\mathbb{R}^{n+1}$ is the semi-direct product

$$G = \mathrm{O}(1,n) \ltimes \mathbb{R}^{n+1},$$

where $\mathrm{O}(1,n)$ is the orthogonal group of the quadratic form

$$[x,y] = x_0 y_0 - x_1 y_1 - \cdots - x_n y_n,$$

where $x = (x_0, x_1, \ldots, x_n)$ and $y = (y_0, y_1, \ldots, y_n)$ are elements of $\mathbb{R}^{n+1}$.

Set $H = \mathrm{O}(1,n)$. Notice that H, and therefore G, is not connected, it has four connected components; the groups G and H are unimodular. Elements of G are written as pairs $g = (h,a)$ with $h \in H$, $a \in \mathbb{R}^{n+1}$. Such a pair has to be viewed as the product of h and the translation over a, considered as operating on $\mathbb{R}^{n+1}$,

$$g \cdot x = h \cdot x + a \qquad (x \in \mathbb{R}^{n+1}). \tag{9.1.1}$$

It is easily checked (see the Euclidean case in Section 7.1) that θ, defined by $\theta(h,a) = (h,-a)$, is a C^∞ involutive automorphism of G that leaves H fixed. Observe that bi-H-invariant distributions on G are in one-to-one correspondence with H-invariant distributions on $\mathbb{R}^{n+1}$ (natural action of H), and that θ induces the action $T \mapsto \check{T}$ for such distributions on $\mathbb{R}^{n+1}$, where symbolically $\check{T}(x) = T(-x)$. Since to positive-definite bi-H-invariant distributions on G correspond H-invariant positive-definite distributions on $\mathbb{R}^{n+1}$, we may conclude that $\overline{T} = T^\theta$ for such distributions and therefore, by Corollary 8.3.4, that (G,H) is a generalized Gelfand pair.

(ii) Extremal measures

Let $\mathcal{S}(\mathbb{R}^{n+1})$ denote the Schwartz space of $\mathbb{R}^{n+1}$, see [41], [43]. For $\varphi \in \mathcal{S}(\mathbb{R}^{n+1})$ we define its Fourier transform by

$$\widehat{\varphi}(x) = \int_{\mathbb{R}^{n+1}} \varphi(y)\, \mathrm{e}^{-2\pi i [x,y]}\, dy \qquad (x \in \mathbb{R}^{n+1}),$$

and, if T is a tempered distribution on $\mathbb{R}^{n+1}$, we set for $\varphi \in \mathcal{S}(\mathbb{R}^n)$, $\langle \widehat{T}, \widehat{\varphi} \rangle = \langle T, \varphi \rangle$, thus defining the Fourier transform $\widehat{T}$ of T. We recall that $\varphi \mapsto \widehat{\varphi}$ is an isomorphism of $\mathcal{S}(\mathbb{R}^{n+1})$ onto itself, so that $\widehat{T}$ is again tempered if T is.

We are going to determine the extremal positive-definite H-invariant distributions T on $\mathbb{R}^{n+1}$. Recall that the Fourier transform of an H-invariant positive-definite distribution T on $\mathbb{R}^{n+1}$ is an H-invariant positive tempered measure on $\mathbb{R}^{n+1}$. For invariant positive measures one has the following general result by L. Schwartz ([42, pp. 75–76]).

Lemma 9.1.1. *Let G be a group acting on a locally compact space X satisfying the second axiom of countability, and let μ be a positive measure on X, invariant under G. If μ is extremal under the measures with the same property, then the support of μ is the closure of a G-orbit.*

We reproduce the proof by L. Schwartz in our words. Let S be the support of μ. Then S is clearly a G-invariant set, since μ is G-invariant. Assume $\mu \neq 0$, so $S \neq \emptyset$. We have to show that S contains a dense G-orbit. Take a point $a \in S$, and let V be an open neighbourhood of a. Set $\widetilde{V} = G \cdot V = \bigcup_{g \in G} g \cdot V$. The open set V has positive measure, because otherwise $a \notin S$. The complement $\complement\widetilde{V}$ of $\widetilde{V}$ has measure zero; if not, the measure μ could be multiplied by the characteristic function of $\complement\widetilde{V}$ and we would obtain a G-invariant measure $\mu_1 \leq \mu$ that is not proportional to μ. Hence $\widetilde{V} = S$ almost everywhere (a.e.). The point a has a countable basis of neighbourhoods V_n. Any $\widetilde{V}_n$ is a.e. equal to S. The intersection $\bigcap \widetilde{V}_n$ is also a.e. equal to S. In other words: $b \in \bigcap \widetilde{V}_n$ for almost all $b \in S$. Let $b \in \bigcap \widetilde{V}_n$, then $b \in \widetilde{V}_n$ for all n. The orbit $\widetilde{b}$ of b intersects every V_n; so $\widetilde{b}$ intersects all neighbourhoods of a. Thus $a \in \overline{\widetilde{b}}$ (closure of $\widetilde{b}$) for almost all b. Let $\{a_k\}$ be a dense sequence in S. For every k fixed, $a_k \in \overline{\widetilde{b}}$ for almost all $b \in S$. Since $\{a_k\}$ is a countable set, $a_k \in \overline{\widetilde{b}}$ for all k, for almost all $b \in S$. Consequently, $\overline{\widetilde{b}} = S$ for almost all $b \in S$.

So, to determine the extremal, positive, H-invariant measures on $\mathbb{R}^{n+1}$, we have to determine first the set of H-orbits. Set for $x \in \mathbb{R}^{n+1}$, $Q(x) = [x, x]$. Then the H-orbits are given by

$$\Gamma_t = \{x : Q(x) = t\} \ (t \in \mathbb{R},\ t \neq 0), \quad \Gamma_0 = \{x : Q(x) = 0,\ x \neq 0\}, \quad \{0\}.$$

Clearly, for $t > 0$ the orbit Γ_t is isomorphic to $\mathrm{O}(1,n)/\mathrm{O}(n)$, whereas for $t < 0$ it is isomorphic to $\mathrm{O}(1,n)/\mathrm{O}(1,n-1)$. In all these cases Γ_t carries an $\mathrm{O}(1,n)$-invariant measure, since $\mathrm{O}(1,n)$, $\mathrm{O}(1,n-1)$ and $\mathrm{O}(n)$ are unimodular. Let us fix, for the time being, such a measure for each $t \neq 0$ and call it μ_t. Since Γ_t is a closed subset of $\mathbb{R}^{n+1}$, we may consider μ_t as a positive measure on $\mathbb{R}^{n+1}$ with support Γ_t. Indeed, define

$$\mu_t(\varphi_0) = \mu_t(\varphi),$$

where φ is the restriction of $\varphi_0 \in C_c(\mathbb{R}^{n+1})$ to Γ_t. We shall denote the extension of μ_t to a measure on all of $\mathbb{R}^{n+1}$ again by μ_t; it is still H-invariant.

With a slight variation on Proposition 7.5.4, one obtains the following invariant measure on Γ_0, which we call μ_0:

$$\mu_0(\psi) = \int_K \int_0^\infty \psi(\lambda\, k \cdot \xi^0)\, \lambda^{n-2}\, d\lambda\, dk \qquad (\psi \in C_c(\Gamma_0)), \tag{9.1.2}$$

where $K = \mathrm{O}(1) \times \mathrm{O}(n)$ and $\xi^0 = (1, 0, \ldots, 0, 1)$. Let now $\psi \in \mathcal{S}(\mathbb{R}^{n+1})$ and set for any natural number N, $\nu_N(\psi) = \sup_{x \in \mathbb{R}^{n+1}} (1 + \|x\|^2)^N\, |\psi(x)|$. Then clearly

$$|\mu_0(\psi)| \leq \nu_N(\psi) \int_0^\infty (1 + 2\lambda^2)^{-N} \lambda^{n-2} d\lambda.$$

Choosing N large enough we see that for $n \geq 2$ the measure μ_0 can be extended to a (tempered) positive H-invariant measure on $\mathbb{R}^{n+1}$ and that, moreover, the support of μ_0 is equal to the set $\{x : Q(x) = 0\}$. Let us call this extension μ_0 again. If $n = 1$ then μ_0 obviously has no extension to $\mathbb{R}^{n+1}$. So we will assume from now on that $n \geq 2$.

Finally, the H-invariant measure with support in $\{0\}$ is clearly the delta-function δ at $x = 0$. We have the following lemma:

Lemma 9.1.2. *Let μ be an extremal, H-invariant, positive measure on the space $\mathbb{R}^{n+1}$. Then μ is either proportional to some μ_t $(t \in \mathbb{R})$ or to δ.*

By Lemma 9.1.1, μ has support equal to the closure of some H-orbit, so equal to one of the sets $\{Q(x) = t\}$ $(t \in \mathbb{R})$ or to $\{0\}$. Restrict μ to the orbit and call this restriction $\widetilde{\mu}$:

$$\widetilde{\mu}(\varphi) = \mu(\varphi_0),$$

where φ is a continuous function with compact support on the orbit and $\varphi_0 \in C_c(\mathbb{R}^{n+1})$ is any extension of φ to $\mathbb{R}^{n+1}$. This is well-defined by Proposition 4.1.3. Then $\widetilde{\mu} = c\,\mu_t$ for some $t \neq 0$, or $\widetilde{\mu} = c\,\mu_0$ or $\widetilde{\mu} = c\,\delta$ for some positive scalar c. But then $\mu = c\,\mu_t$ for some $t \neq 0$, or $\mu = c\,\delta$, or $\mu = c\mu_0 + c'\delta$ where c' is another scalar. This scalar c' must be positive again. Indeed, setting for $\varphi \in C_c(\mathbb{R}^{n+1})$, $\varphi_\lambda(x) = \varphi(x/\lambda)$ $(x \in \mathbb{R}^{n+1}, \lambda \in \mathbb{R}^*)$, we get by (9.1.2)

$$\mu_0(\varphi_\lambda) = |\lambda|^{n-1}\, \mu_0(\varphi)$$

and also $\delta(\varphi_\lambda) = \delta(\varphi) = \varphi(0)$. Hence $\mu(\varphi_\lambda) = c\,|\lambda|^{n-1}\mu_0(\varphi) + c'\varphi(0)$. Let now $\lambda \to 0$. Then we find $c' \geq 0$. Since μ is extremal, we then get that μ is proportional to μ_0 or to δ.

As for the converse of Lemma 9.1.2 we have:

Lemma 9.1.3. *The positive, H-invariant measures μ_t $(t \in \mathbb{R})$ and δ are extremal in the cone of the measures with the same properties.*

Let μ be a positive, H-invariant measure on $\mathbb{R}^{n+1}$ with $\mu \leq \mu_t$ for some $t \in \mathbb{R}$. Then, if $t \neq 0$, we have, as above, that μ is proportional to μ_t. Let now $\mu \leq \mu_0$. Then again, as above, $\mu = c\,\mu_0 + c'\,\delta$ with $0 \leq c \leq 1$, $c' \geq 0$. By the same trick as in the proof of Lemma 9.1.2 we get, for $\lambda \in \mathbb{R}^*$,

$$c\,|\lambda|^{n-1}\mu_0(\varphi) + c'\,\varphi(0) \leq |\lambda|^{n-1}\,\mu_0(\varphi)$$

for all $\varphi \in C_c(\mathbb{R}^{n+1})$, $\varphi \geq 0$. Now let λ tend to zero again and we obtain $c' \leq 0$, hence $c' = 0$, thus $\mu = c\,\mu_0$ and hence μ_0 is extremal. The measure δ is clearly extremal.

(iii) Tempered extremal measures

Let again $n \geq 2$. We have already seen that μ_0 is a tempered measure on $\mathbb{R}^{n+1}$. The same holds for μ_t, $t \neq 0$. This is easily seen from a concrete expression for μ_t, which can be obtained in the same way as in Section 7.5, using hyperbolic coordinates. We can take

$$\mu_t(\varphi) = \int_K \int_0^\infty \varphi(ka_u \cdot t^{1/2}\,e_0)\,\sinh^{n-1} u\,dk\,du \qquad (t > 0), \tag{9.1.3}$$

$$\mu_t(\varphi) = \int_K \int_0^\infty \varphi(ka_u \cdot |t|^{1/2}\,e_n)\,\cosh^{n-1} u\,dk\,du \qquad (t < 0) \tag{9.1.4}$$

for $\varphi \in C_c(\mathbb{R}^{n+1})$. Using the same method as for μ_0 we see that we have to show that for $t > 0$ the following integral is finite for N sufficiently large:

$$\int_0^\infty [1 + t(2\sinh^2 u + 1)]^{-N}\,\sinh^{n-1} u\,du.$$

Similarly for $t < 0$. We get, taking $v = \sinh^2 u$,

$$\begin{aligned}
&\int_0^\infty [1 + t(2\sinh^2 u + 1)]^{-N}\,\sinh^{n-1} u\,du \\
&\quad = \frac{1}{2}\int_0^\infty [1 + t(2v + 1)]^{-N}\,(v+1)^{-1/2}\,v^{\frac{n-2}{2}}\,dv \\
&\quad \leq \int_0^\infty [1 + 2tv]^{-N}\,v^{\frac{n-3}{2}}\,dv \\
&\quad \leq (2t)^{-\frac{n-1}{2}} \int_0^\infty [1 + v]^{-N}\,v^{\frac{n-3}{2}}\,dv < \infty
\end{aligned}$$

if $N > \frac{n-1}{2}$. So all μ_t are tempered, as is δ. We now have

Theorem 9.1.4. *The extremal, positive-definite, H-invariant distributions on $\mathbb{R}^{n+1}$ are, up to a positive scalar, given by the Fourier transform of the positive tempered measures μ_t ($t \in \mathbb{R}$) and δ.*

Indeed, any H-invariant positive tempered measure that is extremal in the cone of all such measures, is also extremal in the cone of H-invariant positive measures.

The G-invariant Hilbert subspaces of $\mathbb{R}^{n+1}$ (G acting by (9.1.1)) are given by the spaces $L^2(\widehat{\mu}_t)$ and the trivial representation. The space $L^2(\widehat{\mu}_t)$ is the closure of the space $D(\mathbb{R}^{n+1})$ with respect to the 'norm' $\int_{\mathbb{R}^{n+1}} |\widehat{\varphi}(x)|^2 d\mu_t(x)$. Notice that G acts irreducibly on these spaces.

Let $\Box$ be the differential operator on $\mathbb{R}^{n+1}$ given by

$$\Box = -\frac{\partial^2}{\partial x_0^2} + \frac{\partial^2}{\partial x_1^2} + \cdots + \frac{\partial^2}{\partial x_n^2}.$$

Then $\Box\,\widehat{\mu}_t = 4\pi^2 t\,\widehat{\mu}_t$ for all t. Any H-invariant differential operator on $\mathbb{R}^{n+1}$ is a polynomial in $\Box$.

(iv) Plancherel formula

Let us normalize the measures μ_t as in (9.1.3) and (9.1.4). Then one has, using hyperbolic coordinates,

$$\int_{\mathbb{R}^{n+1}} \varphi(x)\,dx = 2\frac{\pi^{\frac{n}{2}}}{\Gamma(\frac{n}{2})}\int_{-\infty}^{\infty} |t|^{\frac{n-1}{2}}\,\mu_t(\varphi)\,dt \qquad (\varphi \in \mathcal{S}(\mathbb{R}^{n+1})). \quad (9.1.5)$$

We leave the proof to the reader. So we easily get the following Plancherel formula.

Theorem 9.1.5. *For all $\varphi \in D(\mathbb{R}^{n+1})$ one has*

$$\varphi(0) = \int_{\mathbb{R}^{n+1}} \widehat{\varphi}(x)\,dx = 2\frac{\pi^{\frac{n}{2}}}{\Gamma(\frac{n}{2})}\int_{-\infty}^{\infty} |t|^{\frac{n-1}{2}}\,\widehat{\mu}_t(\varphi)\,dt.$$

(v) Class-one representations

We give here another realization of the irreducible unitary representations of G which can be realized on $D'(\mathbb{R}^{n+1})$. They are, of course, determined up to unitary equivalence. We call such representations *class-one representations*, similar to the case of a classical Gelfand pair. We follow Section 7.1 (vi).

Let for $y \in \mathbb{R}^{n+1}$, $\chi_y(x) = \mathrm{e}^{-2\pi i[x,y]}$ ($x \in \mathbb{R}^{n+1}$) be a character of $\mathbb{R}^{n+1}$. One immediately sees that $\chi_y(h^{-1}\cdot x) = \chi_{h\cdot y}(x)$ for $x \in \mathbb{R}^{n+1}$, $h \in H$. So H acts on the characters of $\mathbb{R}^{n+1}$. Let us denote by H_y the stabilizer of y in H, which is a closed unimodular subgroup of H. Actually $H/H_y \simeq H\cdot y$, the orbit of y in $\mathbb{R}^{n+1}$. Observe that $H\cdot y = \Gamma_t$ with $t = Q(y)$ or $H\cdot y = \{0\}$. Choose the H-invariant measure μ_t on $H\cdot y$ if $Q(y) = t$ (and δ on $\{0\}$) and let it correspond to an H-invariant measure $d\dot{h}$ on H/H_y via the mapping $h \mapsto h\cdot y$ ($h \in H$).

We now define a unitary representation of G by inducing the representation $1 \otimes \chi_y$ from $H_y \ltimes \mathbb{R}^{n+1}$ to G. The Hilbert space V_y consists of the measurable functions f on G satisfying

(a) $f(ga) = \chi_y(-a)\, f(g) \quad (g \in G,\, a \in \mathbb{R}^{n+1})$,

(b) $f(gh) = f(g) \quad (g \in G,\, h \in H_y)$,

(c) $\int_{H/H_y} |f(h)|^2\, d\dot{h} < \infty$.

The squared Hilbert norm is given by $\|f\|^2 = \int_{H/H_y} |f(h)|^2\, d\dot{h}$. The representation π_y is defined by

$$\pi_y(g)\, f(g') = f(g^{-1}g') \qquad (g, g' \in G).$$

One verifies that π_y is a unitary representation of G on V_y, called the representation induced by $1 \otimes \chi_y$. Clearly, π_0 is the trivial representation of G.

Notice that any $f \in V_y$ is completely determined by its values on H. Transferring V_y into a Hilbert space $\widetilde{V}_y$ of functions on H by just restricting the functions in V_y to H, yields a unitary representation on $L^2(H/H_y)$, again denoted by π_y, given by

$$\begin{aligned} \pi_y(h)\, f(h') &= f(h^{-1}h') && (h, h' \in H), \\ \pi_y(a)\, f(h') &= \chi_{h'\cdot y}(a)\, f(h') && (h' \in H,\, a \in \mathbb{R}^{n+1}). \end{aligned}$$

Now define the mapping $j^* : S(\mathbb{R}^{n+1}) \to \widetilde{V}_y$ as follows:

$$j^*(\varphi) = \psi \quad \text{where} \quad \psi(h) = \widehat{\varphi}(h \cdot y) \qquad (h \in H).$$

Then one easily verifies that j^* has a dense image, is continuous and commutes with the G-actions. So $j = j^{**} : \widetilde{V}_y \to S'(\mathbb{R}^{n+1})$ gives the required G-equivariant embedding of V_y into $D'(\mathbb{R}^{n+1})$. Moreover, if $y \neq 0$,

$$\|j^*\varphi\|^2 = \int_{H/H_y} |\widehat{\varphi}(h \cdot y)|^2\, d\dot{h} = \int_{\mathbb{R}^{n+1}} |\widehat{\varphi}(x)|^2\, d\mu_t,$$

so the reproducing distribution of π_y is either $\widehat{\mu}_t$ or the function identically 1. Therefore we have, applying Propositions 8.2.2 and 8.2.3,

Theorem 9.1.6. (i) *The unitary representations π_y ($y \in \mathbb{R}^{n+1}$), induced by $1 \otimes \chi_y$ from $H_y \ltimes \mathbb{R}^{n+1}$ to G, are irreducible.*

(ii) *Two representations π_y and $\pi_{y'}$ are equivalent if and only if either $y \neq 0$, $y' \neq 0$ and $Q(y) = Q(y')$ or $y = y' = 0$.*

(iii) *The representations π_y ($y \in \mathbb{R}^{n+1}$) exhaust the set of irreducible unitary representations of class one (up to equivalence).*

(vi) Exercise

We have excluded the case $n = 1$. Treat this case and examine also what becomes different if we replace H by its connected component $\mathrm{SO}_0(1,n)$.

9.2 Pseudo-Riemannian real hyperbolic spaces

(i) Introduction

For $n \in \mathbb{N}, n \geq 2$, let $G = \mathrm{O}(1,n)$ be the orthogonal group of the quadratic form

$$[x,y] = -x_0 y_0 + x_1 y_1 + \cdots + x_n y_n$$

where $x = (x_0, x_1, \ldots, x_n)$ and $y = (y_0, y_1, \ldots, y_n)$ are elements of $\mathbb{R}^{n+1}$. Observe that we use here another definition of the quadratic form than in Sections 9.1 and 7.5. Let $X = \{x \in \mathbb{R}^{n+1} : [x,x] = 1\}$. It is again a one-sheeted hyperboloid. The group G acts transitively on X. Let $H = \mathrm{O}(1, n-1)$ be the stabilizer of $e_n = (0, \ldots, 0, 1)$ in G and define the subgroups A, N, M as in Section 7.5 (ii):

$$A = \left\{ a_t = \begin{pmatrix} \cosh t & 0 & \sinh t \\ 0 & I_{n-1} & 0 \\ \sinh t & 0 & \cosh t \end{pmatrix} : t \in \mathbb{R} \right\},$$

$$N = \left\{ n = n_z = \begin{pmatrix} 1 + \frac{1}{2}\|z\|^2 & z^t & -\frac{1}{2}\|z\|^2 \\ z & I_{n-1} & -z \\ \frac{1}{2}\|z\|^2 & z^t & 1 - \frac{1}{2}\|z\|^2 \end{pmatrix} : z \in \mathbb{R}^{n-1} \right\},$$

$$M \simeq \mathrm{O}(n-1) = \left\{ \begin{pmatrix} 1 & 0 & 0 \\ 0 & l & 0 \\ 0 & 0 & 1 \end{pmatrix} : l \in \mathrm{O}(n-1) \right\}.$$

Let $x = (x_0, x_1, \ldots, x_n) \in X$. There is an element $k \in K \simeq \mathrm{O}(1) \times \mathrm{O}(n)$ such that $k \cdot x = (x_0, 0, \ldots, 0, \alpha)$ where $\alpha^2 = x_1^2 + \cdots + x_n^2$, $\alpha \geq 1$. Then we can find $t \in \mathbb{R}$ with

$$a_t \cdot e_n = (x_0, 0, \ldots, 0, \alpha).$$

So $x = k^{-1} a_t \cdot e_n$ and therefore

$$G = KAH.$$

Let w be the matrix $\mathrm{diag}(-1, 1, \ldots, 1)$ in $K \cap H$. Then $w a_t w^{-1} = a_{-t}$ for all $t \in \mathbb{R}$. So we actually have

Proposition 9.2.1. *The group $G = \mathrm{O}(1,n)$ can be written as*

$$G = K\overline{A}^+ H,$$

where $\overline{A}^+ = \{a_t \in A : t \geq 0\}$.

Observe that in the expression $x = ka_t \cdot e_n$ the parameter t is unique if we assume $t \geq 0$, and the parameter k is determined up to right multiplication by an element of the subgroup M, provided $t > 0$. To write down an explicit expression for the invariant measure dx on X, we use hyperbolic coordinates again an get

$$\int_X f(x)\,dx = \frac{4\pi^{\frac{n}{2}}}{\Gamma(\frac{n}{2})} \int_K \int_0^\infty f(ka_t \cdot e_n)\cosh^{n-1} t\,dt\,dk \tag{9.2.1}$$

for $f \in C_c(X)$. The measure dk on K is the normalized Haar measure. Notice that the normalizing constant has been chosen in such a way that for $f \in C_c(\mathbb{R}^{n+1})$ with $\operatorname{Supp} f \subset \{x : [x,x] \geq 0\}$,

$$\int_{\mathbb{R}^{n+1}} f(y)dy = \int_0^\infty \int_X f(rx)r^n\,dxdr.$$

Compare these expressions with (7.5.1). Let us fix the Haar measures dg on G and dh on H in such a way that $dg = dx\,dh$. Observe that we still have freedom to select dg (or dh).

(ii) Iwasawa decomposition

Let Ξ be the cone

$$\Xi = \{\xi = (\xi_0, \xi_1, \ldots, \xi_n) : [\xi, \xi] = 0,\ \xi_0 \neq 0\}.$$

Observe that G acts on Ξ. Set $\Xi' = \{\xi \in \Xi : \xi_n \neq 0\}$. Then we clearly have $\Xi' = HA\,\xi^0 \cup HA\,(-w)\,\xi^0$, where $\xi^0 = (1, 0, \ldots, 0, 1)$, both 'cells' being open in Ξ. Of course one also has, similar to Section 7.5 (ii),

$$\Xi = KA\,\xi^0.$$

The stabilizer of ξ^0 in G is the subgroup MN (see Section 7.5 (ii)). We have $G = KAN$ and Theorem 7.5.3 holds. Also Proposition 7.5.4 remains true in our context.

Proposition 9.2.2. *One can normalize the Haar measure dg of G in such a way that, for all $f \in C_c(G)$,*

$$\int_G f(g)\,dg = \int_K \int_{-\infty}^\infty \int_{\mathbb{R}^{n-1}} f(ka_t n_z)\,e^{(n-1)t}\,dzdtdk.$$

Let now $f \in C_c(\Xi)$. Then this proposition implies that

$$\mu : f \mapsto \int_K \int_0^\infty f(\lambda k \cdot \xi^0)\,\lambda^{n-1}\,\frac{d\lambda}{\lambda}\,dk$$

is a G-invariant measure on Ξ. Here dk is again the normalized Haar measure on K. Notice that the integral can actually be taken over K/M instead of K. Setting

$B = K \cdot \xi^0 \simeq K/M$ we might also write for the invariant measure μ:

$$\mu(f) = \int_B \int_0^\infty f(\lambda b)\, \lambda^{n-1} \frac{d\lambda}{\lambda}\, d\sigma(b)$$

with $d\sigma(b)$ the invariant (normalized) measure on B. We also have $B = \{\xi \in \Xi : \xi_0 = \pm 1\}$ and $B \simeq S^0 \times S^{n-1}$ where S^k is the unit sphere in $\mathbb{R}^{k+1}$.

(iii) The Laplace–Beltrami operator and its radial part

We define the Laplace–Beltrami (or Laplace) operator Δ on the space X as in Section 7.5 (iii).

Let now q be the mapping $q(x) = x/r$ defined on $\{x \in \mathbb{R}^{n+1} : [x, x] > 0\}$ where $r = [x, x]^{1/2}$. Then for any C^∞ function F on X one has

$$-\Box\, (F \circ q) = (\Delta F) \circ q$$

where $\Box = -\frac{\partial^2}{\partial x_0^2} + \frac{\partial^2}{\partial x_1^2} + \cdots + \frac{\partial^2}{\partial x_n^2}$.

In a similar way we define the radial part L of Δ. Let $Q : X \to \mathbb{R}$ be defined by

$$Q(x_0, x_1, \ldots, x_n) = x_n.$$

Then for any C^∞ function f on $\mathbb{R}$ we have

$$(Lf) \circ Q = \Delta\, (f \circ Q).$$

We can compute L as before and we obtain

$$L = (1 - u^2)\, \frac{d^2}{du^2} - nu \frac{d}{du}.$$

Let now ω be the Casimir operator on G, see Section 7.5. Considering functions on X as right-H-invariant functions on G, we easily see, similarly to Section 7.5 (vii) (2), that ω acts on $C^\infty(X)$ as a real scalar times Δ. Moreover it follows that Δ is a *symmetric* differential operator. Notice that L is symmetric only if $n = 2$.

Actually we have $\omega = \omega_{\mathfrak{q}}$ on $C^\infty(X)$ where $\omega_{\mathfrak{q}}$ is defined as follows. Let τ be the involution on G (and so on its Lie algebra $\mathfrak{g}$) given by

$$\tau(g) = J_0 g J_0$$

where J_0 is the diagonal matrix with entries $(1, 1, \ldots, 1, -1)$. The ± 1 eigenspaces of τ in $\mathfrak{g}$ are called $\mathfrak{h}$ and $\mathfrak{q}$. Clearly, $\mathfrak{h}$ is the Lie algebra of H and $\mathfrak{q}$ is given by the space of matrices of the form

$$Z = Z(z_1, \dots, z_n) = \begin{pmatrix} 0 & 0 & \dots & 0 & z_1 \\ 0 & 0 & \dots & 0 & -z_2 \\ \vdots & \vdots & & \vdots & \vdots \\ 0 & 0 & \dots & 0 & -z_n \\ z_1 & z_2 & \dots & z_n & 0 \end{pmatrix}$$

with $z_1, \dots, z_n \in \mathbb{R}$. It is easily checked that τ and the Cartan involution θ commute. So $\mathfrak{q} = \mathfrak{q} \cap \mathfrak{k} \oplus \mathfrak{q} \cap \mathfrak{p}$ and $\dim \mathfrak{q} \cap \mathfrak{p} = 1$. Now choose an orthonormal basis $Z_2, \dots, Z_n$ in $\mathfrak{q} \cap \mathfrak{k}$ and Z_1 in $\mathfrak{q} \cap \mathfrak{p}$ with respect to the form

$$\langle Y, Z \rangle = -B(Y, \theta Z) \qquad (Y, Z \in \mathfrak{g})$$

where B is the Killing form of $\mathfrak{g}$, $B(Y, Z) = (n-1) \operatorname{tr} YZ$. Then

$$\omega_{\mathfrak{q}} = -Z_2^2 - \dots - Z_n^2 + Z_1^2.$$

Observe that $\omega_{\mathfrak{q}}$ is not definite. Notice also that $\omega_{\mathfrak{q}}$ is $\operatorname{Ad}(H)$-invariant, so that it commutes with left and right translations over elements of H. It can be considered as a left-invariant differential operator on X. Follow now the reasoning from Section 7.5 (vii) (2).

(iv) Spherical distributions

We start with a definition.

Definition 9.2.3. A *spherical distribution* on X is an H-invariant eigendistribution of the Laplace–Beltrami operator Δ.

If F is a smooth function on $\mathbb{R}$ satisfying $LF = \lambda F$ where L is the radial part of Δ, then $f(x) = F(Q(x))$ $(x \in X)$ satisfies $\Delta f = \lambda f$. We shall obtain the spherical distributions by replacing F by a distribution on $\mathbb{R}$ of a certain type, which we shall make precise. For that purpose we apply results by Méthée, De Rham and Tengstrand, which are discussed in detail in the Appendices. We shall summarize the main results here.

For $x \neq \pm e_n$ the function Q has non-vanishing differential, so we can define the average of a function $f \in D(X)$ over the surfaces $\{Q(x) = t\}$ $(t \in \mathbb{R})$, that we denote by $M_f(t)$. One has the formula

$$\int_X F(Q(x))\, f(x)\, dx = \int_{\mathbb{R}} F(t)\, M_f(t)\, dt \tag{9.2.2}$$

for all continuous function F on $\mathbb{R}$.

The function M_f admits singularities at $t = \pm 1$. One has

$$M_f(t) = \varphi_1(t) + \eta(t-1)\,\varphi_2(t) + \eta(-t-1)\,\varphi_3(t) \qquad (9.2.3)$$

where $\varphi_1, \varphi_2, \varphi_3 \in D(\mathbb{R})$ and

$$\eta(t) = \begin{cases} Y(t)\,|t|^{\frac{n-2}{2}} & \text{if } n \text{ is odd,} \\ \log|t|\,|t|^{\frac{n-2}{2}} & \text{if } n \text{ is even.} \end{cases}$$

Here Y is the Heaviside function: $Y(t) = 1$ if $t \geq 0$, $Y(t) = 0$ if $t < 0$.

For simplicity of the presentation we shall restrict from now on to the case that *n is odd.*

Call $\mathcal{H}_\eta$ the linear space of functions of the form (9.2.3). We define a topology on $\mathcal{H}_\eta$ as in Appendices A and B; $\mathcal{H}_\eta$ becomes a complete locally convex topological vector space. We recall (see Appendix B):

Theorem 9.2.4. (a) *The mapping M from $D(X)$ to $\mathcal{H}_\eta$ is linear, continuous and surjective.*

(b) *The adjoint mapping M' from $\mathcal{H}'_\eta$ to $D'(X)$ is injective, continuous and has as image the space of H-invariant distributions on X.*

(c) *The mapping M' intertwines the operators Δ and L, that is*

$$\Delta \circ M' = M' \circ L.$$

We now come to the main theorem of this subsection.

Theorem 9.2.5. *Let $D'_\lambda(X)^H$ be the space of H-invariant distributions T on X satisfying $\Delta T = \lambda T$. Then* $\dim D'_\lambda(X)^H = 2$ *for all $\lambda \in \mathbb{C}$.*

A fundamental system of solutions of the differential equation $\Delta T = \lambda T$ is given by $M'S_\lambda$ and $M'T_\lambda$, see Section B.4. We shall apply their explicit expressions later on and then write down these expressions. For the moment, Theorem 9.2.5 gives sufficient information.

We are now going to construct a fundamental system in another way, namely from the point of view of representation theory, keeping in mind Section 8.2, Theorem 8.2.1 in particular.

Theorem 9.2.4 also has another consequence, namely

Corollary 9.2.6. *The pair (G, H) is a generalized Gelfand pair.*

See [52]. Fix, as before, Haar measures dg on G and dh on H in such a way that $dg = dx dh$. For $f \in D(G)$ set

$$f^o(x) = \int_H f(gh)\,dh \qquad (x = g \cdot e_n).$$

Given a bi-H-invariant distribution T_0 on G, there is a unique H-invariant distribution T on X satisfying $\langle T_0, f\rangle = \langle T, f^o\rangle$ ($f \in D(G)$), and conversely. This is a well-known fact. Let us extend the function Q from X to G by setting $Q(g) = Q(g\cdot e_n) = [g\cdot e_n, e_n]$.

To show that (G, H) is a generalized Gelfand pair, we apply Proposition 8.3.3 with $JT = \overline{T}$ ($T \in D'(X)$). We have to show that $\overline{T} = T$ for all bi-H-invariant positive-definite distributions T on G. Since $\overline{T} = \check{T}$ for such T, we shall show the following: for any bi-H-invariant distribution T on G one has $T = \check{T}$. We recall the definition of $\check{T}$: $\langle \check{T}, f\rangle = \langle T, \check{f}\rangle$, $\check{f}(g) = f(g^{-1})$ ($g \in G, f \in D(G)$). In view of the relation between bi-H-invariant distributions on G and H-invariant distributions on X, and because of Theorem 9.2.4 (b), this amounts to the relation

$$M_{[(\check{f})^o]} = M_{f^o}$$

for all $f \in D(G)$. This is easily checked: for all $F \in D(\mathbb{R})$ one has

$$\begin{aligned}\int_{-\infty}^{\infty} F(t)\, M_{[(\check{f})^o]}(t)\, dt &= \int_X F(Q(x))\, (\check{f})^o(x)\, dx \\ &= \int_G F(Q(g))\, \check{f}(g)\, dg = \int_G F(Q(g))\, f(g)\, dg,\end{aligned}$$

since $Q(g) = Q(g^{-1})$ ($g \in G$), and we get the result.

(v) Representations associated with the cone

The representations under consideration are also called principal series representations, see also Section 7.5 (vii), where we considered these representations for the group $\mathrm{SO}_0(1,n)$. In our situation the definition is slightly different.

The stabilizer of the line $\mathbb{R}\,\xi^0$ is equal to the subgroup $P = M_1AN$ where $M_1 = M \cup (-M)$. The group P is a so-called (minimal) parabolic subgroup of G. Recall that G acts transitively on Ξ and $\mathrm{Stab}\,\xi^0 = MN$, so $\Xi \simeq G/MN$. In (ii) we have also defined a G-invariant measure μ on Ξ.

The representations we are considering are representations induced by one-dimensional representations of P.

Let $s \in \mathbb{C}$, $\varepsilon = 0, 1$ and define $\mathcal{H}_{s,\varepsilon}$ to be the space of C^∞ functions f on G satisfying

$$f(xma_tn) = \mathrm{sgn}^\varepsilon[m\xi^0, e_n]\, \mathrm{e}^{(s-\rho)t}\, f(x)$$

where $x \in G, m \in M_1, n \in N, t \in \mathbb{R}$ and $\rho = (n-1)/2$ (as usual). Let $\pi_{s,\varepsilon}$ be the representation of G on $\mathcal{H}_{s,\varepsilon}$ defined by

$$\pi_{s,\varepsilon}(g) f(x) = f(g^{-1}x), \qquad g \in G.$$

We may also realize $\pi_{s,\varepsilon}$ on a space of C^∞ functions on Ξ, called $\mathcal{H}_{s,\varepsilon}(\Xi)$, defined by

$$\mathcal{H}_{s,\varepsilon}(\Xi) = \{f \in C^\infty(\Xi) : f(\lambda\xi) = \mathrm{sgn}^\varepsilon\,\lambda\, |\lambda|^{s-\rho}\, f(\xi) \text{ for } \lambda \neq 0,\ \xi \in \Xi\}.$$

In addition, setting $B = K\xi^0$, so $B \simeq K/M$, we can, by restricting functions on Ξ to B consider $\mathcal{H}_{s,\varepsilon}(\Xi)$ as a space of C^∞ functions on B, satisfying $f(b) = f(-b)$ if $\varepsilon = 0$, $f(b) = -f(-b)$ if $\varepsilon = 1$. We shall denote this space by $D_\varepsilon(B)$.

Let us define the linear form ℓ on $\mathcal{H}_{-\rho,0}$ as in Section 7.5 (v):

$$\ell(f) = \int_B f(b)\, d\sigma(b).$$

This form is again G-invariant (same proof).

Denote by $\langle\, ,\rangle_{s,\varepsilon}$ the bilinear form on $\mathcal{H}_{s,\varepsilon} \times \mathcal{H}_{-s,\varepsilon}$ defined by

$$\langle f, g\rangle_{s,\varepsilon} = \int_B f(b)\, g(b)\, d\sigma(b).$$

This form is continuous, non-degenerate and G-invariant:

$$\langle \pi_{s,\varepsilon}(x) f, \pi_{-s,\varepsilon}(x) g\rangle_{s,\varepsilon} = \langle f, g\rangle_{s,\varepsilon} \qquad (x \in G),$$

as follows from the invariance of ℓ.

If s is imaginary, say $s = i\nu$ with $\nu \in \mathbb{R}$, then $\mathcal{H}_{s,\varepsilon} = \mathcal{H}_{i\nu,\varepsilon}$ can be provided with a pre-Hilbert space structure with squared norm

$$\|f\|^2 = \int_B |f(b)|^2\, d\sigma(b),$$

and $\pi_{i\nu,\varepsilon}(x)$ is then unitary for all $x \in G$. Completing the space $\mathcal{H}_{i\nu,\varepsilon}$ and extending $\pi_{i\nu,\varepsilon}(x)$ to this completion gives a unitary representation on a Hilbert space, which we again denote by $\pi_{i\nu,\varepsilon}$.

For general $s \in \mathbb{C}$, $\mathcal{H}_{s,\varepsilon}$ is a Fréchet space with the natural topology induced by $C^\infty(B)$ and $\pi_{s,\varepsilon}$ is a (continuous) representation on this space.

To study the (ir)reducibility of the representations $\pi_{s,\varepsilon}$, one may follow Section 7.5 and derive Corollary 7.5.12 again for $\pi_{s,\varepsilon}$ ($\varepsilon = 0, 1$). For a more detailed analysis of the reducible representations at $s = \pm(\rho + k)$, $k \in \mathbb{N}$, see [32]. We do not need it here. See however (xiv).

The bilinear form $\langle\, ,\rangle_{s,\varepsilon}$ on $\mathcal{H}_{s,\varepsilon} \times \mathcal{H}_{-s,\varepsilon}$ permits to consider $\mathcal{H}_{s,\varepsilon}$ as a subspace of the dual of $\mathcal{H}_{-s,\varepsilon}$; we shall therefore denote by $\mathcal{H}'_{s,\varepsilon}(\Xi)$ the dual of $\mathcal{H}_{-s,\varepsilon}(\Xi)$, so that $\mathcal{H}_{s,\varepsilon}(\Xi) \subset \mathcal{H}'_{s,\varepsilon}(\Xi)$. In particular, a continuous function F on Ξ satisfying

$$F(\lambda\xi) = \operatorname{sgn}^\varepsilon \lambda\, |\lambda|^{s-\rho}\, F(\xi) \qquad (\lambda \neq 0,\ \xi \in \Xi)$$

defines an element of $\mathcal{H}'_{s,\varepsilon}(\Xi)$:

$$f \mapsto \int_B F(b)\, f(b)\, d\sigma(b).$$

A main point is the computation of the action of the Casimir operator ω on $\mathcal{H}_{s,\varepsilon}$. It turns out that ω acts as a scalar.

Proposition 9.2.7. *For all $s \in \mathbb{C}$ and $\varepsilon = 0, 1$ one has*

$$\pi_{s,\varepsilon}(\omega) = \frac{s^2 - \rho^2}{2n-2} I,$$

where I is the identity on $\mathcal{H}_{s,\varepsilon}$.

Recall the definition of ω: $\omega = \sum_{i=1}^k X_i X_i'$ where $X_1, \dots, X_k$ is a basis of $\mathfrak{g}$ and $X_1', \dots, X_k'$ the dual basis with respect to the Killing form $B(X,Y) = (n-1)\operatorname{tr} XY$.

We are going to select a special basis relative to the Lie algebras $\mathfrak{a}$ of A, $\mathfrak{m}$ of M, $\mathfrak{n}$ of N, $\theta\mathfrak{n}$ of θN. Observe that $\mathfrak{g} = \mathfrak{m} + \mathfrak{a} + \mathfrak{n} + \theta\mathfrak{n}$, $\mathfrak{m} = \mathfrak{k} \cap \mathfrak{h}$, $\mathfrak{a} = \mathfrak{q} \cap \mathfrak{p}$, $\mathfrak{a} = \mathbb{R}L$ with

$$L = \begin{pmatrix} 0 & 0 & 1 \\ 0 & 0_{n-1} & 0 \\ 1 & 0 & 0 \end{pmatrix}.$$

The Lie algebra $\mathfrak{n}$ consists of the matrices

$$N(z) = \begin{pmatrix} 0 & z^t & 0 \\ z & 0 & -z \\ 0 & z^t & 0 \end{pmatrix} \qquad (z \in \mathbb{R}^{n-1}).$$

Notice that $\mathfrak{n} \subset \mathfrak{h} \cap \mathfrak{p} + \mathfrak{q} \cap \mathfrak{k}$.

Select a basis $N_1, \dots, N_{n-1}$ of $\mathfrak{n}$ such that

$$B(N_i, \theta N_j) = \begin{cases} 0 & \text{if } i \neq j, \\ -\frac{1}{2} & \text{if } i = j. \end{cases}$$

Take for example $N_i = N(e_i)$ up to a normalizing factor, where e_i is the i-th unit vector in $\mathbb{R}^{n-1}$.

Then we have that

- $-N_1 + \theta N_1, \dots, -N_{n-1} + \theta N_{n-1}$ is an orthonormal basis of $\mathfrak{h} \cap \mathfrak{p}$:

$$B(-N_i + \theta N_i, -N_j + \theta N_j) = \begin{cases} 0 & \text{if } i \neq j, \\ 1 & \text{if } i = j, \end{cases}$$

and

- $N_1 + \theta N_1, \dots, N_{n-1} + \theta N_{n-1}$ is an anti-orthonormal basis of $\mathfrak{q} \cap \mathfrak{k}$:

$$B(N_i + \theta N_i, N_j + \theta N_j) = \begin{cases} 0 & \text{if } i \neq j, \\ -1 & \text{if } i = j. \end{cases}$$

Let $Y_1, \dots, Y_m$ be an anti-orthonormal basis of $\mathfrak{m}$:

$$B(Y_i, Y_j) = \begin{cases} 0 & \text{if } i \neq j, \\ -1 & \text{if } i = j. \end{cases}$$

Then

$$\begin{aligned}\omega &= \frac{L^2}{2(n-1)} + \sum_{j=1}^{n-1}\{-(N_j+\theta N_j)^2 + (-N_j+\theta N_j)^2\} - \sum_{j=1}^{m} Y_j^2 \\ &= \frac{L^2}{2(n-1)} - 2\sum_{j=1}^{n-1}\{\theta N_j \cdot N_j + N_j \cdot \theta N_j\} - \sum_{j=1}^{m} Y_j^2 \\ &= \frac{L^2}{2(n-1)} - 4\sum_{j=1}^{n-1}\theta N_j \cdot N_j - 2\sum_{j=1}^{n-1}[N_j, \theta N_j] - \sum_{j=1}^{m} Y_j^2.\end{aligned}$$

It is easily seen that the Lie algebra bracket $[N_j, \theta N_j]$ equals $-\frac{L}{2(n-1)}$ for all j. So we finally get

$$\omega = \frac{1}{2(n-1)}\{L^2 + 2\rho L\} - 4\sum_{j=1}^{n-1}\theta N_j \cdot N_j - \sum_{j=1}^{m} Y_j^2. \tag{9.2.4}$$

Now it follows, considering $L, N_j, \theta N_j, Y_j$ as left-invariant differential operators, that for $f \in \mathcal{H}_{s,\varepsilon}$ we have

$$\begin{aligned}(\omega f)(g) &= \frac{1}{2(n-1)}\{L^2 + 2\rho L\} f(g) \\ &= \frac{1}{2(n-1)}\left\{\frac{d^2}{dt^2} + 2\rho\frac{d}{dt}\right\} f(ga_t)|_{t=0} \\ &= \frac{1}{2(n-1)}\left\{\frac{d^2}{dt^2} + 2\rho\frac{d}{dt}\right\} e^{(s-\rho)t} f(g)|_{t=0} = \frac{s^2-\rho^2}{2(n-1)} f(g) \qquad (g \in G).\end{aligned}$$

Remark 9.2.8. It is now easy to connect $\omega_{\mathfrak{q}}$ and Δ as differential operators on X. We know that $\Delta = c\,\omega_{\mathfrak{q}}$ for some constant c. To determine it, consider the function $f(x,\xi) = [x,\xi]$ on $X \times \Xi$. This function is G-invariant: $f(gx, g\xi) = f(x,\xi)$ for all $g \in G$. As a function of x we know that $\Delta_X f(x,\xi) = (2\rho + 1) f(x,\xi)$, see Section 7.5. But

$$\Delta_X f(x,\xi) = c\,\omega_X f(x,\xi) = c\,\omega_\xi f(x,\xi) = \frac{c}{2(n-1)}(2\rho+1) f(x,\xi)$$

since $\xi \mapsto f(x,\xi)$ is in $\mathcal{H}_{\rho+1,0}$.

So $c = 2(n-1)$. Here we considered ω as a right-invariant differential operator on X and Ξ.

(vi) Spherical distributions associated with the representations $\pi_{s,\varepsilon}$

For $\operatorname{Re} s > \rho$ we define continuous linear forms $F_{s,\varepsilon}$ on $\mathcal{H}_{-s,\varepsilon}(\Xi)$ by

$$\langle F_{s,\varepsilon}, f\rangle = \int_B \operatorname{sgn}^\varepsilon[b, e_n]\, |[b, e_n]|^{s-\rho} f(b) d\sigma(b).$$

So $F_{\sigma,\varepsilon} \in \mathcal{H}'_{s,\varepsilon}$ and moreover $\pi'_{s,\varepsilon}(h) F_{s,\varepsilon} = F_{s,\varepsilon}$ for all $h \in H$.

For any real number x we shall write from now on $x^{s,\varepsilon} = |x|^s \operatorname{sgn}^\varepsilon x$ where $s \in \mathbb{C}$ and $\varepsilon = 0, 1$.

Proposition 9.2.9. *If f is a function in $\mathcal{H}_{-s,\varepsilon}(\Xi)$, then the integral*

$$Z_{s,\varepsilon}(f) = \int_B [b, e_n]^{s-\rho,\varepsilon} f(b)\, d\sigma(b)$$

is defined for $\operatorname{Re} s > \rho$, *is an analytic function of s and admits a meromorphic extension to $\mathbb{C}$ with at most simple poles at $s = \rho - 2k - 1 - \varepsilon$ ($k \geq 0$).*

One has $Z_{s,\varepsilon}(f) = \int_{S^{n-1}} \sigma_n^{s-\rho,\varepsilon} f_1(\sigma)\, d\sigma$ with $f_1(\sigma) = \frac{1}{2}\{f(1,\sigma) + f(-1,\sigma)\}$. Clearly, σ_n can be taken as coordinate near any point of S^{n-1} with $\sigma_n = 0$. The proposition now follows from a classical result [17].

For $\operatorname{Re} s > \rho$ and for a function f in $\mathcal{H}_{-s,\varepsilon}(\Xi)$ set

$$u_{s,\varepsilon}(f) = \frac{1}{\Gamma(\frac{s-\rho+1+\varepsilon}{2})} \int_B [b, e_n]^{s-\rho,\varepsilon} f(b)\, d\sigma(b).$$

By Proposition 9.2.9 above, the function $s \mapsto u_{s,\varepsilon}(f)$ has an entire analytic extension to all of $\mathbb{C}$ and one shows that for all $s \in \mathbb{C}$, $u_{s.\varepsilon} \in \mathcal{H}'_{s,\varepsilon}(\Xi)$ with, moreover,

$$\pi'_{s,\varepsilon}(h)\, u_{s,\varepsilon} = u_{s,\varepsilon}$$

for all $h \in H$. Observe that

$$[\pi'_{s,\varepsilon}(g)\, u_{s,\varepsilon}](f) = \frac{1}{\Gamma(\frac{s-\rho+1+\varepsilon}{2})} \int_B [g^{-1} \cdot b, e_n]^{s-\rho,\varepsilon} f(b)\, d\sigma(b)$$
$$= \frac{1}{\Gamma(\frac{s-\rho+1+\varepsilon}{2})} \int_B [b, g \cdot e_n]^{s-\rho,\varepsilon} f(b)\, d\sigma(b).$$

Let $\varphi \in D(G)$. Then we have

$$[\pi'_{s,\varepsilon}(\varphi)\, u_{s,\varepsilon}](f) = \int_G [\pi'_{s,\varepsilon}(g)\, u_{s,\varepsilon}]\,(f)\, \varphi(g)\, dg$$
$$= \frac{1}{\Gamma(\frac{s-\rho+1+\varepsilon}{2})} \int_G \int_B [b, g \cdot e_n]^{s-\rho,\varepsilon} f(b)\, \varphi(g) d\sigma(b) dg,$$

which shows that $\pi'_{s,\varepsilon}(\varphi)\, u_{s,\varepsilon}$ is an element of $\mathcal{H}_{s,\varepsilon}(\Xi)$ given by

$$[\pi'_{s,\varepsilon}(\varphi)\, u_{s,\varepsilon}](\xi) = \frac{1}{\Gamma(\frac{s-\rho+1+\varepsilon}{2})} \int_G [\xi, g \cdot e_n]^{s-\rho,\varepsilon} \varphi(g)\, dg.$$

We leave the proof to the reader (for $\operatorname{Re} s \gg \rho$ this is clear; for other s use analytic continuation).

Definition 9.2.10. The distribution $\zeta_{s,\varepsilon}$ is the anti-linear form on $D(G)$ defined by

$$\zeta_{s,\varepsilon}(\varphi) = \langle \pi'_{s,\varepsilon}(\bar{\varphi})\, u_{s,\varepsilon}, u_{-s,\varepsilon} \rangle.$$

Proposition 9.2.11. *The distribution $\zeta_{s,\varepsilon}$ is a spherical distribution:*

- $\zeta_{s,\varepsilon}$ *is bi-H-invariant,*
- $\Delta\, \zeta_{s,\varepsilon} = (s^2 - \rho^2)\, \zeta_{s,\varepsilon}$.

It is instructive to compare the construction of $\zeta_{s,\varepsilon}$ with the construction of the spherical functions in the Riemannian case, see Section 7.5.

We know by Theorem 9.2.5 that $\dim D'_\lambda(X)^H = 2$. Notice that $\zeta_{s,\varepsilon}$ and $\zeta_{-s,\varepsilon}$ ($\varepsilon = 0, 1$) belong to $D'_\lambda(X)^H$ for $\lambda = s^2 - \rho^2$. We shall show later on that $\zeta_{s,\varepsilon} = \zeta_{-s,\varepsilon}$. Notice that $-I_{n+1} \in \mathrm{O}(1,n)$ and $L_{-I_{n+1}}\zeta_{s,\varepsilon} = (-1)^\varepsilon \zeta_{s,\varepsilon}$, so that $\zeta_{s,0}$ and $\zeta_{s,1}$ span $D'_\lambda(X)^H$ provided both do not vanish.

(vii) Intertwining operators

Let f be a function in $\mathcal{H}_{-s,\varepsilon}(\Xi)$ and set

$$W_{s,\varepsilon}(f) = \int_B [\xi^0, b]^{s-\rho,\varepsilon} f(b)\, d\sigma(b).$$

Theorem 9.2.12. *The integral $W_{s,\varepsilon}(f)$ exists for* $\mathrm{Re}\, s > 0$ *and is holomorphic in this region. It can be meromorphically extended to the complex plane with at most simple poles in $-\mathbb{N}$. The mapping $f \mapsto W_{s,\varepsilon}(f)$ is a continuous linear form on $D_\varepsilon(B)$ for $s \notin -\mathbb{N}$.*

In Section 7.5 (vi) we have computed $W_{s,0}(1)$, which exists for $\mathrm{Re}\, s > 0$. Hence $W_{s,\varepsilon}(f)$ exists for $\mathrm{Re}\, s > 0$, and is clearly holomorphic there. Identifying $B = K\xi^0$ with $S^o \times S^{n-1}$, where $S^0 = \{\pm 1\}$, the function $F_+(\sigma) = [\xi^0, \sigma] = 1 - \sigma_n$ ($\sigma \in S^{n-1}$) vanishes at $\sigma_n = 1$, so at ξ^0. This point turns out to be a non-degenerate critical point of F_+ with signature $(0, n-1)$. Similar observations hold for $F_-(\sigma) = -1 - \sigma_n$. Observe that $F_+ \geq 0$, $F_- \leq 0$. Applying now Morse's lemma and [17], the theorem follows.

Recall from Section 7.5 (vi) that

$$W_{s,0}(1) = 2^{s+\rho-1} \frac{\Gamma(\frac{n}{2})}{\sqrt{\pi}} \frac{\Gamma(s)}{\Gamma(s+\rho)}.$$

For $f \in \mathcal{H}_{-s,\varepsilon}(\Xi)$ set

$$(A_{s,\varepsilon} f)(\xi) = \frac{1}{W_{s,0}(1)} \int_B [\xi, b]^{s-\rho,\varepsilon} f(b)\, db.$$

Theorem 9.2.13. (a) *As a function of* s, $(A_{s,\varepsilon} f)(\xi)$ *is a holomorphic function for* $\operatorname{Re} s > 0$, *that admits a meromorphic extension to* $\mathbb{C}$ *with at most simple poles in* $-\rho - \mathbb{N}$.

(b) *If* $s \notin -\rho - \mathbb{N}$, *the function* $A_{s,\varepsilon} f$ *is in* $\mathcal{H}_{s,\varepsilon}(\Xi)$ *and the mapping* $A_{s,\varepsilon} : \mathcal{H}_{-s,\varepsilon} \to \mathcal{H}_{s,\varepsilon}$ *is continuous.*

(c) *The operator* $A_{s,\varepsilon}$ *intertwines the representations* $\pi_{-s,\varepsilon}$ *and* $\pi_{s,\varepsilon}$, *that is*

$$A_{s,\varepsilon} \circ \pi_{-s,\varepsilon} = \pi_{s,\varepsilon} \circ A_{s,\varepsilon}.$$

According to Theorem 9.2.12 the mapping $s \mapsto W_{s,\varepsilon}$ defined for $\operatorname{Re} s > 0$ with values in $D'(B)$ has a meromorphic extension and the operator $A_{s,\varepsilon}$ may be regarded as the convolution with the linear form $W_{s,0}(1)^{-1} W_{s,\varepsilon}$ on the group K, so $A_{s,\varepsilon} f \in \mathcal{H}_{s,\varepsilon}(\Xi)$ for any $f \in \mathcal{H}_{-s,\varepsilon}(\Xi)$ and $A_{s,\varepsilon} : \mathcal{H}_{-s,\varepsilon} \to \mathcal{H}_{s,\varepsilon}$ is continuous. Property (c) is easily checked for $\operatorname{Re} s > 0$, and hence it is valid for all $s \notin -\rho - \mathbb{N}$ by analytic continuation.

(viii) Diagonalization of the intertwining operators

We may consider the intertwining operators $A_{s,\varepsilon}$ as acting on $D_\varepsilon(B)$, the space of C^∞ functions f on B satisfying $f(-b) = (-1)^\varepsilon f(b)$. Observe that $B \simeq S^0 \times S^{n-1}$, where $S^0 = \{-1, 1\}$. The structure of the $\mathrm{O}(n)$-action on $L^2(S^{n-1})$, being a closed subspace of $L^2(B)$, is well known from Section 7.3. Since its decomposition into $\mathrm{O}(n)$-irreducible subspaces is multiplicity free, and since $A_{s,\varepsilon}$ commutes with the action of $\mathrm{O}(n)$, $A_{s,\varepsilon}$ acts as a scalar on each irreducible component of $L^2(S^{n-1})$.

As in Section 7.3, let us write $L^2(S^{n-1}) = \bigoplus_{l=0}^{\infty} \mathcal{H}_l$, where $\mathcal{H}_l$ is the space of harmonic polynomials on $\mathbb{R}^n$, homogeneous of degree l, restricted to S^{n-1}. Clearly, $D_\varepsilon(B)$ is now spanned (in the C^∞ topology) by the spaces $\mathcal{Y}_{l,m} = \sigma_0^m \mathcal{H}_l$ with $l + m \equiv \varepsilon \pmod 2$. Actually, it is sufficient to consider $m = 0$ and $m = 1$ only. To determine the scalar action of $A_{s,\varepsilon}$ on each $\mathcal{Y}_{l,m}$ we introduce a special differential operator on $D_\varepsilon(B)$, namely

$$\pi_{s,\varepsilon}(L) = \frac{d}{dt}\Big|_{t=0} \pi_{s,\varepsilon}(a_t).$$

This operator commutes with the action of the subgroup $M \simeq \mathrm{O}(n-1)$. So if f is a function on $D_\varepsilon(B)$ invariant under the action of $\mathrm{O}(n-1)$, then the same is true for $\pi_{s,\varepsilon}(L) f$. A function in $D(B)$, invariant under the action of M, depends only on the variables σ_0 and σ_n of S^0 and S^{n-1} respectively. Notice that $\sigma_0 = \pm 1$. We shall write down the differential operator associated with $\pi_{s,\varepsilon}(L)$ acting on such functions.

Lemma 9.2.14. *For a function f in $D(B)$, invariant under* $\mathrm{O}(n-1)$, *one has*

$$\pi_{s,\varepsilon}(L)f = (s-\rho)\sigma_0\sigma_n f + \sigma_0(1-\sigma_n^2)\,\frac{\partial f}{\partial \sigma_n}.$$

Write $a_{-t}k = la_u n$ according to the Iwasawa decomposition. Here $k, l \in K$, $n \in N$ and $u, t \in \mathbb{R}$. Set $[ke_n, e_n] = \sigma_n$, $[ke_0, e_0] = \sigma_0$, $[le_n, e_n] = \sigma_n'$, $[le_0, e_0] = \sigma_0'$. We have to determine σ_0' and σ_n' in terms of t, σ_0 and σ_n.

Starting from $a_{-t}k\xi^0 = \mathrm{e}^u l\xi^0$ we get

(1) $[a_{-t}k\xi^0, e_n] = \mathrm{e}^u\,[l\xi^0, e_n]$,

(2) $[a_{-t}k\xi^0, e_0] = \mathrm{e}^u\,[l\xi^0, e_0]$,

and hence

(1)′ $\sinh t\,\sigma_0 + \cosh t\,\sigma_n = \mathrm{e}^u\,\sigma_n'$,

(2)′ $\cosh t\,\sigma_0 + \sinh t\,\sigma_n = \mathrm{e}^u\,\sigma_0'$.

From (2)′ we may conclude that $\sigma_0 = \sigma_0'$, and then

$$\mathrm{e}^u = \sigma_0\,(\cosh t\,\sigma_0 + \sinh t\,\sigma_n).$$

So if f is M-invariant (only depending on σ_0 and σ_n), we have to compute

$$\frac{d}{dt}\bigg|_{t=0} [\sigma_0\,(\cosh t\,\sigma_0 + \sinh t\,\sigma_n)]^{s-\rho}\, f\left(\sigma_0, \frac{\sinh t\,\sigma_0 + \cosh t\,\sigma_n}{\sigma_0(\cosh t\,\sigma_0 + \sinh t\,\sigma_n)}\right).$$

The result then follows easily.

The $\mathrm{O}(n-1)$-invariant functions in $\mathcal{H}_l$ are proportional to the Gegenbauer polynomials $C_l^{\frac{n-2}{2}}$. Set

$$\omega_{l,m}\,(\sigma_0, \sigma_n) = \sigma_0^m\, C_l^{\frac{n-2}{2}}\,(\sigma_n).$$

We recall two relations for Gegenbauer polynomials C_l^λ (see [13, 3.15.2 (27–30)]), that can easily be derived from other relations for the C_l^λ.

(a) $z\,C_l^\lambda(z) = \frac{l+1}{2(\lambda+l)} C_{l+1}^\lambda(z) + \frac{2\lambda+l-1}{2(\lambda+l)} C_{l-1}^\lambda(z)$,

(b) $(1-z^2)\,\frac{d}{dz} C_l^\lambda(z) = (2\lambda+l)zC_l^\lambda(z) - (l+1)C_{l+1}^\lambda(z)$.

One clearly has

$$\pi_{s,\varepsilon}(L)\,\omega_{l,m} = a_{l,m}^{(1)}(s)\,\omega_{l+1,m+1} + a_{l,m}^{(-1)}(s)\,\omega_{l-1,m+1},$$

where

$$a_{l,m}^{(1)}(s) = (s-\rho-l)\,\frac{l+1}{n-2+2l}, \tag{9.2.5}$$

$$a_{l,m}^{(-1)}(s) = (n-2+l+s-\rho)\,\frac{n-3+l}{n-2+2l}. \tag{9.2.6}$$

The operator $A_{s,\varepsilon}$ acts on $\mathcal{Y}_{l,m}$ as a scalar $\alpha_{l,m}(s)$ and one also has

$$A_{s,\varepsilon} \circ \pi_{-s,\varepsilon}(L) = \pi_{s,\varepsilon}(L) \circ A_{s,\varepsilon},$$

which implies

$$a_{l,m}^{(1)}(-s)\,\alpha_{l+1,m+1}(s) = \alpha_{l,m}(s)\,a_{l,m}^{(1)}(s), \tag{9.2.7}$$

$$a_{l,m}^{(-1)}(-s)\,\alpha_{l-1,m+1}(s) = \alpha_{l,m}(s)\,a_{l,m}^{(-1)}(s). \tag{9.2.8}$$

So, if $s-\rho$ is not an integer, (9.2.7) determines the row $\alpha_{l,m}(s)$ (since $\alpha_{0,0}(s) = \alpha_{0,1}(s) = 1$) and hence determines $A_{s,\varepsilon}$. The same holds for (9.2.8). This yields the following theorem.

Theorem 9.2.15. *The eigenvalues $\alpha_{l,m}(s)$ of the operator $A_{s,\varepsilon}$ are given by*

$$\alpha_{l,m}(s) = \frac{Q_l(-s)}{Q_l(s)}$$

where Q_l is the polynomial defined by

$$Q_l(s) = \prod_{j=1}^{l}(\rho + s + j - 1) \qquad (l \geq 1).$$

If $s-\rho$ is not an integer, then

$$\alpha_{l,m}(-s) = \alpha_{l,m}(s)^{-1}$$

and thus $A_{s,\varepsilon} \circ A_{-s,\varepsilon} = I$ on $D_\varepsilon(B)$. This can also be shown directly by using Schur's lemma and applying the irreducibility of $\pi_{s,\varepsilon}$ for such values of s.

Notice that $a_{l,m}^{(1)}, a_{l,m}^{(-1)}, \alpha_{l,m}$ are independent of the parameter m.

(ix) Fourier transform

Let φ_0 be a function in $D(G)$. In paragraph (vi) we have seen that the distribution $\pi'_{s,\varepsilon}(\varphi_0)\,u_{s,\varepsilon}$ is an element of $\mathcal{H}_{s,\varepsilon}(\Xi)$ given by

$$[\pi'_{s,\varepsilon}(\varphi_0)\,u_{s,\varepsilon}](\xi) = \frac{1}{\Gamma(\frac{s-\rho+1+\varepsilon}{2})} \int_G [\xi, g \cdot e_n]^{s-\rho,\varepsilon} \varphi_0(g)\,dg.$$

This function depends only on the function φ in $D(X)$ defined by

$$\varphi(x) = \int_H \varphi_0(gh)\,dh \qquad (x = g \cdot e_n),$$

since

$$[\pi'_{s,\varepsilon}(\varphi_0)\,u_{s,\varepsilon}](\xi) = \frac{1}{\Gamma(\frac{s-\rho+1+\varepsilon}{2})} \int_X [\xi, x]^{s-\rho,\varepsilon} \varphi(x)\,dx.$$

We define:

Definition 9.2.16. Let φ be a function in $D(X)$. The *Fourier transform* of φ is the function $\widehat{\varphi}$ defined on $\Xi \times \mathbb{C} \times \{0,1\}$ by

$$\widehat{\varphi}(\xi, s, \varepsilon) = \frac{1}{\Gamma(\frac{s-\rho+1+\varepsilon}{2})} \int_X [\xi, x]^{s-\rho,\varepsilon} \varphi(x)\, dx.$$

This expression is perfectly well-defined for $\operatorname{Re} s > \rho$; for other s we apply analytic continuation.

Proposition 9.2.17. *The Fourier transform $\widehat{\varphi}$ of $\varphi \in D(X)$ has the following properties:*

(a) *$\widehat{\varphi}(\xi, s, \varepsilon)$ is an entire function of s,*

(b) *$\widehat{\varphi}(\xi, s, \varepsilon)$ is a C^∞ function of ξ and belongs to $\mathcal{H}_{s,\varepsilon}(\Xi)$,*

(c) *the Fourier transform commutes with the G-action, that is*

$$(L_g\varphi)\hat{}(., s, \varepsilon) = \pi_{s,\varepsilon}(g)[\widehat{\varphi}(., s, \varepsilon)]$$

for all $g \in G$,

(d) *the Laplace–Beltrami operator satisfies*

$$\widehat{\Delta\varphi}(\xi, s, \varepsilon) = (s^2 - \rho^2)\,\widehat{\varphi}(\xi, s, \varepsilon).$$

Let us return to the spherical distributions $\zeta_{s,\varepsilon}$, which may be regarded as H-invariant distributions on X. We may write

$$\zeta_{s,\varepsilon}(\varphi) = \langle \widehat{\widehat{\varphi}}(., s, \varepsilon), u_{-s,\varepsilon} \rangle,$$

so for $\operatorname{Re} s < -\rho$,

$$\zeta_{s,\varepsilon}(\varphi) = \frac{1}{\Gamma(\frac{-s-\rho+1+\varepsilon}{2})} \int_B \widehat{\varphi}(b, s, \varepsilon)\, [b, e_n]^{-s-\rho,\varepsilon}\, db.$$

(x) Fourier transform of K-finite functions

According to Proposition 9.2.1 any $x \in X$ can be written as

$$x = k a_t \cdot e_n \qquad (k \in K,\ t \geq 0).$$

This expression depends only on the class kM where M is, as before, the centralizer of A in K, stabilizing e_n (for $t = 0$ we have to choose $\mathrm{O}(1) \times M$).

Because $B \simeq K/M$ we may parametrize X by $[0,\infty) \times B$. In this coordinate system the invariant measure dx on X is given by (see (i))

$$dx = \frac{4\pi^{\frac{n}{2}}}{\Gamma(\frac{n}{2})} A(t)dtdb,$$

with $A(t) = \cosh^{n-1} t$. We shall also need the Laplace–Beltrami operator in these coordinates. Here we use again that $B \simeq S^0 \times S^{n-1}$ and invoke the Laplacian Ω on S^{n-1}, see Section 7.3 (v). One has:

Lemma 9.2.18. *In the coordinates of* $[0,\infty) \times B$ *on* X *the Laplace–Beltrami operator* Δ *is given by*

$$\Delta f = \frac{1}{A(t)}\frac{\partial}{\partial t}\Big(A(t)\frac{\partial f}{\partial t}\Big) - \frac{1}{\cosh^2 t}\Omega f.$$

The proof of this lemma, which the reader may postpone to a later occasion, can be given along two lines, the differential geometric one (see, e.g., [21]) or the Lie algebra theoretic one. We shall follow the Lie algebra approach.

We return to the notations used in the proof of Proposition 9.2.7. First let $Z \in \mathfrak{g}$ and $x \in X$. We define the following directional derivative of $f \in C^\infty(X)$ at x:

$$(Zf)(x) = \frac{d}{ds}\Big|_{s=0} f(ka_t \exp sZ \cdot e_n),$$

if $x = ka_t \cdot e_n$. This definition depends on the choice of k, so we will write Z_k instead of Z for the time being. If also $x = k'a_t \cdot e_n$, then $k' = km$ for some $m \in M$, so $Z_{k'} = (\mathrm{Ad}(m)Z)_k$. We also define Z_k^2 as

$$(Z_k^2 f)(x) = \frac{d^2}{ds^2}\Big|_{s=0} f(ka_t \exp sZ \cdot e_n).$$

Clearly, $Z_{k'}^2 = (\mathrm{Ad}(m)Z)_k^2$.

Write

$$\omega_k = -\sum_{j=1}^{n-1}(N_j + \theta N_j)_k^2 + \frac{L^2}{2(n-1)}.$$

This operator does not depend on the choice of k anymore, since it is M-invariant, and we just write $\omega_k = \omega_x$. It is precisely the operator $\omega_{\mathfrak{q}}$ taken at the point x. Call

$$Y_j = N_j + \theta N_j \quad \text{and} \quad Y_j' = N_j - \theta N_j \qquad (j = 1,\dots,n-1).$$

Observe that the Y_j span $\mathfrak{q} \cap \mathfrak{k}$, whereas the Y_j' span $\mathfrak{h} \cap \mathfrak{p}$. One obviously has

$$\begin{aligned}\mathrm{Ad}(a_t)Y_j &= \cosh t\, Y_j + \sinh t\, Y_j',\\ \mathrm{Ad}(a_t)Y_j' &= \sinh t\, Y_j + \cosh t\, Y_j'.\end{aligned}$$

Notice that

$$\frac{d^2}{ds^2} f(ka_t \exp sY_j \cdot e_n) = \frac{d^2}{ds^2} f(ka_t \exp sY_j \exp c_j sY_j' \cdot e_n)$$

for every choice of $c_j \in \mathbb{R}$. We shall make a special choice for the c_j later on.

We have

$$\begin{aligned}\frac{d^2}{ds^2} & f(k\, a_t \exp sY_j \cdot e_n) \\ &= \frac{d^2}{ds^2} f(k \exp[\cosh t\, sY_j + \sinh t\, sY_j'] \cdot \exp(c_j[\sinh t\, sY_j + \cosh t\, sY_j')]\, a_t \cdot e_n).\end{aligned}$$

Taking $c_j = -\tanh t$ and using the first two terms in the Baker–Campbell–Hausdorff formula (see [55, Theorem 2.15.4]), we obtain

$$\begin{aligned}\exp[\cosh t\, sY_j + \sinh t sY_j'] \cdot \exp\Big(-\Big[\frac{\sinh^2 t}{\cosh t}\, sY_j + \sinh t\, sY_j'\Big]\Big) \\ = \exp(\cosh^{-1} t\, sY_j - \tanh t\, s^2[Y_j, Y_j'] + o\,(s^2)).\end{aligned}$$

So,

$$\begin{aligned}\frac{d^2}{ds^2}\Big|_{s=0} & f(k\, a_t \exp sY_j \cdot e_n) \\ &= \frac{d^2}{ds^2}\Big|_{s=0} f(k \exp(\cosh^{-1} t\, sY_j - \frac{1}{2} \tanh t\, s^2\, [Y_j, Y_j'])\, a_t \cdot e_n).\end{aligned}$$

One has $[Y_j, Y_j'] = -2[N_j, \theta N_j] = \frac{L}{2(n-1)}$ for all j (cf. the proof of Proposition 9.2.7).

In order to formulate in a proper way the result we have obtained so far, we define for $Z \in \mathfrak{k}$ and $x = k \cdot e_n$ the differentiation operator Z_k^2 on S^{n-1} by

$$(Z_k^2 f)(x) = \frac{d^2}{ds^2}\Big|_{s=0} f(k \exp sZ \cdot e_n).$$

As before, $\omega_{S^{n-1}} = \sum_{j=1}^{n-1} (Y_j)_k^2$ is independent of the choice of k and we also have

$$\omega_{S^{n-1}} f(k \cdot e_n) = \sum_{j=1}^{n-1} \frac{d^2}{ds^2}\Big|_{s=0} f(\exp sY_j\, k \cdot e_n).$$

We have now, writing $f(k\, a_t \cdot e_n) = f(t, b) = f(t, \sigma_0, \sigma_n)$,

$$\begin{aligned}\omega_{\mathfrak{q}} &= \Big(\frac{L^2}{2(n-1)} + \frac{L}{2} \tanh t - \frac{1}{\cosh^2 t} \omega_{S^{n-1}}\Big) f \\ &= \frac{1}{2(n-1)} \Big[\frac{1}{A(t)} \frac{\partial}{\partial t} \Big(A(t) \frac{\partial f}{\partial t}\Big)\Big] - \frac{1}{\cosh^2 t}\, \omega_{S^{n-1}} f.\end{aligned}$$

We know that $\Delta = 2(n-1)\,\omega_{\mathfrak{q}}$, so we get

$$\Delta f = \frac{1}{A(t)}\frac{\partial}{\partial t}\left(A(t)\frac{\partial f}{\partial t}\right) - 2(n-1)\frac{1}{\cosh^2 t}\,\omega_{S^{n-1}} f.$$

We finally have to show that $2(n-1)\,\omega_{S^{n-1}}$ coincides with the Laplacian Ω on S^{n-1}. This is easily done: we know that the two differential operators are proportional and that Ω acts on the harmonic polynomial s_1+is_2 on S^{n-1} as the scalar $-(n-1)$, see Section 7.3 (vi) (3); computing now the action of $2(n-1)\,\omega_{S^{n-1}}$ on this polynomial gives the same scalar.

Consider a function $\varphi \in D(X)$, K-finite of type (l,m), i.e.

$$\varphi(x) = \varphi(k\,a_t \cdot e_n) = F(t)\,Y(b) \qquad (k \in K,\ t \geq 0,\ b \in K\xi^0),$$

where $Y \in \mathcal{Y}_{l,m}$. The function F is in $C_c^\infty(\mathbb{R})$ and satisfies the relation $F(t) = (-1)^m F(-t)$. We are going to make explicit the Fourier transform $\widehat{\varphi}(\xi,s,\varepsilon)$ of such a function. Notice that, because of the K-equivariance, we must have $l+m \equiv \varepsilon \pmod 2$ because otherwise the Fourier transform would vanish.

Consider $Y(b)$ as a function on $S^0 \times S^{n-1}$, $b = (\sigma_0, \sigma)$ with $\sigma_0 \in S^0$ ($\sigma_0 = \pm 1$) and $\sigma \in S^{n-1}$, and consider the function $\xi \mapsto \widehat{\varphi}(\xi,s,\varepsilon)$ as a function on $\mathbb{R}_* \times S^0 \times S^{n-1}$:

$$\xi = (\lambda\tau_0,\ \lambda\tau) \qquad (\lambda \in \mathbb{R},\ \lambda \neq 0,\ \tau_0 \in S^0,\ \tau \in S^{n-1}).$$

We obtain

$$\widehat{\varphi}(\xi,s,\varepsilon) = \frac{4\pi^{\frac{n}{2}}}{\Gamma(\frac{n}{2})}\,\frac{\lambda^{s-\rho,\varepsilon}}{\Gamma(\frac{s-\rho+1+\varepsilon}{2})}$$
$$\times \int_0^\infty \int_{S^0}\int_{S^{n-1}} [-\sinh t\ \sigma_0\tau_0 + \cosh t\ \langle\sigma,\tau\rangle]^{s-\rho,\varepsilon} F(t) Y(\sigma_0,\sigma) A(t)\, d\sigma_0 d\sigma,$$

where, for the moment, $\langle\ ,\ \rangle$ denotes the usual scalar product in $\mathbb{R}^n$. As said before, $\widehat{\varphi}(b,s,\varepsilon)$ belongs to $\mathcal{Y}_{l,m}$ again and, because of the invariance under K and the irreducibility of $\mathcal{Y}_{l,m}$ under K, we get

$$\widehat{\varphi}(\lambda\tau_0,\lambda\tau) = \lambda^{s-\rho,\varepsilon}\,\widetilde{F}(s,\varepsilon)\,Y(\tau_0,\tau)$$

where $\widetilde{F}(s,\varepsilon)$ only depends on (l,m) and F (and on s and ε, of course).

To compute $\widetilde{F}(s,\varepsilon)$ we consider a special function Y, namely

$$Y(\sigma_0,\sigma) = \sigma_0^m\, C_l^{\frac{n-2}{2}}(\sigma_n)/C_l^{\frac{n-2}{2}}(1).$$

This function was previously also called $\omega_{l,m}/\omega_{l,m}(\xi^0)$. We obtain

$$\widetilde{F}(s,\varepsilon) = \int_0^\infty \Phi_{l,m}(t,s,\varepsilon)\,F(t)\,A(t)\,dt,$$

where

$$\Phi_{l,m}(t,s,\varepsilon) = \frac{4\pi^{\frac{n-1}{2}}}{\Gamma(\frac{n-1}{2})} [C_l^{\frac{n-2}{2}}(1)]^{-1} \frac{1}{\Gamma(\frac{s-\rho+1+\varepsilon}{2})} \tag{9.2.9}$$

$$\times \int_{S^0} \int_{-1}^{1} [-\sinh t\, \sigma_0 + \cosh t\, \sigma_n]^{s-\rho,\varepsilon}\, \sigma_0^m\, C_l^{\frac{n-2}{2}}(\sigma_n)(1-\sigma_n^2)^{\frac{n-2}{2}}\, d\sigma_0\, d\sigma_n.$$

The integral equals

$$\int_{S^0} \int_{-1}^{1} [-\sinh t + \cosh t\, \sigma_n]^{s-\rho,\varepsilon}\, \sigma_0^{m+l+\varepsilon}\, C_l^{\frac{n-2}{2}}(\sigma_n)\,(1-\sigma_n^2)^{\frac{n-2}{2}}\, d\sigma_0 d\sigma_n$$

$$= \int_{-1}^{1} [-\sinh t + \cosh t\, \sigma_n]^{s-\rho,\varepsilon}\, C_l^{\frac{n-2}{2}}(\sigma_n)\,(1-\sigma_n^2)^{\frac{n-2}{2}}\, d\sigma_n.$$

Notice that this integral is absolutely convergent for $\operatorname{Re} s > \rho$ and all $t \in \mathbb{R}$. Moreover,

$$\Phi_{l,m}(-t,s,\varepsilon) = (-1)^m\, \Phi_{l,m}(t,s,\varepsilon). \tag{9.2.10}$$

The relation

$$\widehat{\Delta\varphi}\,(\xi,s,\varepsilon) = (s^2-\rho^2)\,\widehat{\varphi}(\xi,s,\varepsilon)$$

implies that the function $\Phi_{l,m}(t,s,\varepsilon)$, for $\operatorname{Re} s$ large, is a solution of the differential equation

$$\frac{1}{A(t)}\frac{d}{dt}\left(A(t)\,\frac{du}{dt}\right) + \frac{l(l+n-2)}{\cosh^2 t} u = (s^2-\rho^2)\,u. \tag{9.2.11}$$

The change of variable $z = \tanh^2 t$ leads to the hypergeometric differential equation, and we get the following two, linearly independent, solutions of (9.2.11) on $\mathbb{R}$:

- $\Psi_{l,0}(t,s,\varepsilon) = (\cosh t)^{s-\rho}\, {}_2F_1(\frac{-s+\rho+l}{2}, \frac{-s-\rho-l+1}{2}; \frac{1}{2}; \tanh^2 t)$,
- $\Psi_{l,1}(t,s,\varepsilon) = (\tanh t)\,(\cosh t)^{s-\rho}\, {}_2F_1(\frac{-s+\rho+l+1}{2}, \frac{-s-\rho-l+2}{2}; \frac{3}{2}; \tanh^2 t)$.

Because of relation (9.2.10) we get

$$\Phi_{l,m}(t,s,\varepsilon) = \beta_{l,m}(s,\varepsilon)\,\Psi_{l,m}(t,s,\varepsilon).$$

The numbers $\beta_{l,m}(s,\varepsilon)$ can be calculated and we obtain

$$\beta_{l,m}(s,\varepsilon) = 4\pi^{\frac{n-1}{2}}\, 2^{\frac{-l+|\varepsilon-m|}{2}}$$

$$\times (-1)^m \frac{(s-\rho-\varepsilon)(s-\rho-\varepsilon-2)\cdots(s-\rho-m-l+2)}{\Gamma(\frac{s+\rho+l-m+1}{2})},$$

where for $m+l<2$ the nominator of the fraction has to be taken equal to 1.

Clearly,

$$\beta_{l,m}(s,\varepsilon) = \frac{4\pi^{\frac{n-1}{2}}}{\Gamma(\frac{n-1}{2})} \, [C_l^{\frac{n-2}{2}}(1)]^{-1} \, \frac{1}{\Gamma(\frac{s-\rho+1+\varepsilon}{2})} \tag{9.2.12}$$

$$\times (s-\rho)^m (-1)^m \int_{-1}^{1} |\sigma_n|^{s-\rho-m} \operatorname{sgn} \sigma_n^{\varepsilon+m} \, C_l^{\frac{n-2}{2}}(\sigma_n) \, (1-\sigma_n^2)^{\frac{n-3}{2}} \, d\sigma_n.$$

This integral can be computed, using Rodrigues' formula for Gegenbauer polynomials:

$$C_l^\lambda(x)\,(1-x^2)^{\lambda-\frac{1}{2}} = A_l^\lambda \left(\frac{d}{dx}\right)^l \left[(1-x^2)^{l+\lambda-\frac{1}{2}}\right]$$

with

$$A_l^\lambda = \frac{(-1)^l \, (2\lambda)_l}{2^l \, l! \, (\lambda+\frac{1}{2})_l},$$

where, as usual,

$$(\alpha)_l = \frac{\Gamma(\alpha+l)}{\Gamma(\alpha)} = \alpha(\alpha+1)\cdots(\alpha+l-1).$$

Performing l partial integrations we get the result.

We remark that

$$\Psi_{l,m}(t,-s,\varepsilon) = \Psi_{l,m}(t,s,\varepsilon). \tag{9.2.13}$$

Summarizing we now have

$$\widehat{\varphi}(b,s,\varepsilon) = Y(b)\,\beta_{l,m}(s,\varepsilon) \int_0^\infty \Psi_{l,m}(t,s,\varepsilon)\,F(t)\,A(t)\,dt, \tag{9.2.14}$$

where $\beta_{l,m}(s,\varepsilon)$ and $\Psi_{l,m}(t,s,\varepsilon)$ are equal to, respectively,

$$4\,\pi^{\frac{n-1}{2}}\, 2^{\frac{-l+|\varepsilon+m|}{2}} \, (-1)^m \, \frac{(s-\rho-\varepsilon)(s-\rho-\varepsilon-2)\cdots(s-\rho-m-l+2)}{\Gamma(\frac{s+\rho+l-m+1}{2})}$$

and

$$(\tanh t)^m \, (\cosh t)^{s-\rho} \, {}_2F_1\Big(\frac{-s+\rho+l+1}{2}, \frac{-s-\rho-l+2}{2}; \frac{3}{2}; \tanh^2 t\Big).$$

Denote by $\mathcal{F}_{s,\varepsilon}$ the (continuous) mapping from $D(X)$ to $\mathcal{H}_{s,\varepsilon}(\Xi)$ defined by

$$\mathcal{F}_{s,\varepsilon} : \varphi \mapsto \widehat{\varphi}(.,s,\varepsilon).$$

We will study its image. To do this, we introduce the sets

$$E(s,\varepsilon) = \{(l,m) : l+m \equiv \varepsilon \,(\mathrm{mod}\, 2),\ \beta_{l,m}(s,\varepsilon) \neq 0\},$$

and set

$$\mathcal{I}_{s,\varepsilon} = \sum_{(l,m)\in E(s,\varepsilon)} \mathcal{Y}_{l,m}.$$

Then $\mathcal{I}_{s,\varepsilon} \subset \mathcal{F}_{s,\varepsilon}(D(X)) \subset \overline{\mathcal{I}}_{s,\varepsilon}$.

The study of the numbers $\beta_{l,m}(s,\varepsilon)$ leads to the consideration of the following values of s:

$$\rho + \varepsilon + 2r, \quad r \in \mathbb{N}, \tag{9.2.15}$$

$$-1 - \rho + \varepsilon - 2h, \quad h \in \mathbb{N}. \tag{9.2.16}$$

If s is not equal to one of these values, we have $\overline{\mathcal{I}}_{s,\varepsilon} = D_\varepsilon(B)$, so the image of $\mathcal{F}_{s,\varepsilon}$ is dense for such s. On the other hand, if $r \in \mathbb{N}$, then

$$E(\rho + \varepsilon + 2r, \varepsilon) = \{(l,m) : l + m \equiv \varepsilon (\mathrm{mod}\, 2),\ l + m \le 2r + \varepsilon\},$$

and if $h \in \mathbb{N}$, then

$$E(-1 - \rho + \varepsilon - 2h, \varepsilon) = \{(l,m) : l + m \equiv \varepsilon (\mathrm{mod}\, 2),\ l - m > 2h - \varepsilon\}.$$

Proposition 9.2.19. *If s is not equal to one of the values in* (9.2.15) *or* (9.2.16), *in particular if $s + \rho$ is not an integer, then the set*

$$\{\pi'_{s,\varepsilon}(\varphi)\, u_{s,\varepsilon} : \varphi \in D(G)\}$$

is a dense subspace of $\mathcal{H}_{s,\varepsilon}(\Xi)$.

In fact, this set equals

$$\{\widehat{\varphi}(.,s,\varepsilon) : \varphi \in D(X)\} = \mathcal{F}_{s,\varepsilon}(D(X)).$$

This equality says that $u_{s,\varepsilon}$ is a cyclic vector of $\pi_{s,\varepsilon}$ in a generalized sense.

We end this section with a very useful corollary. Setting as in (vii),

$$\omega_{l,m}(b) = \omega_{l,m}(\sigma_0, \sigma_n) = \sigma_0^m\, C_l^{\frac{n-2}{2}}(\sigma_n),$$

we obtain from (9.2.12), taking $m = 0$:

Corollary 9.2.20. *One has*

$$\frac{1}{\Gamma(\frac{s-\rho+1+\varepsilon}{2})} |x|^{s-\rho,\varepsilon} = \sum_{l\equiv\varepsilon (\mathrm{mod}\, 2)} c_l\, \beta_{l,0}(s,\varepsilon)\, \omega_{l,0}(x)$$

where the numbers c_l are non-zero and the convergence is taken in the space $D'((-1,1))$.

(xi) Expansion of $\zeta_{s,\varepsilon}(\varphi)$ when φ is a K-finite function

Suppose that φ is a function of the form

$$\varphi(x) = F(t)\, Y(b) \qquad (x = ka_t \cdot e_n,\ t \geq 0,\ b = k \cdot \xi^0)$$

where $y \in \mathcal{Y}_{l,m}$ with $l + m \equiv \varepsilon \,(\mathrm{mod}\, 2)$. We apply the expression

$$\zeta_{s,\varepsilon}(\overline{\varphi}) = \frac{1}{\Gamma(\frac{-s-\rho+1+\varepsilon}{2})} \int_B \widehat{\varphi}(b, s, \varepsilon)\, [b, e_n]^{-s-\rho,\varepsilon}\, db,$$

which holds for $\mathrm{Re}\, s < -\rho$, and we obtain

$$\zeta_{s,\varepsilon}(\overline{\varphi}) = \beta_{l,m}(s, \varepsilon) \int_0^\infty \Psi_{l,m}(t, s, \varepsilon)\, F(t)\, A(t)\, dt$$
$$\times \frac{1}{\Gamma(\frac{-s-\rho+1+\varepsilon}{2})} \int_B Y(b)\, [b, e_n]^{-s-\rho,\varepsilon}\, db.$$

Applying Corollary 9.2.20 we get

$$\zeta_{s,\varepsilon}(\overline{\varphi}) = 0 \qquad \text{if } m = 1.$$

So $\zeta_{s,\varepsilon}(\overline{\varphi})$ might not vanish only if $m = 0$, and then

$$\zeta_{s,\varepsilon}(\overline{\varphi}) = c_l\, \beta_{l,0}(s, \varepsilon)\, \beta_{l,0}(-s, \varepsilon) \int_0^\infty \Psi_{l,0}(t, s, \varepsilon)\, F(t)\, A(t)\, dt \tag{9.2.17}$$
$$\times \int_B Y(b)\, \omega_{l,0}(b)\, db \qquad (l \equiv \varepsilon \,(\mathrm{mod}\, 2)).$$

Theorem 9.2.21. (a) *For all complex numbers s one has*

$$\zeta_{s,\varepsilon} = \zeta_{-s,\varepsilon} \qquad (\varepsilon = 0, 1).$$

(b) *There holds $\zeta_{s,\varepsilon} \neq 0$ for all $s \in \mathbb{C}$, $\varepsilon = 0, 1$.*

Since $\zeta_{s,\varepsilon}$ and $\zeta_{-s,\varepsilon}$ are spherical distributions belonging to the same eigenvalue $s^2 - \rho^2$, it follows from Theorem 9.2.5 that one can find complex numbers a and b (depending on s and ε) such that

$$\zeta_{-s,\varepsilon} = a\, \zeta_{s,1} + b\, \zeta_{s,0},$$

since $\zeta_{s,0}$ and $\zeta_{s,1}$ are linearly independent. This latter fact is easily seen by taking a function φ with $Y = 1$. Of course, we have to exclude some isolated values of s, but it holds for s "in general position". Taking functions φ with $Y = 1$ and $Y = \omega_{10}$ we easily get that $\zeta_{-s,\varepsilon}$ is proportional to $\zeta_{s,\varepsilon}$ and then, applying that the expression (9.2.17) is even in s, $\zeta_{-s,\varepsilon} = \zeta_{s,\varepsilon}$. The fact that $\zeta_{s,\varepsilon} \neq 0$ is clear for s in general position. One can easily show that for each s and ε there is a number l with

$$\beta_{l,0}(s, \varepsilon)\, \beta_{l,0}(-s, \varepsilon) \neq 0.$$

(xii) An intertwining relation for the Fourier transform

Theorem 9.2.22. *The Fourier transform $\mathcal{F}_{s,\varepsilon}$ satisfies the relation*

$$A_{s,\varepsilon} \circ \mathcal{F}_{-s,\varepsilon} = \gamma(s,\varepsilon)\, \mathcal{F}_{s,\varepsilon},$$

where $A_{s,\varepsilon}$ is the intertwining operator defined in (vii) *and $\gamma(s,\varepsilon)$ is the following meromorphic function:*

$$\gamma(s,\varepsilon) = \frac{\Gamma(\frac{s+\rho+1-\varepsilon}{2})}{\Gamma(\frac{-s+\rho+1-\varepsilon}{2})}.$$

Set

$$\zeta'_{s,\varepsilon}(\overline{\varphi}) = \langle A_{s,\varepsilon}\,[\widehat{\varphi}\,(.,-s,\varepsilon)], u_{-s,\varepsilon}\rangle.$$

This distribution is again spherical with eigenvalue $\lambda = s^2 - \rho^2$, and one easily shows, as in the previous section, that $\zeta'_{s,\varepsilon}$ is proportional to $\zeta_{s,\varepsilon}$:

$$\zeta'_{s,\varepsilon} = \gamma(s,\varepsilon)\, \zeta_{s,\varepsilon}.$$

To determine the factor $\gamma(s,\varepsilon)$, we consider special functions $\varphi \in D(X)$ of the form $\varphi\,(ka_t \cdot e_n) = F(t)\, Y_\varepsilon(b)$, namely with $Y_0 = 1$ (if $\varepsilon = 0$) and $Y_1 = \omega_{10}$ (if $\varepsilon = 1$). We know by Theorem 9.2.15 that $A_{s,0}\, Y_0 = Y_0$ and $A_{s,1}\, Y_1 = \frac{\rho-s}{\rho+s}\, Y_1$. So we obtain

$$\zeta_{s,\varepsilon}(\overline{\varphi}) = c_0\, \beta_{\varepsilon,0}(s,\varepsilon)\, \beta_{\varepsilon,0}(-s,\varepsilon) \int_0^\infty \Psi_{\varepsilon,0}(t,s,\varepsilon)\, F(t)\, A(t)dt,$$

$$\zeta'_{s,\varepsilon}(\overline{\varphi}) = c_0\, [\beta_{\varepsilon,0}(-s,\varepsilon)]^2 \int_0^\infty \Psi_{\varepsilon,0}(t,s,\varepsilon)\, F(t)\, A(t)dt \times \Big(\frac{\rho-s}{\rho+s}\Big)^\varepsilon.$$

This yields the announced formula for $\gamma(s,\varepsilon)$. We have shown that for all $\varphi \in D(X)$

$$\langle A_{s,\varepsilon}\,[\widehat{\varphi}\,(.,-s,\varepsilon)], u_{-s,\varepsilon}\rangle = \gamma(s,\varepsilon)\,\langle \widehat{\varphi}\,(.,s,\varepsilon), u_{-s,\varepsilon}\rangle.$$

So for any function $\psi \in D(G)$ one has

$$\langle A_{s,\varepsilon}\,[\widehat{\varphi}\,(.,-s,\varepsilon)], \pi'_{-s,\varepsilon}(\psi)\, u_{-s,\varepsilon}\rangle = \gamma(s,\varepsilon)\,\langle \widehat{\varphi}\,(.,s,\varepsilon), \pi'_{-s,\varepsilon}(\psi)\, u_{-s,\varepsilon}\rangle.$$

If $s+\rho$ is not an integer, the functions $\pi'_{-s,\varepsilon}(\psi)\, u_{-s,\varepsilon}$ span a dense subspace of $\mathcal{H}_{-s,\varepsilon}(\Xi)$ when ψ runs over $D(G)$, hence

$$A_{s,\varepsilon}\,[\widehat{\varphi}\,(.,-s,\varepsilon)] = \gamma(s,\varepsilon)\,\widehat{\varphi}\,(.,s,\varepsilon),$$

hence we have shown the announced relation between $A_{s,\varepsilon}$, $\mathcal{F}_{-s,\varepsilon}$ and $\mathcal{F}_{s,\varepsilon}$.

Remark 9.2.23. Observe that $j_1^* : \varphi \mapsto \widehat{\varphi}\,(.,s,\varepsilon)$ and $j_2^* : \varphi \mapsto A_{s,\varepsilon}\,[\widehat{\varphi}\,(.,-s,\varepsilon)]$ are both continuous G-equivariant linear mappings from $D(X)$ to $\mathcal{H}_{s,\varepsilon}(\Xi)$. If s is imaginary and non-zero, we know that $\pi_{s,\varepsilon}$ is irreducible and $\mathcal{H}_{s,\varepsilon}(\Xi)$ carries a unitary structure. It is easily seen that j_1^* and j_2^* are both continuous for this structure as well and hence, since (G, H) is a generalized Gelfand pair, j_2^* is proportional to j_1^*:

$$A_{s,\varepsilon} \circ \mathcal{F}_{-s,\varepsilon} = \gamma(s,\varepsilon)\, \mathcal{F}_{s,\varepsilon},$$

first for $s \in i\mathbb{R}^*$ and hence for all $s \in \mathbb{C}$ by analytic (meromorphic) continuation. The computation of $\gamma(s,\varepsilon)$ has to be done as before. Thus we obtain an alternative proof of Theorem 9.2.22.

(xiii) Asymptotic behaviour of the distributions $\zeta_{s,\varepsilon}$; the c-functions

We define the functions $c(s,\varepsilon)$ by

$$c(s,\varepsilon) = \frac{2^{\rho-s}}{\Gamma(\frac{s-\rho+1+\varepsilon}{2})}\, W_{s,0}(1).$$

We recall, see (vii), that

$$W_{s,0}(1) = \frac{2^{s+\rho-1}\,\Gamma(\frac{n}{2})}{\sqrt{\pi}}\,\frac{\Gamma(s)}{\Gamma(s+\rho)} = \int_B |[b,\xi^0]|^{s-\rho}\, db.$$

Proposition 9.2.24. *For $\varphi \in D(X)$ we set $\varphi_t(x) = \varphi(a_t \cdot x)$ $(x \in X)$. Then we have for* $\operatorname{Re} s > \rho$

$$\lim_{t\to\infty} \mathrm{e}^{-(s-\rho)t}\, \zeta_{s,\varepsilon}(\overline{\varphi_t}) = c(s,\varepsilon)\,\gamma(s,\varepsilon)\,\widehat{\varphi}(\xi^0, s, \varepsilon).$$

Indeed,

$$\zeta_{s,\varepsilon}(\overline{\varphi_t}) = \frac{1}{\Gamma(\frac{s-\rho+1+\varepsilon}{2})}\int_B \widehat{\varphi}(b,-s,\varepsilon)\,[a_{-t}\cdot b, e_n]^{s-\rho,\varepsilon}\, db.$$

Since $\mathrm{e}^{-t}\,[a_{-t}\cdot b, e_n] = \mathrm{e}^{-t}\,[b, \sinh t\; e_0 + \cosh t\; e_n]$ tends to $\frac{1}{2}\,[b,\xi^0]$ when $t \to \infty$, we get, by (xii),

$$\begin{aligned}\lim_{t\to\infty} \mathrm{e}^{-(s-\rho)t}\, \zeta_{s,\varepsilon}(\overline{\varphi_t}) &= \frac{2^{\rho-s}}{\Gamma(\frac{s-\rho+1+\varepsilon}{2})}\int_B \widehat{\varphi}(b,-s,\varepsilon)\,[b,\xi^0]^{s-\rho,\varepsilon}\, db \\ &= \frac{2^{\rho-s}}{\Gamma(\frac{s-\rho+1+\varepsilon}{2})}\, W_{s,0}(1)\,\gamma(s,\varepsilon)\,\widehat{\varphi}(\xi^0,s,\varepsilon).\end{aligned}$$

According to Theorem 9.2.5 the spherical distributions $\zeta_{s,\varepsilon}$ can be expressed as linear combinations of the distributions $M'S_\lambda$ and $M'T_\lambda$ $(\lambda = s^2 - \rho^2)$, defined in Section B.4. So

$$\zeta_{s,\varepsilon} = a(s,\varepsilon)\, M'S_\lambda + b(s,\varepsilon)\, M'T_\lambda.$$

We shall determine a and b by comparing the asymptotic behaviour of the distributions at $\pm\infty$.

Recall the definition of S_λ and T_λ of Section B.4. It is appropriate to collect a few formulae which play an important role in the next computations:

$$\begin{aligned} {}_2F_1(a,b;c;z) &= \frac{\Gamma(c)\,\Gamma(b-a)}{\Gamma(b)\,\Gamma(c-a)}\,(-z)^{-a}\,{}_2F_1\Big(a,1-c+a;1-b+a;\frac{1}{z}\Big) \\ &\quad + \frac{\Gamma(c)\,\Gamma(a-b)}{\Gamma(a)\,\Gamma(c-b)}\,(-z)^{-b}\,{}_2F_1\Big(b,1-c-b;1-a+b;\frac{1}{z}\Big) \end{aligned} \qquad (*)$$

for $|\arg(-z)| < \pi$ (see [13, 2.1.4, formula (17)]),

$$\sqrt{\pi}\,\Gamma(2z) = 2^{2z-1}\Gamma(z)\Gamma\Big(z+\frac{1}{2}\Big), \qquad (**)$$

$$\Gamma(z)\,\Gamma(1-z) = \frac{\pi}{\sin(\pi z)}. \qquad (***)$$

For $\operatorname{Re} s > \rho$ and $s \notin \frac{1}{2}\mathbb{Z}$ we have

- $S_\lambda \sim \dfrac{\Gamma(\frac{n}{2})\,\Gamma(2s)}{\Gamma(s+\rho)\,\Gamma(s+\frac{1}{2})}\,2^{\rho-s}\,t^{s-\rho} \quad (t\to\infty),$

  $T_\lambda = 0 \quad (t\to\infty),$
- $S_\lambda \sim \gamma_1(s)\,\dfrac{\Gamma(\frac{n}{2})\,\Gamma(2s)}{\Gamma(s+\rho)\,\Gamma(s+\frac{1}{2})}\,2^{\rho-s}\,(-t)^{s-\rho} \quad (t\to-\infty),$

  $T_\lambda \sim \dfrac{1-\gamma_1(s)^2}{\gamma_2(s)}\;\dfrac{\Gamma(\frac{n}{2})\,\Gamma(2s)}{\Gamma(s+\rho)\,\Gamma(s+\frac{1}{2})}\,2^{\rho-s}\,(-t)^{s-\rho} \quad (t\to-\infty).$

This follows from formula $(*)$. Recall

$$\gamma_1(s) = \frac{\Gamma(\frac{n}{2})\,\Gamma(1-\frac{n}{2})}{\Gamma(s+\frac{1}{2})\,\Gamma(-s+\frac{1}{2})} = (-1)^{\frac{n-1}{2}}\cos\pi s,$$

$$\gamma_2(s) = \frac{\Gamma(\frac{n}{2})\,\Gamma(1-\frac{n}{2})}{\Gamma(s+\rho)\,\Gamma(-s+\rho)} = \frac{(-1)^{\frac{n-1}{2}}\,\pi}{\Gamma(s+\rho)\,\Gamma(-s+\rho)}.$$

If $\varphi \in D(X)$ and $t \in \mathbb{R}$, define $\varphi_t \in D(X)$ again by $\varphi_t(x) = \varphi(a_t\cdot x)$ $(x\in X)$. If $\operatorname{Supp}\varphi \subset \{x\in X : [x,\xi^0] > 0\}$ then, because

$$[x,\xi^o] = 2\lim_{t\to\infty} \mathrm{e}^{-t}\,Q(a_{-t}\,x) \quad \text{for all } x\in X,$$

we have $\operatorname{Supp}\varphi_t \subset \{x \in X : Q(x) > 1\}$ for t large enough. Hence

$$\begin{aligned}
&\lim_{t\to\infty} \mathrm{e}^{-(s-\rho)t}\, \zeta_{s,\varepsilon}(\varphi_t) \\
&\quad = a(s,\varepsilon) \lim_{t\to\infty} \mathrm{e}^{-(s-\rho)t} \int_X M' S_\lambda(Q(a_{-t}\,x))\, \varphi_t(x)\, dx \\
&\quad = a(s,\varepsilon) \frac{\Gamma(\frac{n}{2}\,\Gamma(2s)}{\Gamma(s+\rho)\,\Gamma(-s+\rho)}\, 2^{2\rho-2s} \int_X |[x,\xi^0]|^{s-\rho}\, \varphi(x)\, dx,
\end{aligned}$$

as follows from the application of the dominated convergence theorem. On the other hand, from Proposition 9.2.24 it follows that the left hand side equals

$$\begin{aligned}
&\frac{2^{\rho-s}}{\Gamma(\frac{s-\rho+1+\varepsilon}{2})}\, \frac{2^{s+\rho+1}\,\Gamma(\frac{n}{2})\,\Gamma(s)}{\Gamma(s+\rho)\,\sqrt{\pi}}\, \frac{\Gamma(\frac{s+\rho+1-\varepsilon}{2})}{\Gamma(\frac{-s+\rho+1-\varepsilon}{2})}\, \frac{1}{\Gamma(s-\rho+1+\varepsilon)} \\
&\qquad \times \int_X |[x,\xi^0]|^{s-\rho}\, \varphi(x)\, dx.
\end{aligned}$$

Applying formulae $(**)$ and $(***)$ several times we obtain

$$a(s,\varepsilon) = \frac{4\,(-1)^{\rho+\varepsilon}}{\Gamma(\frac{s-\rho+1+\varepsilon}{2})\,\Gamma(\frac{-s-\rho+1+\varepsilon}{2})}. \tag{9.2.18}$$

Observe that a is even with respect to s, as it should be.

In a similar way we show that if $\operatorname{Supp}\varphi \subset \{x \in X : [x,\xi^0] < 0\}$ then

$$\begin{aligned}
&\lim_{t\to\infty} \mathrm{e}^{-(s-\rho)t}\, \zeta_{s,\varepsilon}(\varphi_t) \\
&\quad = a(s,\varepsilon)\, \gamma_1(s)\, \frac{\Gamma(\frac{n}{2})\,\Gamma(2s)}{\Gamma(s+\rho)\,\Gamma(s+\frac{1}{2})}\, 2^{2\rho-2s} \int_X |[x,\xi^0]|^{s-\rho}\, \varphi(x)\, dx \\
&\quad + b(s,\varepsilon)\, \frac{1-\gamma_1^2(s)}{\gamma_2(s)}\, \frac{\Gamma(\frac{n}{2})\,\Gamma(2s)}{\Gamma(s+\rho)\,\Gamma(s+\frac{1}{2})}\, 2^{2\rho-2s} \int_X |[x,\xi^0]|^{s-\rho}\, \varphi(x)\, dx.
\end{aligned}$$

On the other hand, the left-hand side equals

$$\begin{aligned}
&(-1)^\varepsilon \frac{2^{\rho-s}}{\Gamma(\frac{s-\rho+1+\varepsilon}{2})^2}\, 2^{s+\rho+1}\, \frac{\Gamma(\frac{n}{2})\,\Gamma(s)}{\Gamma(s+\rho)\,\sqrt{\pi}}\, \frac{\Gamma(\frac{s+\rho+1-\varepsilon}{2})}{\Gamma(\frac{-s+\rho+1-\varepsilon}{2})} \\
&\qquad \times \int_X |[x,\xi^0]|^{s-\rho}\, \varphi(x)\, dx \\
&\quad = (-1)^\varepsilon\, \frac{2^{2\rho-2}\,\Gamma(\frac{n}{2})\,\Gamma(s)}{\Gamma(s+\rho)\,\sqrt{\pi}}\, a(s,\varepsilon) \int_X |[x,\xi^0]|^{s-\rho}\, \varphi(x)\, dx.
\end{aligned}$$

Applying again formulae (∗∗) and (∗∗∗) we obtain

$$b(s,\varepsilon) = \frac{4(-1)^{\rho+\varepsilon}\,\pi}{\Gamma(s+\rho)\,\Gamma(-s+\rho)\,\Gamma(\frac{s-\rho+1+\varepsilon}{2})\,\Gamma(\frac{-s-\rho+1+\varepsilon}{2})\,\{\cos\pi s + (-1)^{\rho+\varepsilon}\}}. \tag{9.2.19}$$

Observe that this expression is well-defined and entire analytic in s.

We have obtained:

Proposition 9.2.25. *If* $\lambda = s^2 - \rho^2$, *then*

$$\zeta_{s,\varepsilon} = a(s,\varepsilon)\, M'S_\lambda + b(s,\varepsilon)\, M'T_\lambda$$

with

$$a(s,\varepsilon) = \frac{4\,(-1)^{\rho+\varepsilon}}{\Gamma(\frac{s-\rho+1+\varepsilon}{2})\,\Gamma(\frac{-s-\rho+1+\varepsilon}{2})},$$

$$b(s,\varepsilon) = \frac{4(-1)^{\rho+\varepsilon}\,\pi}{\Gamma(s+\rho)\,\Gamma(-s+\rho)\,\Gamma(\frac{s-\rho+1+\varepsilon}{2})\,\Gamma(\frac{-s-\rho+1+\varepsilon}{2})\,\{\cos\pi s + (-1)^{\rho+\varepsilon}\}}.$$

Though this holds primarily only for $\operatorname{Re} s > \rho$ and $s \notin \frac{1}{2}\mathbb{Z}$, by analytic continuation it holds for all $s \in \mathbb{C}$.

(xiv) Positive-definite spherical distributions

In this subsection we determine the positive-definite spherical distributions. An important ingredient is the fact that $\dim D'_\lambda(X) = 2$ for each $\lambda = s^2 - \rho^2$. Moreover, $\zeta_{s,0}$ and $\zeta_{s,1}$ form a basis of this space of spherical distributions. We shall first determine which $\zeta_{s,\varepsilon}$ are positive-definite. Taking into account the G-equivariant projection of $D(G)$ onto $D(X)$ given by $\varphi_0 \mapsto \varphi$ where

$$\varphi\,(x) = \varphi\,(g \cdot e_n) = \int_H \varphi_0(gh)\, dh,$$

one easily sees that $\zeta_{s,\varepsilon}(\widetilde{\psi}_0 * \varphi_0)$ corresponds to

$$\zeta_{s,\varepsilon}(\widetilde{\psi}_0 * \varphi_0) = \int_B \overline{\widehat{\varphi}}\,(b,s,\varepsilon)\, \widehat{\psi}\,(b,-s,\varepsilon)\, db.$$

Let us assume that $\zeta_{s,\varepsilon}$ is positive-definite. Then clearly $\zeta_{s,\varepsilon} = \widetilde{\zeta}_{s,\varepsilon}$ and moreover

$$\begin{aligned}\zeta_{s,\varepsilon}(\widetilde{\psi}_0 * \omega\varphi_0) &= \zeta_{s,\varepsilon}(\widetilde{\omega\psi_0} * \varphi_0) = \zeta_{s,\varepsilon}(\omega(\widetilde{\psi}_0 * \varphi_0))\\ &= \omega\,\zeta_{s,\varepsilon}(\widetilde{\psi}_0 * \varphi_0) = \frac{s^2-\rho^2}{2(n-1)}\,\zeta_{s,\varepsilon}(\widetilde{\psi}_0 * \varphi_0),\end{aligned}$$

where ω is the Casimir operator. Taking now $\psi_0 = \varphi_0$ we get $s^2 - \rho^2$, hence s^2 must be real, so $s \in i\mathbb{R}$ or $s \in \mathbb{R}$.

(1) Spherical distributions associated with the unitary principal series and the complementary series

(a) Unitary principal series. In this case where $s = i\nu$ with $\nu \in \mathbb{R}$ one has

$$\zeta_{i\nu,\varepsilon}(\widetilde{\psi}_0 * \varphi_0) = \int_B |\widehat{\varphi}(b, i\nu, \varepsilon)|^2 \, db \geq 0.$$

So we may conclude that

- $\zeta_{i\nu,\varepsilon}$ $(\nu \in \mathbb{R},\ \varepsilon = 0, 1)$ *is positive-definite.*

(b) Complementary series. Consider now the case with $s \in \mathbb{R},\ s - \rho \notin \mathbb{Z}$. According to Theorem 9.2.22 we have

$$\begin{aligned}\zeta_{s,\varepsilon}(\widetilde{\varphi}_0 * \varphi_0) &= \int_B \widehat{\overline{\varphi}}(b, s, \varepsilon)\, \widehat{\varphi}(b, -s, \varepsilon)\, db \\ &= \gamma(-s, \varepsilon) \int_B \overline{A_{s,\varepsilon}\, [\widehat{\varphi}\, (., -s, \varepsilon)](b)}\, \widehat{\varphi}(b, -s, \varepsilon)\, db\end{aligned}$$

and if, in addition, φ is K-finite of the form

$$\varphi_0(x) = F(t)\, Y(b) \qquad (x = ka_t \cdot e_n)$$

with $Y \in \mathcal{Y}_{l,m}$, $l + m \equiv \varepsilon \pmod 2$, then

$$\begin{aligned}\zeta_{s,\varepsilon}(\widetilde{\varphi}_0 * \varphi_0) = {}& \gamma(-s, \varepsilon)\, \alpha_{l,m}(s)\, [\beta_{l,m}(-s, \varepsilon)]^2 \\ & \times \left| \int_0^\infty F(t)\, \Psi_{l,m}(t, s, \varepsilon)\, A(t)\, dt \right|^2 \int_B |Y(b)|^2 \, db.\end{aligned} \tag{9.2.20}$$

To obtain this formula, apply (9.2.14) and (viii). In order that $\zeta_{s,\varepsilon}$ (or $-\zeta_{s,\varepsilon}$) is positive-definite, it is necessary and sufficient that the sesqui-linear form on $\mathcal{H}_{-s,\varepsilon}(\Xi)$ given by

$$(f, g) \mapsto \langle A_{s,\varepsilon} f, \overline{g} \rangle$$

(or its negative) is positive-definite. Equivalently, since $\beta_{l,m}(-s, \varepsilon) \neq 0$ and $\gamma(-s, \varepsilon) > 0$ for $-\rho < s < \rho$, if and only if the eigenvalues $\alpha_{l,m}(s)$ of $A_{s,\varepsilon}$ all have the same sign when $l + m \equiv \varepsilon \pmod 2$. Notice that the K-finite functions in $D_\varepsilon(B)$ span a dense subspace, see Section 8.1. The expression for $\alpha_{l,m}(s)$ is given by Theorem 9.2.15. We see:

- $\zeta s, \varepsilon$ *is positive-definite for* $-\rho < s < \rho$.
- *If* $s > \rho$ *and* $s - \rho \notin \mathbb{Z}$ *then neither* $\zeta_{s,\varepsilon}$ *nor* $-\zeta_{s,\varepsilon}$ *is positive-definite.*

(2) Spherical distributions associated with a discrete series representation and the trivial representation

We now consider the cases where $s - \rho$ is a non-negative integer.

For $\varepsilon = 0$ and $s = \rho$ we have

- $\zeta_{\rho,0}(\varphi) = \frac{\Gamma(\frac{n}{2})}{\pi\sqrt{\pi}} \int_X \varphi(x)\,dx.$

This is easily computed and $\zeta_{\rho,0}$ is clearly positive-definite.

We are now turning to the cases $s = \rho + 2r + \varepsilon$ and $s = \rho + 2r + 1 - \varepsilon$ for $\varepsilon = 0, 1$ respectively and r a non-negative integer. Thus we have actually to distinguish between four cases.

Case $\varepsilon = 0$, $s_r = \rho + 2r \quad$ ($r \in \mathbb{N}$, *r positive*)

We apply formula (9.2.20) again and notice that $\beta_{l,m}(-s_r, 0) \neq 0$ for all (l, m) with $l + m \equiv 0 \pmod 2$. Let us consider the expression

$$\lim_{s \to s_r} \gamma(-s, 0)\, \alpha_{l,m}(s) = \alpha'_{l,m}(s_r).$$

It follows that

$$\alpha'_{l,m}(s_r) = \alpha_{l,m}(s_r)\, \gamma(-s_r, 0),$$

$\alpha'_{l,m}(s_r) = 0$ for $l \geq 2r + 1$ and $l \mapsto \alpha'_{l,m}(s_r)$ is non-zero and alternates in sign for $l < 2r + 1$, hence:

- *Neither* $\zeta_{\rho+2r,0}$ *nor* $-\zeta_{\rho+2r,0}$ *is positive-definite.*

Case $\varepsilon = 0, s_r = \rho + 2r + 1 \quad (r \in \mathbb{N})$

We have

$$\begin{aligned} E(-s_r, 0) &= \{(l, m) : l + m \equiv 0 \pmod 2,\ \beta_{l,m}(-s_r, 0) \neq 0\} \\ &= \{(l, m) : l + m \equiv 0 \pmod 2,\ l - m \geq 2r + 2\}. \end{aligned}$$

Furthermore,

$$\begin{aligned} \alpha'_{l,m}(s_r) &= \lim_{s \to s_r} \gamma(-s, 0)\, \alpha_{l,m}(s) \\ &= \frac{(-1)^{r+1}\, 2^{2r+2}\, \Gamma(\frac{1}{2} + r)\, \Gamma(-2r - 1 + l)}{\sqrt{\pi}\, \Gamma(\rho + r + 1)\, (2\rho + 2r + 1) \cdots (2\rho + 2r + l)}. \end{aligned}$$

So $(-1)^{r+1} \alpha'_{l,m}(s_r) > 0$ for $(l, m) \in E(-s_r, 0)$. Hence:

- $(-1)^{r+1} \zeta_{\rho+2r+1,0}$ *is positive-definite.*

We now turn to the two cases where $\varepsilon = 1$ and obtain in a completely similar way:

- *Neither* $\zeta_{\rho+2r+1,1}$ *nor* $-\zeta_{\rho+2r+1,1}$ *is positive-definite.*
- $(-1)^r \zeta_{\rho+2r,1}$ *is positive-definite (also for* $r = 0$*).*

Since $\dim D'_\lambda(X)^H = 2$ and $\zeta_{s,0}$ and $\zeta_{s,1}$ form a basis of $D'_\lambda(X)^H$ for $\lambda = s^2 - \rho^2$, we now easily see that the positive-definite spherical distributions associated with the discrete points $s \in \rho + \mathbb{Z}_+$ are, up to a positive scalar, given by

$$(-1)^{r+1}\zeta_{\rho+2r+1,0}, \quad (-1)^r \zeta_{\rho+2r,1}, \quad \zeta_{\rho,0}.$$

In particular it follows that these distributions are extremal.

Let T be one of these distributions, $\mathcal{H}$ the corresponding invariant Hilbert subspace of $D'(X)$, π the unitary representation and j the equivariant injection $\mathcal{H} \to D'(X)$. Since T is spherical we have $\omega T = \lambda T$ for some $\lambda \in \mathbb{C}$, where ω is the Casimir operator. Then $\pi_{-\infty}(\omega) = \lambda I$ on $\mathcal{H}_{-\infty}$. If T_1 is a positive-definite H-invariant distribution with $T_1 \le T$, and if $(\pi_1, j_1, \mathcal{H}_1)$ is the associated triple as above, then the mapping A given by $A(j^*\varphi) = j_1^*\varphi$ ($\varphi \in D(X)$) is well-defined and continuous. Extending A to $\mathcal{H}$ we get a continuous linear mapping $\mathcal{H} \to \mathcal{H}_1$ intertwining π and π_1. This implies, since A has dense image, that $\pi_1(\omega) = \lambda I$ on $\mathcal{H}_{1,-\infty}$ and thus T_1 is spherical again: $\omega T_1 = \lambda T_1$. So T_1 is a linear combination of T and a second distribution in $D'_\lambda(X)$. To see that T_1 is proportional to T, it is sufficient to consider the action on K-finite functions of type (l, m) with $l + m \equiv \varepsilon \pmod 2$ for $\varepsilon = 0$ and $\varepsilon = 1$ respectively. Observe that $\zeta_{s,\varepsilon}(\widetilde{\varphi_0} * \varphi_0) = 0$ if φ is of type (l, m) with $l + m \equiv (1 - \varepsilon) \pmod 2$. Hence T is extremal.

(3) Extremal positive-definite distributions

We summarize:

Theorem 9.2.26. *The extremal positive-definite H-invariant distributions are the following (up to positive scalars):*

- $\zeta_{i\nu,\varepsilon}$ $\quad(\nu \in \mathbb{R};\ \varepsilon = 0, 1)$,
- $\zeta_{s,\varepsilon}$ $\quad(0 < s < \rho;\ \varepsilon = 0, 1)$,
- $\zeta_{\rho,0}$,
- $(-1)^{r+1}\zeta_{\rho+2r+1,0}$ *and* $(-1)^r \zeta_{\rho+2r,1}$ $\quad(r = 0, 1, 2, \ldots)$.

Since, by Proposition 8.2.6, any extremal positive-definite H-invariant distribution is spherical, the result is a direct consequence of (1) and (2).

(4) Associated class-one representations

The representations associated with the extremal positive-definite H-invariant distributions can be described as follows.

• *Unitary principal series*

The distribution $\zeta_{i\nu,\varepsilon}$ ($\nu \in \mathbb{R}$, $\varepsilon = 0, 1$) is the reproducing distribution of $\pi_{i\nu,\varepsilon}$ on the Hilbert space completion of $\mathcal{H}_{-i\nu,\varepsilon}$ with respect to the scalar product

$$\langle f \mid g \rangle = \int_B f(b)\, \overline{g(b)}\, db.$$

• *Complementary series*

The distribution $\zeta_{s,\varepsilon}$ ($0 < s < \rho$; $\varepsilon = 0, 1$) is the reproducing distribution of $\pi_{s,\varepsilon}$ on the Hilbert space completion with respect to the scalar product

$$(f, g) = \langle A_{s,\varepsilon} f \mid g \rangle.$$

• *Trivial representation*

The distribution $\zeta_{\rho,0}$ *corresponds to the trivial representation.*

• *Discrete series*

The distribution $(-1)^{r+1}\,\zeta_{\rho+2r+1,0}$ ($r = 0, 1, 2, \ldots$) corresponds to the restriction of $\pi_{-(\rho+2r+1),0}$ to the space $\overline{\mathcal{I}}_{-(\rho+2r+1),0}$, being the closure in $D_0(B)$ of $\mathcal{I}_{-(\rho+2r+1),0}$. See (x) for the definition of this set. We provide $\overline{\mathcal{I}}_{-(\rho+2r+1),0}$ with the G-invariant scalar product

$$(f, g) = (-1)^{r+1} \lim_{s \to \rho+2r+1} \gamma(-s, 0)\, \langle A_{s,0} f \mid g \rangle,$$

and take the Hilbert space completion, so that $(-1)^{r+1}\,\zeta_{\rho+2r+1,0}$ is now the reproducing distribution of the unitary extension of $\pi_{-(\rho+2r+1),0}$ to this completion.

In a similar way we proceed with the distributions $(-1)^r\,\zeta_{\rho+2r,1}$. Each one corresponds to the restriction of $\pi_{-\rho-2r,1}$ to the space $\overline{\mathcal{I}}_{-\rho-2r,1}$, being the closure in $D_1(B)$ of $\mathcal{I}_{-\rho-2r,1}$. See again (x) for definition of this set. We provide $\overline{\mathcal{I}}_{-\rho-2r,1}$ with the G-invariant scalar product

$$(f, g) = (-1)^r \lim_{s \to \rho+2r} \gamma(-s, 1)\, \langle A_{s,1} f \mid g \rangle.$$

(xv) Plancherel formula

We are now ready to determine the Plancherel formula for X: the expansion of the delta-function δ at the origin e_n of X into extremal positive-definite H-invariant distributions on X, so a formula like

$$\delta = \sum_{\varepsilon=0}^{1} \int_0^{\infty} \zeta_{i\nu,\varepsilon}\, d\mu_\varepsilon(\nu) + \sum_{\varepsilon=0}^{1} \int_0^{\rho} \zeta_{s,\varepsilon}\, dm_\varepsilon(s)$$
$$+ \sum_{r=0}^{\infty} a_r\, (-1)^{r+1}\, \zeta_{\rho+2r+1,0} + \sum_{r=0}^{\infty} b_r\, (-1)^r\, \zeta_{\rho+2r,1},$$

where μ_0, μ_1 and m_0, m_1 are positive (Radon) measures on $[0,\infty)$ and $[0,\rho]$ respectively and a_r, b_r are positive numbers for all r. Since (G,H) is a generalized Gelfand pair, such an expansion must be unique, hence the data μ_0, μ_1, m_0, m_1, a_r, b_r are uniquely determined by δ in the given parametrization of the extremal positive-definite H-invariant distributions on X. We are going to determine μ_0, $\mu_1, m_0, m_1, a_r, b_r$.

Theorem 9.2.27. *One has*

$$\frac{\Gamma(\frac{n}{2})\,2^{2\rho-2}}{\pi^{\frac{n}{2}}}\,\delta = \frac{1}{2\pi}\sum_{\varepsilon=0}^{1}\int_0^\infty \zeta_{i\nu,\varepsilon}\,\frac{d\nu}{|c(i\nu,\varepsilon)|^2}$$
$$+\sum_{\rho+2r+1>0}\zeta_{\rho+2r+1,0}\,\mathrm{Res}_{s=\rho+2r+1}\left(\frac{1}{c(s,0)c(-s,0)}\right)$$
$$+\sum_{\rho+2r>0}\zeta_{\rho+2r,1}\,\mathrm{Res}_{s=\rho+2r}\left(\frac{1}{c(s,1)c(-s,1)}\right).$$

This formula has been inspired by the Riemannian case, and the proof is also at several points similar to the case $\mathrm{SO}_0(1,n)/\mathrm{SO}(n)$, see Section 7.5 (vii). We shall show that the formula is correct, but refrain from long computations. Of course, Res stands for residue.

(1) Method of proof. A first remark is that the Laplace–Beltrami operator Δ is not only symmetric on its domain $D(X)\subset L^2(X)$, but in addition essentially self-adjoint. The proof of Proposition 7.5.13 applies to this case as well, the reader may easily check it.

Denote by $\overline{\Delta}$ the closure of Δ and by R_λ the resolvent of $\overline{\Delta}$. We are going to determine R_λ, because we can then derive the spectral function $\lambda\mapsto E_\lambda$ of $\overline{\Delta}$ by means of the formula

$$\int_{-\infty}^{\infty}\varphi(\lambda)\,d(E_\lambda f\mid f) = -\frac{1}{\pi}\lim_{\varepsilon\downarrow 0}\int_{-\infty}^{\infty}\mathrm{Im}(R_{\lambda+i\varepsilon}f\mid f)\,\varphi(\lambda)\,d\lambda \qquad (9.2.21)$$

for all $\varphi\in C_c(\mathbb{R})$ and $f\in L^2(X)$. We shall use the following two properties of R_λ:

(a) *For all $f,g\in L^2(X)$ and all $\lambda\in\mathbb{C}$ with* $\mathrm{Im}\,\lambda\neq 0$ *one has*

$$|(R_\lambda f\mid g)|\leq\frac{1}{|\mathrm{Im}\,\lambda|}\|f\|_2\|g\|_2.$$

(b) *If f and g are in $D(X)$, then*

$$(R_\lambda\Delta f\mid g) = (R_\lambda f\mid \Delta g).$$

(2) Computation of R_λ. Let $\psi\in D(X)$ and $\varphi_0\in D(G)$ and let, as usual, $\varphi\in D(X)$ be defined by

$$\varphi(x) = \varphi(g\cdot e_n) = \int_H \varphi_0(gh)\,dh.$$

Set

$$\widetilde{\varphi}_0 * \psi(x) = \int_G \overline{\varphi_0(g)}\, \psi(g \cdot x)\, dg.$$

Clearly, $\widetilde{\varphi}_0 * \psi \in D(X)$. Consider now the function $M(\widetilde{\varphi}_0 * \psi)$ in $\mathcal{H}_\eta$. If we take a right-translate of φ_0 over $h \in H$, this function does not change, hence we get a continuous mapping of $D(X) \times D(X)$ into $\mathcal{H}_\eta$ which we denote by

$$(\varphi, \psi) \mapsto \varphi \# \psi.$$

So $\varphi\#\psi = M(\widetilde{\varphi}_0 * \psi)$. This mapping is anti-linear in φ, linear in ψ.

Applying Schwartz' kernel theorem, see, e.g., Section 4.5 in [50] and Theorem B.3.2, one has

Proposition 9.2.28. *Let B be a continuous G-invariant mapping of $D(X) \times D(X)$ to $\mathbb{C}$, linear in the first and anti-linear in the second variable. There exists a unique element $K \in \mathcal{H}'_\eta$ such that*

$$B(\varphi, \psi) = K(\varphi \# \psi).$$

For example, from the relation $M' B_0^{(1)} = \frac{(-1)^{\frac{n-1}{2}} \pi^{\frac{n}{2}}}{\Gamma(\frac{n}{2})} \delta$ (see Remark A.3.7), where $B_0^{(1)}$ is defined on $\mathcal{H}_\eta$ by

$$B_0^{(1)}(\varphi) = \overline{\varphi(1)}$$

and δ is the delta-function at the origin e_n of X, we conclude

$$\int_X \varphi(x)\, \overline{\psi(x)}\, dx = \frac{(-1)^{\frac{n-1}{2}} \Gamma(\frac{n}{2})}{\pi^{\frac{n}{2}}} \langle B_0^{(1)}, \varphi \# \psi \rangle.$$

Let us apply the proposition to the kernel

$$(\varphi, \psi) \mapsto (R_\lambda \varphi \,|\, \psi).$$

If $\operatorname{Im} \lambda \neq 0$, this kernel is continuous (see (b)) and G-invariant, so there exists a unique element $K_\lambda \in \mathcal{H}'_\eta$ with

$$(R_\lambda \varphi \,|\, \psi) = \langle K_\lambda, \varphi \# \psi \rangle$$

for all $\varphi, \psi \in D(X)$.

Clearly, K_λ satisfies the following equation in $\mathcal{H}'_\eta$

$$\lambda K_\lambda - L K_\lambda = \frac{(-1)^{\frac{n-1}{2}} \Gamma(\frac{n}{2})}{\pi^{\frac{n}{2}}} B_0^{(1)}.$$

By Theorem 9.2.5 and similar to the computations in Section B.4, we have

$$K_\lambda = \frac{1}{4(n-2)} \frac{(-1)^{\frac{n-3}{2}} \Gamma(\frac{n}{2})}{\pi^{\frac{n}{2}}} T_+^{(1)} + A\, S_\lambda + B\, T_\lambda$$

for some complex numbers A and B.

To determine A and B we look at the behaviour of the anti-linear forms at $\pm\infty$ again and take $\operatorname{Re} s > 0$ in $\lambda = s^2 - \rho^2$, $s \notin \mathbb{R}$.

(3) Asymptotics. For a function $\varphi \in D(X)$ let us set again

$$\varphi_\tau(x) = \varphi(a_\tau \cdot x) \qquad (\tau \in \mathbb{R},\, x \in X).$$

Let φ and ψ be two functions in $D(X)$ and F a continuous function on $\mathbb{R}$. Then we have

$$\int_{-\infty}^{\infty} F(t)\,(\varphi\#\psi)(t)\,dt = \int_X \int_X F([x,y])\,\overline{\varphi(x)}\,\psi(y)\,dxdy.$$

Hence, if $\operatorname{Supp}\varphi \times \operatorname{Supp}\psi \subset \{(x,y) : [x,y] > 1\}$, then

$$\langle S_\lambda, \varphi\#\psi\rangle = \int_X \int_X \Phi^{(1)}([x,y])\,\overline{\varphi(x)}\,\psi(y)\,dxdy$$

(see Section B.4 for the definition of $\Phi^{(1)}$), and

$$\begin{aligned}\langle S_\lambda, \varphi_\tau\#\psi\rangle &= \int_X \int_X \Phi^{(1)}([x,y])\,\overline{\varphi(a_\tau \cdot x)}\,\psi(y)\,dxdy \\ &= \int_X \int_X \Phi^{(1)}([a_{-\tau} \cdot x, y])\,\overline{\varphi(x)}\,\psi(y)\,dxdy.\end{aligned}$$

We have

$$\begin{aligned}[a_{-\tau} \cdot x, y] = -y_0\,(\cosh\tau \cdot x_0 - \sinh\tau \cdot x_n) &+ x_1 y_1 + \cdots + x_{n-1} y_{n-1} \\ &+ y_n\,(-\sinh\tau \cdot x_0 + \cosh\tau \cdot x_n),\end{aligned}$$

and thus

$$\lim_{\tau\to\infty} e^{-\tau}\,[a_{-\tau} \cdot x, y] = \frac{1}{2}\,(y_0 + y_n)(x_n - x_0) = \frac{1}{2}[x,\xi^0]\,[y, w \cdot \xi^0],$$

where w is the matrix $w = \operatorname{diag}(-1, 1, \ldots, 1) \in G$. Suppose

$$\begin{aligned}\operatorname{Supp}\varphi &\subset \{x : [x,\xi^0] > 0\}, \\ \operatorname{Supp}\psi &\subset \{x : [x, w \cdot \xi^0] > 0\}.\end{aligned}$$

We have for $\operatorname{Re} s > \rho$

$$\Phi^{(1)}(t) \sim \frac{\Gamma(\frac{n}{2})\,\Gamma(2s)}{\Gamma(s+\rho)\,\Gamma(s+\frac{1}{2})}\,2^{\rho-s}\,t^{s-\rho} \qquad (t \to \infty),$$

and therefore

$$\Phi^{(1)}\,([a_{-\tau}x, y]) \sim \frac{\Gamma(\frac{n}{2})\,\Gamma(2s)}{\Gamma(s+\rho)\,\Gamma(s+\frac{1}{2})}\,2^{2\rho-2s}\,e^{(s-\rho)t}\,[x,\xi^0]^{s-\rho}\,[x, w\xi^0]^{s-\rho} \quad (\tau \to \infty),$$

and thus, applying the dominated convergence theorem,

$$\begin{aligned}&\lim_{\tau\to\infty} e^{-(s-\rho)\tau}\,\langle S_\lambda, \varphi\#\psi\rangle \\ &= \frac{\Gamma(\frac{n}{2})\,\Gamma(2s)}{\Gamma(s+\rho)\,\Gamma(s+\frac{1}{2})}\,2^{2\rho-2s} \int_X [x,\xi^0]^{s-\rho}\,\overline{\varphi(x)}\,dx \int_X [y, w\xi^0]^{s-\rho}\,\psi(y)\,dy.\end{aligned}$$

Similar observations hold for T_λ and $T_+^{(1)}$:

$$\lim_{\tau\to\infty} \mathrm{e}^{-(s-\rho)\tau} \langle T_\lambda, \varphi_\tau \# \psi\rangle = 0$$

and

$$\lim_{\tau\to\infty} \mathrm{e}^{-(s-\rho)\tau} \langle T_+^{(1)}, \varphi_\tau \# \psi\rangle$$
$$= \frac{\Gamma(2-\frac{n}{2})\,\Gamma(2s)}{\Gamma(s+\rho)\,\Gamma(s+\frac{1}{2})}\, 2^{2\rho-2s} \int_X [x,\xi^0]^{s-\rho}\,\overline{\varphi(x)}\,dx \int_X [y, w\xi^0]^{s-\rho}\,\psi(y)\,dy.$$

We also have

$$|\langle K_\lambda, \varphi_\tau \# \psi\rangle| \le \frac{1}{|\operatorname{Im}\lambda|}\,\|\varphi\|_2\,\|\psi\|_2,$$

since $\|\varphi_\tau\|_2 = \|\varphi\|_2$. This yields the condition

$$\frac{(-1)^{\frac{n-3}{2}}\,\Gamma(\frac{n-2}{2})\,\Gamma(2-\frac{n}{2})\,\Gamma(2s)}{8\,\pi^{\frac{n}{2}}\,\Gamma(s+\frac{1}{2})\,\Gamma(s-\rho+1)} + A\,\frac{\Gamma(\frac{n}{2})\,\Gamma(2s)}{\Gamma(s+\rho)\,\Gamma(s+\frac{1}{2})} = 0,$$

hence $A = -\frac{1}{8\,\pi^{\frac{n-2}{2}}\,\Gamma(\frac{n}{2})}\,\frac{\Gamma(s+\rho)}{\Gamma(s-\rho+1)}$.

In a similar way, applying asymptotics at $-\infty$, we obtain in the notation of Section B.4,

$$\frac{\Gamma(\frac{n}{2})\,\Gamma(2s)}{\Gamma(s+\rho)\,\Gamma(s+\frac{1}{2})}\left\{\gamma_1(s)\,A + \frac{1-\gamma_1^2(s)}{\gamma_2(s)}\,B\right\} = 0,$$

hence

$$\begin{aligned} B &= -\frac{\gamma_1(s)\gamma_2(s)}{1-\gamma_1^2(s)}\,A \\ &= -\frac{\pi\,\cos\pi s}{\sin^2\pi s\,\Gamma(s+\rho)\,\Gamma(-s+\rho)}\,A \\ &= \frac{\pi}{8\,\pi^{\frac{n-2}{2}}\,\Gamma(\frac{n}{2})}\,\frac{\cos\pi s}{\sin^2\pi s}\,\frac{1}{\Gamma(s-\rho+1)\,\Gamma(-s+\rho)} \\ &= \frac{(-1)^{\rho+1}}{8\,\pi^{\frac{n-2}{2}}\,\Gamma(\frac{n}{2})\,\operatorname{tg}\pi s}. \end{aligned}$$

So we have now

$$K_\lambda = \frac{(-1)^{\frac{n-3}{2}}\,\Gamma(\frac{n-2}{2})}{8\,\pi^{\frac{n}{2}}}\,T_+^{(1)} - \frac{1}{8\,\pi^{\frac{n-3}{2}}\,\Gamma(\frac{n}{2})}\left\{\frac{\Gamma(s+\rho)}{\Gamma(s-\rho+1)}\,S_\lambda + \frac{(-1)^\rho}{\operatorname{tg}\pi s}\,T_\lambda\right\}. \quad (9.2.22)$$

(4) The c-functions. The c-functions were introduced in (xiii). For the formulation of the Plancherel measure, it is convenient to have a simple expression for $c(s,\varepsilon)\,c(-s,\varepsilon)$. One has, using the formulae for Γ-functions in (xiii)

$$c(s,\varepsilon)\,c(-s,\varepsilon) = -\frac{\Gamma(\frac{n}{2})^2}{\pi}\,\frac{\cos(\frac{s+\rho+\varepsilon}{2})\pi\cdot\cos(\frac{-s+\rho+\varepsilon}{2})\pi}{\Gamma(\frac{s+\rho+\varepsilon}{2})\,\Gamma(\frac{-s+\rho+\varepsilon}{2})\,s\,\sin\pi s}.$$

(5) The Plancherel measure. We are going to rewrite formula (9.2.21) in terms of the variable s, $\lambda = s^2 - \rho^2$. We take $\operatorname{Re} s > 0$ and $\operatorname{Re} s = 0$, $\operatorname{Im} s > 0$ as unique solutions of the latter equation. So $\operatorname{Im} \lambda > 0$ corresponds to $\operatorname{Re} s > 0$, $\operatorname{Im} s > 0$ and $\operatorname{Im} \lambda < 0$ to $\operatorname{Re} s > 0$, $\operatorname{Im} s < 0$.

We start, as said, with formula (9.2.21), written as

$$\int_{-\infty}^{\infty} \varphi(\lambda)\, d(E_\lambda f \mid f) = -\frac{1}{2\pi i} \lim_{\varepsilon \downarrow 0} \int_{-\infty}^{\infty} [(R_{\lambda+i\varepsilon} f \mid f) - (R_{\lambda-i\varepsilon} f \mid f)]\, \varphi(\lambda)\, d\lambda$$

with $\varphi \in C_c(\mathbb{R})$ and $f \in L^2(X)$, in particular $f \in D(X)$. Changing to the new variables, $\lambda = s^2 - \rho^2$ gives two branches, namely $\lambda \in (-\infty, -\rho^2)$, so $s \in i\mathbb{R}$ and $\lambda > -\rho^2$, so $s \in \mathbb{R}$. Let K_s correspond to K_λ (as usual). The line $\lambda + i\varepsilon$ ($\lambda < -\rho^2$) transforms into the curve κ_ε, the line $\lambda + i\varepsilon$ ($\lambda > -\rho^2$) into $\widetilde{\kappa}_\varepsilon$. Similarly we obtain for $\lambda - i\varepsilon$ the curves $\kappa_{-\varepsilon}$ and $\widetilde{\kappa}_{-\varepsilon}$ (see Figure 9.1).

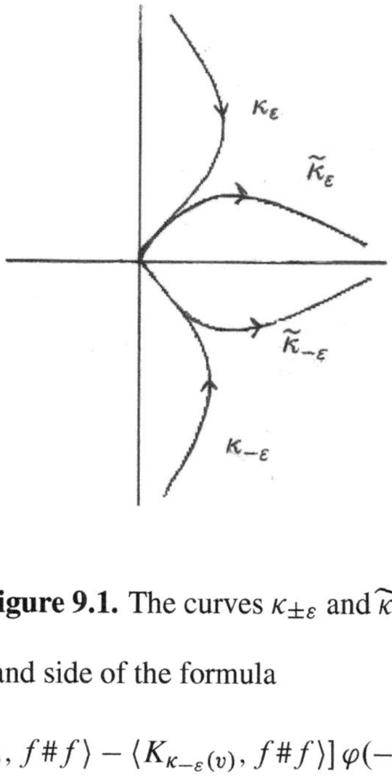

Figure 9.1. The curves $\kappa_{\pm\varepsilon}$ and $\widetilde{\kappa}_{\pm\varepsilon}$

We obtain for the right-hand side of the formula

$$\frac{1}{2\pi i} \lim_{\varepsilon \downarrow 0} \int_{-\infty}^{0} [\langle K_{\kappa_\varepsilon(v)}, f\#f\rangle - \langle K_{\kappa_{-\varepsilon}(v)}, f\#f\rangle]\, \varphi(-v^2 - \rho^2)\, 2v\, dv$$
$$- \frac{1}{2\pi i} \lim_{\varepsilon \downarrow 0} \int_{0}^{\infty} [\langle K_{\widetilde{\kappa}_\varepsilon(s)}, f\#f\rangle - \langle K_{\widetilde{\kappa}_{-\varepsilon}(s)}, f\#f\rangle]\, \varphi(-s^2 - \rho^2)\, 2s\, ds.$$

Taking into account the expression for K_s, this gives

$$-\frac{1}{\pi} \int_0^\infty \operatorname{Im}\langle K_{iv}, f\#f\rangle\, \varphi(-v^2 - \rho^2)\, 2v\, dv + \sum_r \operatorname{Res}_{s=s_r} \langle K_s, f\#f\rangle\, \varphi(s_r^2 - \rho^2)\, 2s_r,$$

where s_r, $r = 0, 1, 2, \ldots$ are simple poles of $s \mapsto \langle K_s, f\#f\rangle$ in $(0, \infty)$. This second term is easily obtained by writing in a neighbourhood of a pole $s_r > 0$ of K_s, $\widetilde{\kappa}_\varepsilon(s) =$

$s + \frac{1}{2}i\varepsilon + O(\varepsilon^2)$ and similarly $\widetilde{\kappa}_{-\varepsilon}(s) = s - \frac{1}{2}i\varepsilon + O(\varepsilon^2)$. Finally, letting φ tend to the function 1, this gives the Plancherel formula in a more or less concrete form. What remains to show can be formulated as follows. In the first place we have to prove that

$$-\frac{1}{\pi}\operatorname{Im}\langle K_{i\nu}, f\#f\rangle\, 2\nu = \frac{1}{2\pi c}\Big[\langle \zeta_{i\nu,0}, \widetilde{f}_0\#f\rangle \frac{1}{|c(i\nu,o)|^2} + \langle \zeta_{i\nu,1}, \widetilde{f}_0\#f\rangle \frac{1}{|c(i\nu,1)|^2}\Big].$$

This can be seen in the following way. From formula (9.2.22) we derive for $s = i\nu$ and $\lambda = s^2 - \rho^2$,

$$\operatorname{Im}\langle K_s, f\#f\rangle = -\frac{1}{8\pi^{\frac{n-2}{2}}\Gamma(\frac{n}{2})}\Big[\operatorname{Im}\Big\{\frac{\Gamma(s+\rho)}{\Gamma(s-\rho+1)}\Big\}\langle S_\lambda, f\#f\rangle$$
$$+ \operatorname{Im}\Big\{\frac{(-1)^\rho}{\operatorname{tg}\pi s}\Big\}\langle T_\lambda, f\#f\rangle\Big].$$

Making this formula explicit leads to a routine computation, where one applies Proposition 9.2.25. We leave this part to the reader.

In the second place, we have to show that the terms

$$\operatorname{Res}_{s=s_r}\langle K_s, f\#f\rangle\, 2s_r$$

give the discrete terms with the right weights in the Plancherel formula. From the form of K_s we see that poles in $(0,\infty)$ occur if $\operatorname{tg}\pi s$ has zeros there, so for $s_r = 1, 2, 3, \ldots$ and $\operatorname{Res}_{s=s_r} K_s = \frac{(-1)^{\rho+1}}{8\pi^{\frac{n-2}{2}}\Gamma(\frac{n}{2})} T_{s_r}$ ($T_{s_r} = T_\lambda, \lambda = s_r^2 - \rho^2$). By Proposition 9.2.25 we see that $M'T_{\rho+2r}$ is a multiple of $\zeta_{\rho+2r,1}$ and $M'T_{\rho+2r+1}$ is a multiple of $\zeta_{\rho+2r+1,0}$. This multiple is easily computed, applying the transformation formulae for the Γ-function from (xiii), and we finally obtain the weights described in the Plancherel formula.

Remark 9.2.29. The *relative discrete series* of the group G is easily read off from the Plancherel formula. The corresponding distributions are

$$(-1)^{r+1}\,\zeta_{\rho+2r+1,0} \quad \text{and} \quad (-1)^r\,\zeta_{\rho+2r,1}$$

with $r = 0, 1, 2, \ldots$, and

$$\zeta_{\rho+2r+1,0} \quad \text{and} \quad \zeta_{\rho+2r,1}$$

with $0 < \rho + 2r + 1 < \rho$ and $0 < \rho + 2r < \rho$ respectively, where r is an integer.

Notice that if ζ is one of these distributions, then the Plancherel formula implies

$$|\langle \zeta, f\#f\rangle| \le \text{const.}\|f\|_2^2 \qquad \text{for all } f \in D(X),$$

so ζ belongs to a relative discrete series representation, by Proposition 8.4.1. For all such ζ one has: ζ is a multiple of $M'T_{s_r}$ for some $s_r = 0, 1, 2, \ldots$.

It is known that the group G itself has no discrete series.

Remark 9.2.30. For more (recent) examples of generalized Gelfand pairs we refer to [53, Chapter 8].

Appendix A
The Averaging Mapping on the Space $\mathbb{R}^{n+1}$

A.1 Special case of a theorem of Harish-Chandra

Let Q be the quadratic form on $\mathbb{R}^{n+1}$ given by

$$Q(x) = x_0^2 - x_1^2 - \cdots - x_n^2.$$

As in Chapter 9, set $\Gamma_t = \{x \in \mathbb{R}^{n+1} : Q(x) = t\}$ where $t \in \mathbb{R}$. Write $\mathbb{R}^{n+1}\backslash\{0\} = \mathbb{R}^{n+1}_*$. One has the following result.

Theorem A.1.1. *There exists a mapping $f \mapsto M_f$ from $D(\mathbb{R}^{n+1}_*)$ onto $D(\mathbb{R})$ such that for all continuous functions Φ on $\mathbb{R}$ one has*

$$\int_{\mathbb{R}^{n+1}} f(x)\, \Phi(Q(x))\, dx = \int_{\mathbb{R}} M_f(t)\, \Phi(t)\, dt.$$

Moreover, Supp $M_f \subset Q$(Supp f) *and $f \mapsto M_f$ is continuous.*

This follows easily from the fact that $Q : \mathbb{R}^{n+1}_* \to \mathbb{R}$ is everywhere submersive, i.e. the differential of Q is non-zero. It is also a particular case of a general theorem of Harish-Chandra, see [20, Theorem 1]. We shall sketch the proof of the theorem.

Take $x_0 \in \mathbb{R}^{n+1}_*$ and set $t_0 = Q(x_0)$. Since Q is submersive, we can choose an open neighbourhood V of x_0 in $\mathbb{R}^{n+1}_*$, a neighbourhood W of 0 in $\mathbb{R}^n$ and a neighbourhood A of t_0 in $\mathbb{R}$ such that there is a diffeomorphism $\theta : V \to W \times A$ satisfying

$$Q(\theta^{-1}(y,t)) = t \quad \text{for } y \in W,\, t \in A.$$

Let $\omega(t,y)dydt$ be the image under θ of $dx|_V$. Then one defines for $f \in D(V)$

$$M_f^V(t) = \int_W (f \circ \theta^{-1})(y,t)\, \omega(y,t)\, dy.$$

Obviously, $M_f^V \in D(A)$, Supp $M_f^V \subset Q$(Supp f) and $f \mapsto M_f^V$ is linear and continuous. Moreover, M_f^V is well-defined, i.e. the definition does not depend on the coordinate system (y,t), provided $Q(\theta^{-1}(y,t)) = t$. To define M_f for $f \in D(\mathbb{R}^{n+1}_*)$ we now use partition of unity.

Lemma A.1.2 (Partition of unity). *Let $O_1, \dots, O_k$ be open sets in $\mathbb{R}^m$ and let K be a compact set satisfying $K \subset \bigcup_{i=1}^k O_i$. Then there exist functions $\varphi_i \in D(\mathbb{R}^m)$ with* Supp $\varphi_i \subset O_i$, $\varphi_i \geq 0$, $\sum_{i=1}^k \varphi_i \leq 1$ *and* $\sum_{i=1}^k \varphi_i = 1$ *in a neighbourhood of K.*

The proof of this lemma is similar to that of Lemma 4.1.2. The counterpart of Theorem 3.2.1 is:

Lemma A.1.3. *Let $K \subset \mathbb{R}^m$ be a compact set and let $K \subset O$ with O open. There is a function $\varphi \in D(\mathbb{R}^m)$ such that $0 \le \varphi \le 1$, $\varphi = 1$ on a neighbourhood of K and* $\operatorname{Supp}\varphi \subset O$.

Let us prove this lemma. Let $0 < a < b$ and consider the function f on $\mathbb{R}$ defined by $f(x) = \exp(\frac{1}{x-a} - \frac{1}{x-b})$ if $a < x < b$ and $f(x) = 0$ otherwise. Then f is a C^∞ function and the same holds for $F(x) = \int_x^b f(t)\,dt / \int_a^b f(t)\,dt$. Notice that $F(x) = 1$ if $x \le a$, $F(x) = 0$ if $x \ge b$. The function φ on $\mathbb{R}^m$ given by $\varphi(x_1, \dots, x_m) = F(x_1^2 + \dots + x_m^2)$ is C^∞, is equal to 1 for $r^2 \le a$ and equal to 0 for $r^2 \ge b$, where $r^2 = x_1^2 + \dots + x_m^2$. If $B' \subset B$ are different concentric open balls in $\mathbb{R}^m$, then we can easily construct a function $\psi \in D(\mathbb{R}^m)$ that is equal to 1 on B' and zero outside B. We can find open balls B_i and B_i' $(i = 1, \dots, k)$ such that B_i and B_i' are concentric, the radius of B_i is strictly larger then that of B_i', and such that the B_i' cover K while the $\overline{B_i}$ are contained in O. Let $\psi_i \in D(\mathbb{R}^m)$ be such that $\psi_i = 1$ on $\overline{B_i'}$ and $\psi_i = 0$ outside B_i. Then the function $\varphi = 1 - (1-\psi_1)(1-\psi_2)\cdots(1-\psi_k)$ is in $D(\mathbb{R}^m)$ and it is equal to 1 in a neighbourhood of K, and it satisfies $\operatorname{Supp}\varphi \subset O$ and $0 \le \varphi \le 1$.

Let us now define M_f for $f \in D(\mathbb{R}_*^{n+1})$. Cover $\operatorname{Supp} f$ with finitely many sets V_i, $i = 1, \dots, k$, and choose a partition of unity $\varphi_1, \dots, \varphi_k$ relative to this covering. Then $f = \sum_{i=1}^k \varphi_i f$ and $\varphi_i f \in D(V_i)$ for all i. We define

$$M_f = \sum_{i=1}^{k} M_{\varphi_i f}^{V_i}.$$

Clearly, M_f is well-defined. It remains to show that $f \mapsto M_f$ is a mapping of $D(\mathbb{R}_*^{n+1})$ onto $D(\mathbb{R})$.

Let $t_0 \in \mathbb{R}$ and choose $x_0 \in \Gamma_{t_0}$. As before, choose neighbourhoods V of x_0, U of $y = 0$ in $\mathbb{R}^n$, A of $t = t_0$ in $\mathbb{R}$ such that there is a diffeomorphism $\theta : V \to W \times A$ satisfying $Q(\theta^{-1}(y,t)) = t$ for $y \in W$, $t \in A$. Let $\omega(y,t)dydt$ be the image of $dx|_V$ under θ. Then we have for $f \in D(V)$

$$M_f(t) = \int_W (f \circ \theta^{-1})(y,t)\,\omega(y,t)\,dy.$$

There exist a function $g \in D(W \times A)$ and a neighbourhood $B(t_0)$ of t_0 such that

$$\int_W g(y,t)\,\omega(y,t)\,dy = 1 \quad \text{for } t \in B(t_0).$$

Set $f_{t_0} = g \circ \theta$. Then f_{t_0} belongs to $D(V)$ and $M_{f_{t_0}} = 1$ for $t \in B(t_0)$.

Let now $u \in D(\mathbb{R})$. Associate to every point t in $\operatorname{Supp} u$ a function f_t and a neighbourhood $B(t)$ of t as above. There are finitely many $t_1, \dots, t_l$ such that $\operatorname{Supp} u$ is covered by the $B(t_i)$.

Let $\varphi_1, \dots, \varphi_l$ be a partition of unity relative to this covering. Then $u = \varphi_1 u + \dots + \varphi_l u$, $\operatorname{Supp} \varphi_i u \subset B(t_i)$. The function $f = (\varphi_1 \circ Q) f_{t_1} + \dots + (\varphi_l \circ Q) f_{t_l}$ satisfies the conditions $f \in D(\mathbb{R}_*^{n+1})$ and $M_f = u$. Hence $f \mapsto M_f$ is a surjective mapping.

It is clear that $f \mapsto M_f(t)$ $(t \neq 0)$ is a positive measure on $\mathbb{R}^{n+1}$, since its support Γ_t is closed in $\mathbb{R}_*^{n+1}$; moreover it is H-invariant, so $f \mapsto M_f(t)$ is proportional to the measure μ_t as given by (9.1.3) and (9.1.4), for instance. For $t = 0$ a similar statement holds, see (9.1.2).

Show that one actually has for $t \neq 0$ and $f \in D(\mathbb{R}^{n+1})$

$$M_f(t) = \frac{2\pi^{\frac{n}{2}}}{\Gamma(\frac{n}{2})} |t|^{\frac{n-1}{2}} \mu_t(f),$$

if μ_t is defined by (9.1.3) and (9.1.4).

A.2 Results of Méthée

Let again $G = \mathrm{O}(1, n)$. By Theorem A.1.1 there exists a continuous mapping $f \mapsto M_f$ of $D(\mathbb{R}_*^{n+1})$ onto $D(\mathbb{R})$. If S is a distribution on $\mathbb{R}$ then T defined by $\langle T, f\rangle = \langle S, M_f\rangle$ is a G-invariant distribution on $\mathbb{R}_*^{n+1}$. We shall show that the converse is also true. This result is due to Méthée [31].

Theorem A.2.1. *Let M be the continuous linear mapping from $D(\mathbb{R}_*^{n+1})$ onto $D(\mathbb{R})$ defined in* Theorem A.1.1 *and let M' be the dual mapping from the space $D'(\mathbb{R})$ into the space $D'(\mathbb{R}_*^{n+1})$. Then M' maps $D'(\mathbb{R})$ onto the space of G-invariant distributions in $D'(\mathbb{R}_*^{n+1})$.*

The proof is rather involved, we follow the arguments in [37]. The structure of the proof is as follows.

1. *There is a well-defined continuous linear mapping from $D(\mathbb{R})$ into $D(\mathbb{R}_*^{n+1})$, $\varphi \mapsto \alpha_\varphi$, such that $M_{\alpha_\varphi} = \varphi$.*

Consequently, any distribution T on $\mathbb{R}_*^{n+1}$ gives rise to a distribution S on $\mathbb{R}$ by $\langle S, \varphi\rangle = \langle T, \alpha_\varphi\rangle$ $(\varphi \in D(\mathbb{R}))$.

2. *Let T be a G-invariant distribution on $\mathbb{R}_*^{n+1}$. The restriction T_0 of T to $\{x : Q(x) \neq 0\}$ is of the form $T_0 = M' S_0$ for some $S_0 \in D'(\mathbb{R}^*)$.*

Hence, if S is as in 1., then $S = S_0$ on $\mathbb{R}^*$, so S_0 can be continued to $\mathbb{R}$ and the distribution $T - M'S$ is G-invariant with support in $\{x : Q(x) = 0\}$.

3. *Any G-invariant distribution T on $\mathbb{R}_*^{n+1}$ with support in $Q(x) = 0$ is of the form $p(\frac{d}{dt}) M_{\overline{f}}(t)|_{t=0}$, where p is a polynomial in one variable.*

So $T = M'S$ for $S \in D'(\mathbb{R})$, $S = p(\frac{d}{dt})\,\delta$.

ad 1. We recall that M is surjective: to any $v \in D(\mathbb{R})$ there is $f \in D(\mathbb{R}^{n+1}_*)$ with $M_f = v$ (Theorem A.1.1). We are now going to construct a C^∞ function α on $\mathbb{R}^{n+1}_*$ with the following properties:

(a) For any compact set $K \subset \mathbb{R}$, $Q^{-1}(K) \cap \operatorname{Supp}\alpha$ is compact.

(b) $M_\alpha(t) = 1$ for all $t \in \mathbb{R}$.

Set $\mathbb{R}^{n+1}_+ = \{x \in \mathbb{R}^{n+1} : Q(x) > 0\}$. There is $\psi \in D(\Gamma_1)$ with

$$\int_{\Gamma_1} \psi(x)\, d\mu_1(x) = 1.$$

Set $\psi_+(x) = \psi(x\, Q(x)^{-1/2})\, Q(x)^{-\frac{n-1}{2}}$ for $x \in \mathbb{R}^{n+1}_+$. Then ψ_+ is C^∞ on $\mathbb{R}^{n+1}_+$. If $K \subset R^*_+$ is compact, then $Q^{-1}(K) \cap \operatorname{Supp}\psi_+$ is compact. Moreover $M_{\psi_+}(t) = 1$ for $t \in \mathbb{R}^*_+$.

In a similar way we construct ψ_- on $\mathbb{R}^{n+1}_- = \{x \in \mathbb{R}^{n+1} : Q(x) < 0\}$.

Let now $v \in D(\mathbb{R})$ be such that $v = 1$ in a neighbourhood of $t = 0$. We know that there is $f \in D(\mathbb{R}^{n+1}_*)$ with $M_f = v$. Define

$$\alpha(x) = \begin{cases} f(x) + [1 - v\, Q(x)]\, \psi_+(x) & \text{if } x \in \mathbb{R}^{n+1}_+, \\ f(x) & \text{if } Q(x) = 0, \\ f(x) + [1 - v\, Q(x)]\, \psi_-(x) & \text{if } x \in \mathbb{R}^{n+1}_-. \end{cases}$$

Then α has the required properties.

Let now $\varphi \in D(\mathbb{R})$ and set $\alpha_\varphi(x) = \alpha(x)\,\varphi(Q(x))$. Then α_φ has compact support and $M_{\alpha_\varphi} = \varphi$. Observe that the mapping $\varphi \mapsto \alpha_\varphi$ is continuous.

ad 2. Let T be an invariant distribution on $\mathbb{R}^{n+1}_*$. The mapping

$$(t, y) \mapsto t.y$$

of $\mathbb{R}^*_+ \times \Gamma_1$ onto $\mathbb{R}^{n+1}_+$ is a diffeomorphism. Consider now functions f of the form

$$f(t, y) = \beta(y)\,\alpha(t) \qquad (\alpha \in D(\mathbb{R}^*_+),\ \beta \in D(\Gamma_1)).$$

These functions span a dense subspace of $D(\mathbb{R}^{n+1}_+)$. Fix α and consider $\beta \mapsto \langle T, f\rangle$.

There is a constant $\sigma(\alpha)$ such that

$$\langle T, \alpha \otimes \beta\rangle = \sigma(\alpha) \int_{\Gamma_1} \overline{\beta(y)}\, d\mu_1(y)$$

and σ is a distribution on $\mathbb{R}^*_+$.

We obtain

$$\langle T, f\rangle = \langle \sigma, |t|^{-(n-1)}\, M_f(t^2)\rangle.$$

So there exists a distribution S_+ on $\mathbb{R}^*_+$ with

$$\langle T, f\rangle = \langle S_+, M_f\rangle \qquad (f \in D(\mathbb{R}^{n+1}_+)).$$

Similarly there is a distribution S_- on $\mathbb{R}^*_-$ with

$$\langle T, f \rangle = \langle S_-, M_f \rangle \qquad (f \in D(\mathbb{R}^{n+1}_-)).$$

So $S_0 = (S_+, S_-)$ is a distribution on $\mathbb{R}^*$ with

$$\langle T, f \rangle = \langle S_0, M_f \rangle$$

for $f \in D(\mathbb{R}^{n+1}_+ \cup \mathbb{R}^{n+1}_-)$. Therefore the restriction T_0 of T to $\{x : Q(x) \neq 0\}$ is equal to $M' S_0$.

ad 3. Let T be an invariant distribution on $\mathbb{R}^{n+1}_*$ with support in Γ_0. For every $x_0 \in \Gamma_0$ there is a neighbourhood V where we may use a coordinate system (t, y) $(t \in \mathbb{R},\ y \in \mathbb{R}^n)$ as before in (A.1), so with $t = Q(x)$. We can write T in V as

$$T = \sum_{i=0}^{l} \frac{d^i \delta}{dt^i} \otimes S_i \tag{A.2.1}$$

with $S_1, \dots, S_l$ distributions on $V \cap \Gamma_0$. We call $l = l(x_0)$ the transversal order of T at x_0. Since T is invariant, $l(x_0)$ does not depend on $x_0 \in \Gamma_0$; we write l for the global transversal order. We are going to show by induction on l that there is a polynomial in one variable of order l such that

$$\langle T, f \rangle = p\Big(\frac{d}{dt}\Big)\, M_{\overline{f}}(t)|_{t=0}$$

for $f \in D(\mathbb{R}^{n+1}_*)$.

If $l = 0$, then T is an invariant distribution on Γ_0, so equal to a scalar multiple of μ_0.

Let our statement be true for $0, 1, \dots, l-1$. We shall prove it for l. Write near $x_0 \in \Gamma_0$ the distribution T as in (A.2.1). So

$$\langle T, f \rangle = \sum_{i=0}^{l} \langle S_i, \frac{\partial^i f}{\partial t^i}(0, y) \rangle \qquad (f \in D(V)).$$

We shall show that S_l is (locally) G-invariant.

Let $\beta \in D(V \cap \Gamma_0)$ and take f of the form

$$f(t, y) = \alpha(t)\, \beta(y)$$

with $\alpha(0) = \alpha'(0) = \cdots = \alpha^{(l-1)}(0) = 0$ and $\alpha^{(l)}(0) = 1$. Then $\langle T, f \rangle = \langle S_l, \beta \rangle$. Let $g \in G$ be such that $\operatorname{Supp} L_g(\beta) \subset V \cap \Gamma_0$. If $x \in V$, $x = (t, y)$, we set $g^{-1} \cdot x = (t, \xi(g, y.t))$. We then have

$$(L_g f)(t, y) = \alpha(t)\, \beta(\xi(g, y, t)),$$

and taking into account the conditions imposed on α,

$$\langle T, L_g f \rangle = \langle S_l, \beta(\xi(g, y, t)) \rangle = \langle S_l, L_g \beta \rangle.$$

Since T is invariant, we see that so is S_l, hence $\langle S_l, \beta\rangle = c\, M_{\overline{\beta}}(0)$, where c is a constant. Hence $T - c\,\frac{d^l}{dt^l}\, M_{\overline{f}}(0)$ is a G-invariant distribution with locally and hence, by the G-invariance, also globally, transversal order at most $l-1$. Now apply induction to get the result.

The next step is to extend Theorems A.1.1 and A.2.1 from $\mathbb{R}^{n+1}_*$ to $\mathbb{R}^{n+1}$. This will be done in the next section.

A.3 Results of Tengstrand

We start with the extension of Theorem A.1.1. Let $n \geq 2$ and let $f \in D(\mathbb{R}^{n+1})$. Observe that $M_f(t)$ is defined for all t, see Section 9.1. The crucial point is the determination of the behaviour of $M_f(t)$ near $t = 0$. We begin with a lemma.

Lemma A.3.1. *Let $f \in D(\mathbb{R}^{n+1})$ and $n \geq 2$. Then $M_f(t)$ is continuous at $t = 0$ and C^∞ for $t \neq 0$.*

Let us apply formulae (9.1.3) and (9.1.4). We take f K-invariant, which can be done without loss of generality. We then have (up to a positive constant equal to $\frac{2\pi^{n/2}}{\Gamma(n/2)}$)

$$M_f(t) = \begin{cases} |t|^{\frac{n-1}{2}} \displaystyle\int_0^\infty f(|t|^{1/2}\sinh u, 0, \dots, 0, |t|^{1/2}\cosh u)\, \cosh^{n-1} u\, du & (t<0), \\ t^{\frac{n-1}{2}} \displaystyle\int_0^\infty f(t^{1/2}\cosh u, 0, \dots, 0, t^{1/2}\sinh u)\, \sinh^{n-1} u\, du & (t>0). \end{cases}$$

For simplicity we considered f as a function of two variables, C^∞ and of compact support. Moreover, since f is even in the first and in the second variable, we may write $f(x, y) = g(x^2, y^2)$ for some $g \in D(\mathbb{R}^2)$.

So, for $t < 0$,

$$M_f(t) = |t|^{\frac{n-1}{2}} \int_0^\infty g(|t|\,\sinh^2 u,\ |t|\,\cosh^2 u)\, \cosh^{n-1} u\, du.$$

Substituting $|t|\cosh^2 u = w$, we get

$$\begin{aligned} M_f(t) &= \frac{1}{2}\int_0^\infty g(w - |t|, w)\, w^{\frac{n-2}{2}}\, (w - |t|)^{-1/2}\, dw \qquad \text{(A.3.1)} \\ &= \frac{1}{2}\int_0^\infty g(w, w + |t|)\, (w + |t|)^{\frac{n-2}{2}}\, w^{-1/2}\, dw \end{aligned}$$

and similarly, for $t > 0$,

$$M_f(t) = \frac{1}{2}\int_0^\infty g(w + t, w)\, w^{\frac{n-2}{2}}\, (w + t)^{-1/2}\, dw. \qquad \text{(A.3.2)}$$

Hence $\lim_{t\to 0} M_f(t) = c\, M_f(0)$ by (9.1.2), since $n \geq 2$. Because $M_f(t)$ is already continuous at $t = 0$ if $f \in D(\mathbb{R}^{n+1}_*)$ (by Theorem A.1.1), we get $c = 1$ and $M_f(t)$ is continuous at $t = 0$ for $f \in D(\mathbb{R}^{n+1})$. Clearly, formulae (A.3.1) and (A.3.2) also imply that M_f is C^∞ for $t \neq 0$.

Notice that (A.3.1) and (A.3.2) can be combined to one formula valid for all t (even $t = 0$), namely

$$M_f(t) = \frac{1}{2}\int_{|\tau|}^{\infty} g(z-\tau, z+\tau)\,(z+\tau)^{\frac{n-2}{2}}\,(z-\tau)^{-1/2}\,dz \tag{A.3.3}$$

where $t = -2\tau$ (again up to the positive constant equal to $\frac{2\pi^{n/2}}{\Gamma(n/2)}$). Though $M_f(t)$ is a few times differentiable at $t = 0$, there is obviously a singularity at $t = 0$. The next theorem gives a precise result. Recall the definition of the Heaviside function Y: $Y(t) = 1$ for $t \geq 0$, $Y(t) = 0$ otherwise.

Theorem A.3.2 (Tengstrand). *The image of $D(\mathbb{R}^{n+1})$ under the mapping M is the space $\mathcal{H}_\eta$ consisting of functions on $\mathbb{R}$ of the form*

$$\varphi_1 + \eta\,\varphi_2$$

where $\varphi_1, \varphi_2 \in D(\mathbb{R})$ and

$$\eta(t) = \begin{cases} Y(t)\,|t|^{\frac{n-1}{2}} & \text{if } n \text{ is even,} \\ \log|t|\,t^{\frac{n-1}{2}} & \text{if } n \text{ is odd.} \end{cases} \tag{A.3.4}$$

We follow [47]. Let $f \in D(\mathbb{R}^{n+1})$. Then clearly, M_f has compact support and is C^∞ for $t \neq 0$. We shall examine $M_f(t)$ in $\mathcal{O} = \{\tau : |\tau| \leq \frac{1}{2}\}, -2\tau = t$. It is easily seen that

$$\int_1^{\infty} g(z-\tau, z+\tau)\,(z+\tau)^{\frac{n-1}{2}}\,(z-\tau)^{-1/2}\,dz$$

is C^∞ in τ, so in $C^\infty(\mathcal{O})$.

Let first n be *even*. By expanding g in a Taylor series we get

$$2M_f(\tau) = \sum_{\beta+\gamma\leq v} g_{\beta,\gamma}(0,0)\int_{|\tau|}^{1}(z+\tau)^{\frac{n-2}{2}+\beta}\,(z-\tau)^{\gamma-\frac{1}{2}}\,dz + w, \tag{A.3.5}$$

where $w \in C^{v+\frac{n-2}{2}}$ and $g_{\beta,\gamma}(0,0) = \frac{\partial^{\beta+\gamma} g}{\partial x^\beta \partial y^\gamma}(0,0)$.

Integration by parts (to move the $(z+\tau)$-term to the $(z-\tau)$-term, to get finally $(z-\tau)^{-1/2+\frac{n-2}{2}+\beta+\gamma}$) now gives

$$\begin{aligned} 2M_f(\tau) = Y(-\tau)\,(-2\tau)^{\frac{n-1}{2}} &\sum_{\beta+\gamma\leq v} a_{\beta,\gamma}\,(-\tau)^{\beta+\gamma}\,g_{\beta,\gamma}(0,0) \\ &+ \sum_{\beta+\gamma\leq v} g_{\beta,\gamma}(0,0)\,w_{\beta,\gamma} + w, \end{aligned}$$

where $w \in C^{\frac{n-2}{2}+v}(\mathcal{O})$ and $w_{\beta,\gamma} \in C^\infty(\mathcal{O})$. The $w_{\beta,\gamma}$ are independent of g. Furthermore $a_{\beta,\gamma} = (-1)^{\frac{n}{2}+\beta}\, b_{\beta,\gamma}$ with

$$b_{\beta,\gamma} = 2^{\frac{n-1}{2}+\beta+\gamma} \binom{\beta+\gamma}{\beta} \frac{\Gamma(\frac{n}{2}+\beta)\,\Gamma(\gamma+\frac{1}{2})}{\Gamma(\frac{n+1}{2}+\beta+\gamma)}.$$

By a classical result, due to E. Borel, we can find a function φ in $D(\mathbb{R})$ with

$$\frac{d^v}{d\tau^v}\varphi(0,0) = \sum_{\beta+\gamma=v} a_{\beta,\gamma}\, g_{\beta,\gamma}(0,0).$$

Then we have

$$2M_f(\tau) = Y(-\tau)\,(-2\tau)^{\frac{n-1}{2}}\,\varphi(-\tau) \in C^{\frac{n-2}{2}+v}(\mathcal{O})$$

for all v, so $M_f(\tau) = \varphi_1(\tau) + \eta(\tau)\,\varphi_2(\tau)$ with $\varphi_1, \varphi_2 \in D(\mathbb{R})$.

To see that M maps $D(\mathbb{R}^{n+1})$ onto $\mathcal{H}_\eta$, take $\varphi \in \mathcal{H}_\eta$ of the form $\varphi = \varphi_1 + \eta\,\varphi_2$ and choose g in $D(\mathbb{R}^2)$ of the form

$$g(x,y) = \alpha(x)\,\beta(y)$$

with $\beta = 1$ near $y = 0$, $\beta \in D(\mathbb{R})$ and $\alpha \in D(\mathbb{R})$ such that $2^{\frac{n-1}{2}}\,(-1)^v\,\alpha^{(v)}(0) = \varphi_2^{(v)}(0)$ $(v = 0,1,2,\ldots)$.

Set $f(x,y) = g(x_0^2, x_1^2 + \cdots + x_n^2)$. Then f can be seen as a K-invariant C^∞ function on $\mathbb{R}^{n+1}$ with compact support. Moreover $2\,M_f(\tau) - \varphi(\tau)$ is in $C^\infty(\mathbb{R})$, so in $D(\mathbb{R})$, and thus, applying Theorem A.1.1, M is surjective.

The case n *odd* is a little more difficult, but the proof goes along the same lines. To handle the analog of (A.3.5), i.e. the integrals

$$\int_{|\tau|}^1 (z+\tau)^{\frac{n-2}{2}+\beta}\,(z-\tau)^{\gamma-\frac{1}{2}}\,dz,$$

we consider separately $\tau > 0$ and $\tau < 0$. Let $\tau > 0$. By partial integration we reduce it to

$$\int_\tau^1 (z+\tau)^{\frac{n-3}{2}+\beta+\gamma+\frac{1}{2}}\,(z-\tau)^{-1/2}\,dz$$

(up to a well-defined constant).

Now write

$$\left(\frac{z-\tau}{z+\tau}\right)^{-1/2} = \left(1-\frac{2\tau}{z+\tau}\right)^{-1/2} = \sum_{k=0}^N \binom{-1/2}{k} \frac{2^k\tau^k}{(z+\tau)^k} + \frac{\tau^{N+1}}{(z+\tau)^{N+1}}\,\psi_N(z.\tau).$$

Similarly for $\tau < 0$. We finally get

$$\begin{aligned} 2\,M_f(\tau) = \pi^{-1}\log|\tau| &\sum_{\beta+\gamma\le v} a_{\beta,\gamma}\,\tau^{\frac{n-1}{2}+\beta+\gamma}\,g_{\beta,\gamma}(0,0) \\ &+ \sum_{\beta+\gamma\le v} g_{\beta,\gamma}(0,0)\,w_{\beta,\gamma} + w, \end{aligned}$$

where $w \in C^{\frac{n-1}{2}+v+1}(\mathcal{O})$ and $w_{\beta,\gamma} \in C^\infty(\mathcal{O})$. The $w_{\beta,\gamma}$ are independent of g. Furthermore, $a_{\beta,\gamma} = (-1)^{\frac{n-1}{2}+\beta}\, b_{\beta,\gamma}$, where $b_{\beta,\gamma}$ is as above. The proof is now easily completed.

A careful look at the proof of Theorem A.3.2 also reveals the following. Define the functionals A_k and B_k ($k = 0, 1, 2, \ldots$) on $\mathcal{H}_\eta$ by

$$A_k(\varphi_1 + \eta\,\varphi_2) = \frac{d^k\overline{\varphi_1}}{dt^k}(0), \quad B_k(\varphi_1 + \eta\,\varphi_2) = \frac{d^k\overline{\varphi_2}}{dt^k}(0).$$

Observe that A_k and B_k are well-defined: if $\varphi_1 = \eta\,\varphi_2 \in D(\mathbb{R})$, then all derivatives of φ_1 and φ_2 vanish at $t = 0$. We see that for $f \in D(\mathbb{R}^{n+1})$

$$B_k(M_f) = \sum_{\beta+\gamma=k} c'_{\beta,\gamma}\, \overline{g_{\beta,\gamma}(0,0)}$$

with $c'_{\beta,\gamma}$ well-defined constants, independent of f. About the relation between derivatives of f and g, observe the following.

For any $(n+1)$-tuple $\alpha = (\alpha_0, \alpha_1, \ldots, \alpha_n)$ of non-negative integers we set $|\alpha| = \alpha_0 + \alpha_1 + \cdots + \alpha_n$ and $D^\alpha = \frac{\partial^{|\alpha|}}{\partial x_0^{\alpha_1}\partial x_1^{\alpha_1}\cdots\partial x_n^{\alpha_n}}$. Let us call α even if all α_i are even.

For $f \in D(\mathbb{R}^{n+1})$ set $F(x) = \int_K f(k \cdot x)\,dk$ ($x \in \mathbb{R}^{n+1}$). Notice that F is in $D(\mathbb{R}^{n+1})$ and is even in each variable $x_0, x_1, \ldots, x_n$. So only for even $\alpha = (\alpha_0, \alpha_1, \ldots, \alpha_n)$, $D^\alpha F(0)$ does not vanish and one has

$$(D^\alpha F)(0) = \sum_{|\delta|=|\alpha|} e_\delta\,(D^\delta f)(0),$$

where the e_δ are constants independent of f such that $e_\alpha \neq 0$. In addition we have

$$D^{\beta,\gamma} g(0,0) = g_{\beta,\gamma}(0,0) = \frac{\beta!\gamma!}{(2\beta)!\,(2\gamma)!}\,(D^{2\beta,0,\ldots,0,2\gamma}F)(0),$$

where we have set $F(x, 0, \ldots, 0, y) = g(x^2, y^2)$. The result follows by using the Taylor expansion of g at $(0,0)$. So combining these results we have

$$B_k(M_f) = \sum_{|\delta|=2k} c_\delta\,(D^\delta f)(0) \tag{A.3.6}$$

for all $f \in D(\mathbb{R}^{n+1})$, the c_δ being constants not depending on f and not all equal to zero.

We shall now provide the vector space $\mathcal{H}_\eta$ with a topology, that makes it a locally convex topological space in a natural way. Put on $\mathcal{H}_\eta$ the strongest locally convex topology such that the maps $\varphi \mapsto \varphi$ and $\varphi \mapsto \eta\varphi$ from $D(\mathbb{R})$ to $\mathcal{H}_\eta$ are continuous: the "*topologie localement convexe finale*" (see [4, Chapter II, §4, no. 4]). The restriction of this topology on $\mathcal{H}_\eta$ to $D(\mathbb{R})$ and $\eta\, D(\mathbb{R})$ coincides with the original topologies on $D(\mathbb{R})$ and $\eta\, D(\mathbb{R})$ respectively. So in particular, by Hahn–Banach's theorem, any distribution on $D(\mathbb{R})$ can be extended to an element of $\mathcal{H}'_\eta$, the anti-dual of $\mathcal{H}_\eta$. Let $T \in \mathcal{H}'_\eta$. Then the restriction of T to $D(\mathbb{R})$ is a distribution on $\mathbb{R}$, that can be extended to an element S on $\mathcal{H}'_\eta$. Then $T - S$ vanishes on $D(\mathbb{R})$. Now consider the functional

$$\varphi \mapsto (T - S)(\eta\,\varphi),$$

which is a distribution on $\mathbb{R}$ with support in $\{0\}$, so a finite linear combination of derivatives of the δ-function. Otherwise stated:

$$(T - S)(\eta\,\varphi) = \sum_{k=0}^{m} c_k\, B_k(\varphi)$$

for some positive integer m and constants c_k, independent of φ. Thus $T = S + \sum_{k=0}^{m} c_k\, B_k$. This gives a good insight into the structure of elements of $\mathcal{H}'_\eta$. Notice that $B_k \in \mathcal{H}'_\eta$. Summarizing we have:

Proposition A.3.3. (a) *Any distribution T on $\mathbb{R}$ can be extended into an element of $\mathcal{H}'_\eta$.*
(b) *Two extensions of the same distribution T differ by a finite linear combination of the functionals B_k.*

It is not difficult to show that $f \mapsto M_f$ is now continuous on $D(\mathbb{R}^{n+1})$. We will not go in detail here, but just refer to the proof in [47].

Notice that both $D(\mathbb{R}^{n+1})$ and $\mathcal{H}_\eta$ are inductive limits of Fréchet spaces, so that the closed graph theorem applies both to these spaces and their duals.

Let $\Box$ be the second order differential operator on $\mathbb{R}^{n+1}$ that we defined in Section 9.1 (iii).

Lemma A.3.4. *Any G-invariant distribution T on $\mathbb{R}^{n+1}$ with support in $\{0\}$ is of the form $T = p(\Box)\,\delta$ where p is a polynomial in one variable.*

In fact, every such T has the form $q(\partial)\,\delta$ where q is an invariant polynomial, so $q(\partial) = p(\Box)$.

Lemma A.3.5. *Let $T \in D'(\mathbb{R}^{n+1})$ be G-invariant and with support contained in $\{0\}$. Then T is a finite linear combination of the distributions $M' B_k$.*

Clearly, this follows from (A.3.6) and Lemma A.3.4, since $\delta, \Box\,\delta, \dots, \Box^m\delta$ span the same $(m+1)$-dimensional space as $M' B_0, M' B_1, \dots, M' B_m$.

Theorem A.3.6. *The adjoint M' of the mapping M from $\mathcal{H}'_\eta$ into the space of G-invariant distributions on $\mathbb{R}^{n+1}$ is a continuous surjective mapping, hence an isomorphism.*

The latter statement follows from the closed graph theorem. If $T \in \mathcal{H}'_\eta$ then clearly $M'T$ is a G-invariant distribution on $\mathbb{R}^{n+1}$, since M is continuous. It remains to show that M' is surjective, since M' is clearly injective. So let $T_0 \in D'(\mathbb{R}^{n+1})$ be G-invariant. Then the restriction S_0 of T_0 to $\mathbb{R}^{n+1}_*$ is of the form $M'S$ for some $S \in D'(\mathbb{R})$, by the results of Méthée. Extend S to $\mathcal{H}_\eta$ and call it S again. Then $T - M'S$ is in $D'(\mathbb{R}^{n+1})$, is G-invariant and has support in $\{0\}$. So it is of the form $\sum_{k=0}^m c_k\, M'B_k$ by Lemma A.3.5, for some constants $c_0, \ldots, c_m$. This proves the surjectivity of M'.

Remark A.3.7. From the proof of Theorem A.3.2 we get

$$M'B_0 = \begin{cases} \dfrac{(-1)^{\frac{n}{2}}\, \pi^{\frac{n+1}{2}}}{\Gamma(\frac{n+1}{2})}\,\delta & \text{if } n \text{ is even,} \\ \dfrac{(-1)^{\frac{n+1}{2}}\, \pi^{\frac{n-1}{2}}}{\Gamma(\frac{n+1}{2})}\,\delta & \text{if } n \text{ is odd.} \end{cases}$$

These relations have been obtained earlier by De Rham. Notice that we also have: $M'B_k$ is proportional to $\Box^k\delta$, so $M'B_k = \alpha_k \Box^k \delta$. We leave it as an exercise to compute α_k (notice that $\alpha_k \neq 0$).

A.4 Solutions in $\mathcal{H}'_\eta$ of a singular second order differential equation

For this section we rely on [27].

We begin with the radial part of $\Box = -\frac{\partial^2}{\partial x_0^2} + \frac{\partial^2}{\partial x_1^2} + \cdots + \frac{\partial^2}{\partial x_n^2}$. If $\Phi \in C^\infty(\mathbb{R})$, then we have

$$(L\Phi) \circ Q = \Box\,(\Phi \circ Q)$$

where $L = 4t\,\frac{d^2}{dt^2} + 2(n+1)\,\frac{d}{dt}$.

Now consider the relation

$$\int_{\mathbb{R}^{n+1}} f(x)\,\Phi(Q(x))\,dx = \int_{-\infty}^{\infty} M_f(t)\,\Phi(t)\,dt \tag{A.4.1}$$

for $f \in D(\mathbb{R}^{n+1})$, $\Phi \in C^\infty(\mathbb{R})$.

Let L^* be the differential operator $4t\,\frac{d^2}{dt^2} - 2(n-3)\,\frac{d}{dt}$. Then one has, for $f \in D(\mathbb{R}^{n+1}_*)$, $M_{\Box f}(t) = L^* M_f(t)$ for all $t \in \mathbb{R}$. It follows immediately that for $f \in D(\mathbb{R}^{n+1})$ one has $M_{\Box f}(t) = L^* M_f(t)$ for $t \neq 0$, so L^* leaves $\mathcal{H}_\eta$ invariant. This can also be verified directly, using the explicit expression of L^*.

Moreover, L^* is a continuous operator. Also this fact can be shown straightforwardly applying the explicit expression of L^*, but it also follows from the fact that $f \mapsto M_f$ is not only continuous but also open, by the closed graph theorem. Now we can extend the action of L to elements of $\mathcal{H}'_\eta$: if $T \in \mathcal{H}'_\eta$ then $\langle LT, \varphi\rangle = \langle T, L^*\varphi\rangle$ $(\varphi \in \mathcal{H}_\eta)$. If T is a distribution, this definition agrees with the classical one if one takes $\varphi \in D(\mathbb{R})$. If $\Phi \in C^2(\mathbb{R})$, then $L\Phi$ is the same as usual, a continuous function again. Indeed, if $T_\Phi \in \mathcal{H}'_\eta$ is the element of $\mathcal{H}'_\eta$ defined by Φ, then

$$\begin{aligned}
\langle LT_\Phi, M_f\rangle &= \langle T_\Phi, L^* M_f\rangle \qquad \text{(by definition)} \qquad\qquad \text{(A.4.2)}\\
&= \int_{-\infty}^{\infty} \Phi(t)\,\overline{L^* M_f(t)}\,dt\\
&= \int_{-\infty}^{\infty} \Phi(t)\,\overline{M_{\Box f}(t)}\,dt\\
&= \int_{\mathbb{R}^{n+1}} \overline{\Box f(x)}\,\Phi(Q(x))\,dx\\
&= \int_{-\infty}^{\infty} \overline{M_f(t)}\,(L\Phi)(t)\,dt.
\end{aligned}$$

We shall now determine the solutions of $LT = \lambda T$ in $\mathcal{H}'_\eta$, which is the same problem as the determination of all G-invariant distributions T_0 on $\mathbb{R}^{n+1}$ satisfying $\Box\, T_0 = \lambda T_0$.

The second order differential equation

$$4t\,\frac{d^2u}{dt^2} + 2(n+1)\,\frac{du}{dt} - \lambda u = 0$$

is singular at $t = 0$.

For simplicity of the presentation we shall restrict from now on to the case n *even*.

Set $\mu = \frac{n-1}{2}$, so that $L = 4(t\frac{d^2}{dt^2} + (\mu+1)\frac{d}{dt})$. Since μ is not an integer, we have the following two fundamental *classical* solutions on $\mathbb{R}_*$: $\Phi(t,\lambda,\mu)$ and $|t|^\mu\,\Phi(t,\lambda,-\mu)$. Here Φ is an entire analytic function, namely $\Phi(t,\lambda,\mu) = \sum_{k=0}^{\infty} a_k(\lambda,\mu)t^k$ with

$$a_k(\lambda,\mu) = \frac{(\frac{\lambda}{4})^k}{k!\,(1+\mu)_k},$$

where $(\alpha)_k = \frac{\Gamma(\alpha+k)}{\Gamma(\alpha)} = \alpha(\alpha+1)\cdots(\alpha+k-1)$. The function Φ can be expressed in terms of Bessel functions, see Section 7.1 (ii).

To proceed, let us recall the definition of the Partie finie Pf (see [43, p. 42]). One defines

$$\mathrm{Pf}\int_0^\infty t^{-\mu}\,\varphi(t)\,dt = \lim_{\varepsilon\to 0}\Big[\int_\varepsilon^\infty t^{-\mu}\,\varphi(t)\,dt + \varphi(0)\,\frac{\varepsilon^{-\mu+1}}{-\mu+1} + \varphi'(0)\,\frac{\varepsilon^{-\mu+2}}{-\mu+2} + \cdots + \frac{\varphi^{(k)}(0)}{k!}\,\frac{\varepsilon^{-\mu+k+1}}{-\mu+k+1}\Big],$$

where k is such that $-\mu + k + 2 > 0$, or, writing

$$\int_\varepsilon^\infty t^{-\mu}\,\varphi(t)\,dt = \sum_{k=1}^{[\mu]} c_k\,\varepsilon^{-\mu+k} + c_0 + o\,(1),$$

one has $\mathrm{Pf} = c_0$.

Now define the following distributions on $\mathbb{R}$:

$$\langle S_\lambda^+, \varphi\rangle = \int_0^\infty \Phi(t,\lambda,\mu)\,\overline{\varphi(t)}\,dt, \tag{A.4.3}$$

$$\langle S_\lambda^-, \varphi\rangle = \int_{-\infty}^0 \Phi(t,\lambda,\mu)\,\overline{\varphi(t)}\,dt, \tag{A.4.4}$$

$$\langle T_\lambda^+, \varphi\rangle = \mathrm{Pf}\int_0^\infty t^{-\mu}\,\Phi(t,\lambda,-\mu)\,\overline{\varphi(t)}\,dt, \tag{A.4.5}$$

$$\langle T_\lambda^-, \varphi\rangle = \mathrm{Pf}\int_{-\infty}^0 |t|^{-\mu}\,\Phi(t,\lambda,-\mu)\,\overline{\varphi(t)}\,dt \tag{A.4.6}$$

for $\varphi \in D(\mathbb{R})$.

If $T \in D(\mathbb{R})$ is a distribution solution of $LT = \lambda T$, then

$$T = a\,S_\lambda^+ + b\,S_\lambda^- + c\,T_\lambda^+ + d\,T_\lambda^- + T_0$$

for constants a, b, c, d and T_0 a distribution with support in $\{0\}$. For the following relations we apply the property

$$uL^*\varphi - (Lu)\varphi = [\varphi, u]',$$

where $[\varphi, u] = 4t\,(\varphi' u - \varphi u') - 4\mu\varphi\,u$.

We have:

(a) $(L-\lambda)\,S_\lambda^+ = 4\delta,\ (L-\lambda)\,S_\lambda^- = -4\delta,$

(b) $(L-\lambda)\,T_\lambda^+ = 0,\ (L-\lambda)\,T_\lambda^- = 0,$

(c) $L\,\delta^{(k)} = (\mu - k - 1)\,\delta^{(k+1)},$

hence a fundamental system of *distribution solutions* is given by the three distributions

$$S_\lambda^+ + S_\lambda^-, \quad T_\lambda^+, \quad T_\lambda^-. \tag{A.4.7}$$

Now we determine fundamental solutions in $\mathcal{H}'_\eta$ where $\eta(t) = Y(t)\,|t|^\mu$. Clearly, S_λ^+, S_λ^-, T_λ^+, T_λ^- can be extended to $\mathcal{H}_\eta$ using the expressions (A.4.3)–(A.4.6). We then get on $\mathcal{H}_\eta$:

(a) $(L - \lambda)\, S_\lambda^\pm = \pm\, 4\mu A_0$,

(b) $(L - \lambda)\, T_\lambda^- = 0,\ (L - \lambda)\, T_\lambda^+ = -4\mu B_0$,

(c) $L B_k = c_k\, B_{k+1}$ for $c_k \neq 0$.

As to (c), apply Remark A.3.7. Indeed, the relation $M' B_k = \alpha_k\, \square^k\, \delta$ implies the statement, since $\square\, M' B_k = M'\, L B_k$. Applying Proposition A.3.3 we now get

Proposition A.4.1. *A fundamental system of solutions of $Lu = \lambda u$ in $\mathcal{H}'_\eta$, where $\eta(t) = Y(t)\,|t|^\mu$, $\mu = \frac{n-1}{2}$, n even, is given by $S_\lambda^+ + S_\lambda^-$ and T_λ^-.*

A.5 Expression of $M_{\widehat{f}}(\lambda)$ in terms of Bessel functions

According to Section 9.1 (iii), the distribution $u : f \mapsto M_{\widehat{f}}(\lambda)$ satisfies the differential equation $\square\, u = 4\pi^2 \lambda u$. So we may write

$$M_{\widehat{f}}\Big(\frac{\lambda}{4\pi^2}\Big) = a(\lambda)\,(S_\lambda^+ + S_\lambda^-)(M_f) + b(\lambda)\, T_\lambda^-(M_f)$$

for all $f \in D(\mathbb{R}^{n+1})$. Here $a(\lambda)$ and $b(\lambda)$ are constants, depending on λ only. We are going to determine a and b. The relation $M_{f_u}(\lambda) = |u|^{n-1} M_f(\frac{\lambda}{u^2})$ for $u \neq 0$, where $f_u(x) = f(\frac{x}{u})$ $(x \in \mathbb{R}^{n+1})$, easily implies the following form of a and b:

- $a(\lambda) = a_+\, |\lambda|^{\frac{n-1}{2}}$ $(\lambda > 0)$, $a(\lambda) = a_-\, |\lambda|^{\frac{n-1}{2}}$ $(\lambda < 0)$,
- $b(\lambda) = b_+$ $(\lambda > 0)$, $b(\lambda) = b_-$ $(\lambda < 0)$.

Here a_+, a_-, b_+, b_- are constants, independent of λ. Since $M_{\widehat{f}}$ is continuous at $\lambda = 0$, we get $b_+ = b_- = b$. Moreover, from the general form of M_f it follows that $a_- = 0$. Furthermore,

$$(2\pi)^{n-1}\, a_+ \int_{-\infty}^{\infty} M_f(t)\, dt = c_0 \widehat{f}(0)$$

with $c_0 = \frac{(-1)^{\frac{n}{2}}\, \pi^{\frac{n+1}{2}}}{\Gamma(\frac{n+1}{2})}$ (see Section A.3), hence $(2\pi)^{n-1}\, a_+ = c_0$.

Thus it remains to determine the constant b. For $\lambda < 0$ we have

$$M_{\widehat{f}}\Big(\frac{\lambda}{4\pi^2}\Big) = b\, T_\lambda^-(M_f),$$

for all $f \in \mathcal{S}(\mathbb{R}^{n+1})$. We are going to compute both sides of this equation with $f(x) = f_0(x) = \mathrm{e}^{-\pi\|x\|^2}$. Observe that $f = \widehat{f}$ for this choice of f.

We apply the expression for $M_{f_0}(t)$, previously derived in Section A.3 for $t < 0$,

$$M_{f_0}(t) = \mathrm{e}^{\pi t} \int_0^\infty \mathrm{e}^{-2\pi w}\, w^{-1/2}\,(w-t)^{\frac{n-2}{2}}\, dw,$$

up to a positive constant. This integral can be computed. We obtain

$$M_{f_0}(t) = \mathrm{e}^{\pi t}\, p(-t) \qquad (t < 0)$$

where $p(t)$ is a polynomial of degree $\frac{n-2}{2}$, $p(t) = \sum_{l=0}^{\frac{n-2}{2}} a_l\, t^l$ with

$$a_l = \binom{\frac{n-2}{2}}{l} (2\pi)^{-\frac{n+1}{2}+l}\, \Gamma(\frac{n+1}{2} - l).$$

So $M_{f_0}(0) = a_0 = (2\pi)^{-\frac{n+1}{2}}\, \Gamma(\frac{n+1}{2})$.

Now we can easily compute the right-hand side at $\lambda = 0$:

$$T_0^-(M_{f_0}) = \int_{-\infty}^0 |t|^{-\mu}\, M_{f_0}(t)\, dt = \int_0^\infty t^{-\mu}\, p(t)\, \mathrm{e}^{-\pi t}\, dt.$$

We obtain $T_0^-(M_{f_0}) = 2^{-\frac{n+1}{2}}\pi^{-\frac{n+1}{2}}\,(\frac{n-1}{2} - \pi)\,(1-2\pi)^{\frac{n-4}{2}} = d_0$. Hence $b = a_0/d_0$.

Appendix B
The Averaging Mapping on the Space X

B.1 Special case of a theorem of Harish-Chandra

Let X be the hyperboloid in $\mathbb{R}^{n+1}$ given by

$$-x_0^2 + x_1^2 + \cdots + x_n^2 = 1.$$

It is a C^∞ manifold. The group $G = \mathrm{O}(1, n)$ acts transitively on X, and the stabilizer H of e_n is isomorphic to the group $H = \mathrm{O}(1, n-1)$. As in Appendix A, set $K = \mathrm{O}(1) \times \mathrm{O}(n)$. Let Q be the mapping on X given by

$$Q(x) = x_n.$$

The mapping Q gives a parametrization of the H-orbits on X: $Q(x) = t$ is an H-orbit for all $t \neq \pm 1$, while for $t = \pm 1$ the set splits into two orbits.

Let X_* be the space X without the points e_n and $-e_n$. It easily follows that Q is submersive on X_*: if $Q(x) = 0$, then x_n can be taken as one of the local coordinates near x, whereas if $Q(x) \neq 0$ we can express x_n in the coordinates $x_0, x_1, \ldots, x_{n-1}$ and determine $dQ(x)$. In all cases we easily get $dQ(x) \neq 0$ if $x \in X_*$.

We have the following immediate analog of Theorem A.1.1:

Theorem B.1.1. *There exists a mapping $f \mapsto M_f$ from $D(X_*)$ onto $D(\mathbb{R})$ such that for all continuous functions Φ on $\mathbb{R}$ one has*

$$\int_X f(x)\, \Phi(Q(x))\, dx = \int_{\mathbb{R}} M_f(t)\, \Phi(t)\, dt.$$

Moreover Supp $M_f \subset Q($Supp $f)$ *and $f \mapsto M_f$ is continuous.*

The proof is completely similar to that of Theorem A.1.1.

B.2 Analog of Méthée's results

Our goal is to prove the analog of Theorem A.2.1 for the space X.

Theorem B.2.1. *Let M be the continuous linear mapping from $D(X_*)$ into $D(\mathbb{R})$, defined in Theorem B.1.1, and let M' be the dual mapping from $D'(\mathbb{R})$ into $D'(X_*)$. Then M' maps $D'(\mathbb{R})$ onto the space of H-invariant distributions in $D'(X_*)$.*

The proof is along the same lines as the proof of Theorem A.2.1; only part 1. needs to be reconsidered. It reads:

There is a well-defined continuous mapping from $D(\mathbb{R})$ *to* $D(X_*)$, $\varphi \mapsto \alpha_\varphi$ *such that* $M_{\alpha_\varphi} = \varphi$.

By abuse of notation, let M be the subgroup of H isomorphic to $\mathrm{O}(n-1)$, consisting of the matrices of the form

$$\begin{pmatrix} 1 & 0 & 0 \\ 0 & m & 0 \\ 0 & 0 & 1 \end{pmatrix}$$

with $m \in \mathrm{O}(n-1)$, and let A be the (Cartan) subgroup of the matrices

$$a_u = \begin{pmatrix} \cosh u & 0 & \sinh u \\ 0 & I_{n-1} & 0 \\ \sinh u & 0 & \cosh u \end{pmatrix}$$

with $u \in \mathbb{R}$. Set $X^+ = \{x \in X : Q(x) > 1\}$. Then the mapping $H/M \times R_+^* \to X^+$ given by $(h,u) \mapsto h \cdot a_u \cdot e_n$ is a diffeomorphism and the Haar measure dx on X is given by $c_+ \sinh^{n-1} u \, d\dot{h} du$, where $d\dot{h}$ is an H-invariant measure on H/M and c_+ is a positive constant. This follows easily by taking hyperbolic coordinates on X^+.

Similar considerations hold for the space $X^- = \{x \in X : Q(x) < -1\}$: the mapping $H/M \times \mathbb{R}_+^* \to X^-$ given by $(h,u) \mapsto h \cdot a_u \cdot (-e_n)$ is a diffeomorphism and we have $dx = c_- \sinh^{n-1} u \, d\dot{h} du$.

For the remaining part $X^0 = \{x \in X : -1 < Q(x) < 1\}$ we take another, compact, Cartan subgroup B consisting of the matrices

$$b_\theta = \begin{pmatrix} I_{n-1} & 0 & 0 \\ 0 & \cos\theta & \sin\theta \\ 0 & -\sin\theta & \cos\theta \end{pmatrix} \qquad (0 \le \theta < 2\pi),$$

and the subgroup $M_1 \simeq \mathrm{O}(1, n-2)$, embedded into H by

$$m_1 \mapsto \begin{pmatrix} m_1 & 0 & 0 \\ 0 & 1 & 0 \\ 0 & 0 & 1 \end{pmatrix} \qquad (m_1 \in \mathrm{O}(1, n-2)).$$

The mapping $H/M_1 \times (0,\pi) \to X^0$ given by $(h,\theta) \mapsto h \cdot b_\theta \cdot e_n$ is a diffeomorphism and $dx = c_0 \sin^{n-1} \theta \, d\dot{h}_1 d\theta$, where $d\dot{h}_1$ is now an invariant measure on H/M_1.

Notice that $Q(h \cdot a_u \cdot e_n) = \cosh u$, $Q(h \cdot a_u \cdot (-e_n)) = -\cosh u$ and $Q(h \cdot b_\theta \cdot e_n) = \cos\theta$.

We are going to construct the analogs of the functions ψ_+, ψ_- and a new function ψ_0.

After the change of variables $\cosh u = t$, $-\cosh u = t$, $\cos\theta = t$ respectively, the Haar measure is equal to, respectively

$$c_+ (t^2 - 1)^{\frac{n-2}{2}} dt \, d\dot{h}, \quad c_- (t^2 - 1)^{\frac{n-2}{2}} dt \, d\dot{h}, \quad c_0 (1 - t^2)^{\frac{n-2}{2}} dt \, d\dot{h}_1.$$

The definition of ψ_+ is as follows. Take a function ψ in the space $C_c^\infty(H/M)$ such that $\int_{H/M} \psi(\dot{h}) \, d\dot{h} = 1$. Using for $x \in X^+$ the decomposition $x = h \cdot a_u \cdot e_n$, where $u = u(t)$

is a function of t, we set for $x \in X^+$

$$\begin{aligned}\psi_+(x) &= \psi_+(h \cdot a_{u(t)} \cdot e_n) = c_+^{-1}\, \psi(\dot{h})\, (t^2-1)^{-\frac{n-2}{2}} \\ &= c_+^{-1}\, \psi(\dot{h})\, (Q(x)^2-1)^{-\frac{n-2}{2}}.\end{aligned}$$

Then ψ_+ is C^∞ on X^+ and $M_{\psi_+}(t) = 1$ for $t > 1$. In a similar way we define ψ_- (on X^-) and ψ_0 (on X^0). Let now $v_\pm \in D(\mathbb{R})$ be such that $v_\pm = 1$ in some small neighbourhood of $t = \pm 1$, and let $f_\pm \in D(X_*)$ be such that $M_{f_\pm} = v_\pm$. Then, following the proof of Theorem A.2.1, we construct a function α with the required properties by

$$\alpha(x) = \begin{cases} f_+(x) + [1 - v_+(Q(x))]\, \psi_+(x) & \text{if } x \in X^+, \\ f_+(x) & \text{if } Q(x) = 1, \\ f_+(x) + [1 - v_+(Q(x))]\, \psi_0(x) & \text{if } -1/2 < Q(x) < 1, \\ f_-(x) + [1 - v_-(Q(x))]\, \psi_0(x) & \text{if } -1 < Q(x) < 1/2, \\ f_-(x) & \text{if } Q(x) = -1, \\ f_-(x) + [1 - v_-(Q(x))]\, \psi_-(x) & \text{if } x \in X^-. \end{cases}$$

Let now $\varphi \in D(\mathbb{R})$ and set $\alpha_\varphi(x) = \alpha(x)\, \varphi(Q(x))$. Then α_φ has compact support in X_* and $M_{\alpha_\varphi} = \varphi$. The mapping $\varphi \mapsto \alpha_\varphi$ is continuous.

For the other parts of the proof (2. and 3.), we can just copy the relevant parts of the proof of Theorem A.2.1.

B.3 Tengstrand's results for X

To find the analog of Section A.3, we may choose two ways: either proceed as in Section A.3, using explicit expressions for $M_f(t)$, e.g. for $t > 1$,

$$\begin{aligned}M_f(t) &= \text{const.}(t^2-1)^{\frac{n-1}{2}} \\ &\quad \times \int_0^\infty f(\sqrt{t^2-1}\cosh u, 0, \dots, 0, \sqrt{t^2-1}\sinh u, t)\, \sinh^{n-2} u\, du\end{aligned}$$

if f is a $H \cap K$-invariant function, or applying Morse's lemma. We prefer the latter way, because it is applicable in more general situations where an explicit expression for $M_f(t)$ fails. That lemma reads as follows. Let Y be a real analytic manifold and $f : Y \to \mathbb{R}$ an analytic function. Let $y_0 \in Y$ be a non-degenerate critical point of f, i.e. $df(y_0) = 0$ and the matrix $(\frac{\partial^2 f}{\partial y_i \partial y_j}(y_0))$ is non-degenerate, where $y_1, \dots, y_n$ are local coordinates on Y near y_0. Let (p, q) be the signature of this matrix, which is independent of the local coordinate system. Then we have:

Morse's lemma. *There are local coordinates $y_1, \dots, y_n$ near y_0 such that f can be written as*

$$f(y) = f(y_0) + y_1^2 + \dots + y_p^2 - y_{p+1}^2 - \dots - y_n^2.$$

For a proof, and ample explanation, we refer to [23]. Applying Morse's lemma with $Y = X$ and $f = Q$, we find that $dQ(x) = 0$ if and only if $x = e_n$ or $x = -e_n$; moreover, both points are non-degenerate critical points with signatures $(1, n-1)$ and $(n-1, 1)$ respectively. Indeed, this is easily seen by taking near e_n the coordinates $x_0, x_1, \dots, x_{n-1}$ and setting

$$Q(x) = \sqrt{1 + x_0^2 - x_1^2 - \cdots x_{n-1}^2}.$$

Similarly for $x = -e_n$. We thus obtain the following theorem, similar to Theorem A.2.1.

Theorem B.3.1. *The image of $D(X)$ under the mapping M is the space $\mathcal{H}_\eta$ consisting of functions on $\mathbb{R}$ of the form*

$$\varphi_1(t) + \eta(t-1)\,\varphi_2(t) + \eta(-t-1)\,\varphi_3(t) \qquad (t \in \mathbb{R}),$$

where $\varphi_1, \varphi_2, \varphi_3 \in D(\mathbb{R})$ and

$$\eta(t) = \begin{cases} Y(t)\,|t|^{\frac{n-2}{2}} & \text{if } n \text{ is odd,} \\ \log|t|\,t^{\frac{n-2}{2}} & \text{if } n \text{ is even.} \end{cases}$$

Similar to Appendix A, this section can now be completed. Of course, we have now, instead of the two functionals A_k and B_k, four functionals with obvious notations $A_k^{(1)}$, $A_k^{(-1)}$, $B_k^{(1)}$ and $B_k^{(-1)}$. All these considerations result in an analog of Theorem A.3.6:

Theorem B.3.2. *The adjoint M' of M from $\mathcal{H}'_\eta$ into $D'(X)^H$, the space of H-invariant distributions on X, is a (continuous) isomorphism.*

B.4 Solutions in $\mathcal{H}'_\eta$ of a singular second order differential equation

We follow Section A.4 step by step. The operator L we are considering here is the radial part of the Laplace–Beltrami operator Δ on X (see 9.2 (iii)), that is

$$L = (1 - t^2)\,\frac{d^2}{dt^2} - nt\,\frac{d}{dt}.$$

Here $L^* = (1 - t^2)\,\frac{d^2}{dt^2} + (n-4)\,t\,\frac{d}{dt} + (n-2)$ and

$$u\,(L^*\varphi) - (Lu)\,\varphi = [\varphi, u]',$$

where $[\varphi, u] = (1 - t^2)\,(u\varphi' - u'\varphi) + (n-1)t\,\varphi\,u$, for $u, \varphi \in C^{(2)}(\mathbb{R})$.

We are going to determine classical fundamental solutions of the equation $Lu = \lambda u$ ($\lambda \in \mathbb{C}$) first. Observe that this second order differential equation is singular at $t = \pm 1$.

For simplicity of the presentation, we restrict ourselves to the case that n is *odd*. Substituting $1 - t = 2z$ we get the differential equation

$$z(z-1)\frac{d^2u}{dz^2} + \left(nz - \frac{n}{2}\right)\frac{du}{dz} + \lambda\, u = 0.$$

The same equation is obtained after substitution $1 + t = 2z$. This hypergeometric differential equation has two fundamental solutions on $(-\infty, 1)$, namely

$$ {}_2F_1\left(s+\rho; -s+\rho; \frac{n}{2}; z\right) \quad \text{and} \quad |z|^{-\frac{n-2}{2}}\, {}_2F_1\left(s+\frac{1}{2}; -s+\frac{1}{2}; 2-\frac{n}{2}; z\right).$$

Here we have written $\lambda = s^2 - \rho^2$, $\rho = \frac{n-1}{2}$.

So we have the following solutions of $Lu = \lambda u$:

- on $(-1, \infty)$:

$$\Phi^{(1)}(t) = {}_2F_1\left(s+\rho; -s+\rho; \frac{n}{2}; \frac{1-t}{2}\right),$$

$$\Psi^{(1)}(t) = \left(\frac{|1-t|}{2}\right)^{-\frac{n-2}{2}} {}_2F_1\left(s+\frac{1}{2}; -s+\frac{1}{2}; 2-\frac{n}{2}; \frac{1-t}{2}\right),$$

- on $(-\infty, 1)$:

$$\Phi^{(-1)}(t) = {}_2F_1\left(s+\rho; -s+\rho; \frac{n}{2}; \frac{1+t}{2}\right),$$

$$\Psi^{(-1)}(t) = \left(\frac{|1+t|}{2}\right)^{-\frac{n-2}{2}} {}_2F_1\left(s+\frac{1}{2}; -s+\frac{1}{2}; 2-\frac{n}{2}; \frac{1+t}{2}\right).$$

These solutions meet on $(-1, 1)$; one has (see [13])

$$\Phi^{(1)}(t) = \gamma_1(s)\,\Phi^{(-1)}(t) + \gamma_2(s)\,\Psi^{(-1)}(t), \qquad \text{(B.4.1)}$$

and

$$\Psi^{(1)}(t) = \frac{1-\gamma_1^2(s)}{\gamma_2(s)}\,\Phi^{(-1)}(t) - \gamma_1(s)\,\Psi^{(-1)}(t), \qquad \text{(B.4.2)}$$

where $\gamma_1(s) = \frac{\Gamma(\frac{n}{2})\Gamma(1-\frac{n}{2})}{\Gamma(s+\frac{1}{2})\Gamma(-s+\frac{1}{2})}$ and $\gamma_2(s) = \frac{\Gamma(\frac{n}{2})\Gamma(1-\frac{n}{2})}{\Gamma(s+\rho)\Gamma(-s+\rho)}$.

We now define the following distributions on $(-1, \infty)$:

$$\langle S_+^{(1)}, \varphi\rangle = \int_1^\infty \Phi^{(1)}(t)\,\overline{\varphi(t)}\,dt,$$

$$\langle S_-^{(1)}, \varphi\rangle = \mathrm{Pf}\int_{-1}^1 \Phi^{(1)}(t)\,\overline{\varphi(t)}\,dt,$$

$$\langle T_+^{(1)}, \varphi\rangle = \mathrm{Pf}\int_1^\infty \Psi^{(1)}(t)\,\overline{\varphi(t)}\,dt,$$

$$\langle T_-^{(1)}, \varphi\rangle = \mathrm{Pf}\int_{-1}^1 \Psi^{(1)}(t)\,\overline{\varphi(t)}\,dt.$$

In a similar way we define $S_+^{(-1)}$, $S_-^{(-1)}$, $T_+^{(-1)}$, $T_-^{(-1)}$ on $(-\infty, 1)$. For instance,

$$\langle S_-^{(-1)}, \varphi\rangle = \int_{-\infty}^{-1} \Phi^{(-1)}(t)\,\overline{\varphi(t)}\,dt.$$

Any distribution solution of $LS = \lambda S$ can be written as

$$S = \alpha_1 S_+^{(1)} + \alpha_2 T_+^{(1)} + \alpha_3 S_-^{(1)} + \alpha_4 T_-^{(1)} + \alpha_5 S_-^{(-1)} + \alpha_6 T_-^{(-1)} + T_0$$

on $(-\infty, 1)$ and $(-1, \infty)$ respectively, where T_0 is a distribution on $\mathbb{R}$ with support contained in $\{-1, 1\}$.

Observe that in the computation of $LS_-^{(1)} - \lambda S_-^{(1)}$ and $LT_-^{(1)} - \lambda T_-^{(1)}$ near $t = -1$, we have to apply the expressions (B.4.1) and (B.4.2).

We now proceed as in Appendix A and obtain, finally:

Proposition B.4.1. *The space of solutions of $LS = \lambda S$ in $\mathcal{H}'_\eta$ is two-dimensional. Fundamental solutions are given by*

- $S_+^{(1)} + S_-^{(1)}$ *with its natural extension as* $\gamma_1(s)\,(S_+^{(-1)} + S_-^{(-1)}) + \gamma_2(s)\,T_+^{(-1)}$ *to* $(-\infty, 1)$.
- $T_-^{(1)}$ *with its natural extension as* $\frac{1-\gamma_1(s)^2}{\gamma_2(s)}\,(S_+^{(-1)} + S_-^{(-1)}) - \gamma_1(s)\,T_+^{(-1)}$ *to* $(-\infty, 1)$.

The proof is left to the reader as an exercise. We shall call these solutions S_λ and T_λ respectively.

One might simplify the formulae (B.4.1) and (B.4.2) slightly by using

- $\Gamma(\frac{n}{2})\,\Gamma(1 - \frac{n}{2}) = (-1)^{\frac{n-1}{2}}\,\pi \quad (n \text{ even}),$
- $\Gamma(s + \frac{1}{2})\,\Gamma(-s + \frac{1}{2}) = \frac{\pi}{\cos \pi s}.$

And herewith we finish Appendix B and the book. Further reading? We recommend [53].

Bibliography

[1] S. Bochner, *Lectures on Fourier Integrals*, Ann. of Math. Studies **42**, Princeton Univ. Press, Princeton (1959).

[2] A. Borel, *Représentations de Groupes Localement Compacts*, Springer Lecture Notes **276**, Springer-Verlag, Berlin etc. (1972).

[3] N. Bourbaki, *Eléments de Mathématique, Topologie Générale*, Hermann, Paris (1961).

[4] N. Bourbaki, *Eléments de Mathématique, Espaces Vectoriels Topologiques*, Hermann, Paris (1964).

[5] N. Bourbaki, *Eléments de Mathématique, Intégration*, Hermann, Paris (1965).

[6] T. Carleman, *L'intégrale de Fourier et questions qui s'y rattachent*, Institut Mittag-Leffler, Publications Scientifiques, Uppsala (lectures given at Mittag-Leffler Institute in 1935).

[7] H. Cartan and R. Godement, Théorie de la dualité et analyse harmonique dans les groupes abéliens localement compacts, *Ann. Sci. École Norm. Sup.* **64** (1947), pp. 79–99.

[8] J. Dixmier, *Les C^*-Algèbres et Leurs Représentations*, Gauthier-Villars, Paris (1964).

[9] J. Dixmier and P. Malliavin, Factorisations de fonctions et de vecteurs indéfiniment différentiables, *Bull. Sci. Math.* **102** (1978), pp. 307–330.

[10] N. Dunford and J. T. Schwartz, *Linear Operators I*, Wiley Classics Library Edition, John Wiley & Sons, New York etc. (1988).

[11] N. Dunford and J. T. Schwartz, *Linear Operators II*, Wiley Classics Library Edition, Interscience Publishers, New York etc. (1988).

[12] R. E. Edwards, *Fourier Series*, Vol. 1, Holt Rinehart and Winston Inc., New York (1967).

[13] A. Erdélyi a.o., *Higher Transcendental Functions I*, McGraw-Hill Inc. (1953).

[14] A. Erdélyi a.o., *Higher Transcendental Functions II*, McGraw-Hill Inc. (1953).

[15] J. Faraut, *Analyse harmonique sur les paires de Guelfand et les espaces hyperboliques*, CIMPA lecture notes (1980).

[16] C. Fefferman, Pointwise convergence of Fourier series, *Ann. of Math.* **98** (1973), pp. 551–571.

[17] I. M. Gelfand and G. E. Shilov, *Generalized Functions*, Vol. 1, Academic Press, New York (1964).

[18] R. Godement, Introduction aux travaux de A. Selberg, *Sém. Bourbaki* **9** (1956–1957), exposé 144.

[19] R. R. Goldberg, *Fourier Transforms*, Cambridge At the University Press (1961).

[20] Harish-Chandra, Invariant distributions on Lie algebras, *Amer. J. of Math.* **86** (1964), pp. 271–309.

[21] S. Helgason, *Differential Geometry and Symmetric Spaces*, Academic Press, New York etc. (1962).

[22] E. Hewitt and K. Stromberg, *Real and Abstract Analysis*, Springer-Verlag, New York (1965).

[23] M. W. Hirsch, *Differential Topology*, Springer-Verlag, New York etc. (1976).

[24] J. E. Humphreys, *Introduction to Lie Algebras and Representation Theory*, 2nd edition, Springer-Verlag, New York etc. (1978).

[25] Y. Katznelson, *An Introduction to Harmonic Analysis*, John Wiley, New York (1968).

[26] F. J. M. Klamer, *Group representations in Hilbert subspaces of a locally convex space*, PhD thesis, University of Groningen (1979).

[27] M. T. Kosters, *Spherical distributions on rank one symmetric spaces*, PhD thesis, Leiden University (1983).

[28] M. T. Kosters and G. van Dijk, Spherical distributions on the pseudo-Riemannian space $\mathrm{SL}(n,\mathbb{R})/\mathrm{GL}(n-1,\mathbb{R})$, *J. Funct. Anal.* **68** (1986), pp. 168–213.

[29] L. H. Loomis, *An Introduction to Abstract Harmonic Analysis*, Van Nostrand Company Inc., Princeton etc. (1953).

[30] I. G. Macdonald, *Spherical Functions on a Group of p-adic Type*, Ramanujan Institute Lecture Notes, University of Madras, Madras, India (1971).

[31] P. D. Méthée, Sur les distributions invariantes dans le groupe des rotations de Lorentz, *Comment. Math. Helvetici* **28** (1954), pp. 225–269.

[32] V. F. Molchanov, Representations of the pseudo-orthogonal group associated with a cone, *Mat. Sb.* **81** (1970), pp. 358–375. English translation: *Math. USSR Sb.* **10** (1970), pp. 333–347.

[33] D. Montgomery and L. Zippin, *Topological Transformation Groups*, Interscience, New York (1955).

[34] M. A. Neumark, *Normierte Algebren*, VEB Deutscher Verlag der Wissenschaften, Berlin (1959).

[35] L. S. Pontryagin, *Topological Groups*, 2nd edition, Gordon and Breach, New York (1966).

[36] L. Pukanszky, *Leçons sur les Représentations des Groupes*, Dunod, Paris (1967).

[37] S. Rallis and G. Schiffmann, Distributions invariantes par le groupe orthogonal, *Springer Lecture Notes* **497** (1975), pp. 494–642.

[38] H. Reiter and J. D. Stegeman, *Classical Harmonic Analysis and Locally Compact Groups*, LMS Monographs, new series **22**, Clarendon Press, Oxford (2000).

[39] W. Rudin, *Fourier Analysis on Groups*, Interscience, New York (1962).

[40] W. Rudin, *Real and Complex Analysis*, McGraw-Hill, New York etc. (1995).

[41] L. Schwartz, *Méthodes Mathématiques pour les Sciences Physiques*, Hermann, Paris (1965).

[42] L. Schwartz, *Applications of Distributions to the Theory of Elementary Particles in Quantum Mechanics*, Gordon and Breach, New York (1968).

[43] L. Schwartz, *Théorie des Distributions*, nouvelle édition, Hermann, Paris (1978).

[44] E. M. Stein and G. Weiss, *Introduction to Fourier Analysis on Euclidean Spaces*, Princeton University Press, Princeton (1971).

[45] M. Sugiura, *Unitary Representations and Harmonic Analysis*, 2nd edition, Elsevier, Amsterdam etc. (1990).

[46] A. E. Taylor and D. C. Lay, *Introduction to Functional Analysis*, 2nd edition, John Wiley, New York (1980).

[47] A. Tengstrand, Distributions invariant under an orthogonal group of arbitrary signature, *Math. Scand.* **8** (1969), pp. 201–218.

[48] C. Thiele, *Wave Packet Analysis*, CMRS Regional Conf. Series in Math. **105**, AMS (2006).

[49] E. G. F. Thomas, The theorem of Bochner–Schwartz–Godement for generalized Gelfand pairs, in: *Functional Analysis: Surveys and Results III*, pp. 291–304, edited by K. D. Bierstedt and B. Fuchtsteiner, Elsevier Science Publishers, Amsterdam (1984).

[50] F. Trèves, *Topological Vector Spaces, Distributions and Kernels*, Academic Press, New York etc. (1967).

[51] G. van Dijk, *Spherical functions on the p-adic group* PGL(2), thesis (1969). See also: *Proc. KNAW, Ser. A* **72** (1969), pp. 213–241.

[52] G. van Dijk, On a class of generalized Gelfand pairs, *Math. Z.* **193** (1986), pp. 581–593.

[53] G. van Dijk, *Gelfand Pairs and Beyond*, COE Math-for-Industry Lecture Note **11**, Kyushu University (2008).

[54] E. R. van Kampen, Locally bicompact abelian groups and their character groups, *Annals of Math.* **36** (1935), pp. 448–463.

[55] V. S. Varadarajan, *Lie Groups, Lie Algebras, and Their Representations*, Prentice Hall, Englewood Cliffs, New Jersey (1974).

[56] N. J. Vilenkin, *Special Functions and the Theory of Group Representations*, Translations of Math. Monographs, no. **22**, Amer. Math. Soc., Providence (1968).

[57] G. Warner, *Harmonic Aanlysis on Semi-Simple Lie Groups I*, Springer-Verlag, Berlin etc. (1972).

[58] A. Weil, *L'Intégration dans les Groupes Topologiques et ses Applications*, 2ième édition, Hermann, Paris (1953).

[59] D. V. Widder, *The Laplace Transform*, Princeton Univ. Press, Princeton (1965).

[60] A. C. Zaanen, *Linear Analysis*, North-Holland Publ. Co., Amsterdam (1953).

[61] A. Zygmund, *Trigonometrical Series*, 2nd edition, Cambridge Univ. Press (1959).

Index

I

K

L

M

N

O

P

R

Bei Fragen zur Produktsicherheit wenden Sie sich bitte an:
If you have any questions regarding product safety,
please contact:

Walter de Gruyter GmbH
Genthiner Straße 13
10785 Berlin
productsafety@degruyterbrill.com